POLYMERIC STABILIZATION OF COLLOIDAL DISPERSIONS

COLLOID SCIENCE

Editors
R. H. Ottewill and R. L. Rowell

In recent years colloid science has developed rapidly and now frequently involves both sophisticated mathematical theories and advanced experimental techniques. However, many of the applications in this field require simple ideas and simple measurements. The breadth and the interdisciplinary nature of the subject have made it virtually impossible for a single individual to distill the subject for all to understand. The need for understanding suggests that the approach to an interdisciplinary subject should be through the perspectives gained by individuals.

The series consists of separate monographs, each written by a single author or by collaborating authors. It is the aim that each book will be written at a research level but will be readable by the average graduate student in chemistry, chemical engineering or physics. Theory, experiment and methodology, where necessary, are arranged to stress clarity so that the reader may gain in understanding, insight and predictive capability. It is hoped that this approach will also make the volumes useful for non-specialists and advanced undergraduates.

The author's role is regarded as paramount, and emphasis is placed on individual interpretation rather than on collecting together specialist articles.

The editors simply regard themselves as initiators and catalysts.

1. J. Mahanty and B. W. Ninham: Dispersion Forces.
2. R. J. Hunter: Zeta Potential in Colloid Science.
3. D. H. Napper: Polymeric Stabilization of Colloidal Dispersions.

POLYMERIC STABILIZATION OF COLLOIDAL DISPERSIONS

DONALD H. NAPPER

Department of Physical Chemistry,
University of Sydney, NSW
Australia

1983

ACADEMIC PRESS
A Subsidiary of Harcourt Brace Jovanovich, Publishers
London New York Paris San Diego San Francisco
São Paulo Sydney Tokyo Toronto

ACADEMIC PRESS INC. (LONDON) LTD.
24/28 Oval Road
London NW1

United States Edition published by
ACADEMIC PRESS INC.
111 Fifth Avenue
New York, New York 10003

British Library Cataloguing in Publication Data

Napper, D. H.
Polymeric stabilization of colloidal dispersions.
1. Colloids
I. Title
541.3′541 QD549

ISBN 0-12-513980-2

LCCCN 83-71086

Typeset in Monophoto Times New Roman by
Latimer Trend & Company Ltd, Plymouth
Printed in Great Britain by Whitstable Litho Ltd

Preface

Jacob Israelachvili has recently written that "the interaction of adsorbed macromolecules is arguably the most important fundamental process in modern colloid science". Yet, although our understanding of polymeric stabilization in its many manifestations has now come of age, no treatise has previously been published which deals with the subject in an up-to-date, comprehensive and unified fashion. This book aims to fill that gap. It is directed as much towards senior undergraduate and postgraduate students in universities as it is towards the practitioners of the art of colloid stabilization working in industry. The text has also been written with the needs of biologists, especially biochemists and physiologists, in mind because the stability of so many natural systems is influenced by macromolecules. No prior knowledge of colloid or polymer science is assumed. All the relevant modern theories (and there has been a dramatic proliferation of these) are developed *ab initio* as required. The assumed prerequisite knowledge is little more than that covered in a freshman chemistry course.

This text is an elaboration of a short course of lectures that I presented to MSc students in the winter of 1981 whilst Visiting Professor of Colloid Science and SRC Senior Fellow at the University of Bristol. As John Ziman FRS, my former tutor at King's College, Cambridge, has so aptly stated, it is the purpose of a research monograph to provide a state-of-the-art survey: to stake out the territory that has been recently won, to adjust priorities as to precedence and to place each theory or fact in its appropriate place. I have, in addition, endeavoured to anticipate future advances: first, by preparing the ground that they seem likely to cover and second, by pointing to where current deficiencies lie. This book, however, was never intended to be a bland catalogue of publications in the field. Rather, it expounds a personal perspective. Hopefully, where conflict exists, and it is right that it should exist in a rapidly expanding scientific discipline, the competing viewpoints are treated in an evenhanded, if duly critical, manner.

My grateful thanks go out to my many teachers, especially three who profoundly influenced my thinking in this area: the late A. E. Alexander, P. J. Flory, Nobel Laureate, and R. H. Ottewill FRS. The last named, together with my many friends at ICI Paints Division, introduced me to steric stabilization over a decade ago. Ron Ottewill also provided both the stimulus and the means whereby this book could be written. I include among my teachers three former research students: Robert Evans, Jim Smitham and Robert Feigin; all made significant contributions to my understanding of the area.

I have had the singular good fortune to be blessed with an understanding and supportive family: first, my parents, who allowed me the luxury of doing my own "thing", and subsequently my wife and children, Katherine, Jonathon and Meredith, who have continued that tradition. Their response to the many deprivations suffered by an author's family were words of encouragement.

June, 1983 *Don Napper*

Contents

"*stand still you*
ever moving spheres . . ."

Christopher Marlowe (1564–1593)

The Tragicall History of Dr Faustus

to G.M.N

Chapter 1

Basic Concepts of Colloid Stability

1.1. General introduction

Colloidal dispersions play a pivotal role in our everyday lives: from blood, that is vital to the human organism, through polymer latices, which are generated for use in the manufacture of plastics and paints, to agricultural emulsions and dispersions, exploited in the efficient production of food. Considering the prominent position of colloid science in modern-day technology, it is surprising that the scientific effort expended in the past fifty years investigating colloidal systems has been in no way commensurate with their technological importance. Yet, even though colloidal systems remain to some degree the "world of neglected dimensions" (to borrow Wolfgang Ostwald's picturesque phrase), such is the recent progress in our understanding of colloidal phenomena that Everett (1980) has been led to assert that the designation "*colloid science*" is now fully justified.

Colloid science is currently undergoing a veritable transformation. One primary stimulus for this at the experimental level is the application of modern scattering methods and spectroscopic techniques (e.g. small angle neutron scattering, photon correlation spectroscopy, pulsed NMR) to the study of colloidal phenomena. As we shall see, such experimental methods provide detailed information about the properties of colloidal systems that were mere theoretical conjectures several years ago. Investigations of the role played by polymers in stabilizing and flocculating colloidal dispersions have not been immune to the recent developments in experimental techniques. At the theoretical level, a major impetus to recent progress has come from the deployment of some of the powerful techniques of modern chemical physics into areas of colloid and polymer science. Examples of these new mathematical technologies (e.g. the self-consistent field theory of depletion flocculation) will be displayed later in this text. The successful exploitation of many of these methods has depended critically upon the advent of high speed digital computers.

1.2. The nature of colloidal dispersions

We will be concerned in this book primarily with the stability of *colloidal dispersions*. A colloidal dispersion consists of a *dispersion medium* (the continuous phase) through which are dispersed the colloidal particles. These particles constitute the *dispersed phase*, although the elided term *disperse phase* is also commonly used (Freundlich, 1926). Particles are said to be *colloidal* in character if they possess at least one dimension in the size range 1 to 10^3 nm.

1.2.1. Particulate dispersions

The dispersed phase in colloidal systems may be solid, liquid or gas. If the dispersed phase is solid, a distinction is made according to whether or not the particles are composed of polymer. A colloidal dispersion whose dispersed phase is non-polymeric in nature (e.g. a metal, such as gold, or an inorganic salt, such as silver iodide) is usually referred to as a *sol*. Sols are sometimes differentiated according to the nature of their dispersion medium: if aqueous, they are termed *hydrosols*; whereas dispersions in organic liquids are referred to as *organosols*. If the dispersed phase is polymeric in character, the dispersion is called a *latex*. This terminology has been borrowed from its original use for the sap of the rubber tree *Hevea braziliensis*. The term "latex" is actually the Latin word for "liquid" (and also for wine). Its plural is strictly *latices* (from the Latin *laticis*) but *latexes* is also commonly used (especially to avoid confusion with *lattices*).

A dispersion of particles whose size is greater than, say, 1 μm is usually referred to as a *suspension*.

1.2.2. Emulsions and foams

The dispersion of one liquid in another is called an *emulsion* (strictly, a *liquid emulsion*). The dispersion of a gas in a liquid is a *foam*. In what follows, the discussion will centre on particulate dispersions (i.e. sols and latices). However, passing reference will be made, wherever appropriate, to emulsions. Both emulsions and foams obey the same general principles that will be elaborated hereunder, provided that due allowance is made for the differences in the nature of the dispersed phases.

1.3. Colloid stability

A colloidal dispersion in which the dispersed phase remains essentially as discrete, single particles on a long time scale (e.g. months or even years) is said to be *stable*. Such a dispersion may be stable for thermodynamic or for kinetic reasons. In the latter case, typified by electrostatic stabilization, the system is said to be *thermodynamically metastable*.

1.3.1 Coagulation and flocculation

The dispersed particles of a colloidal dispersion can usually be induced to form doublets and higher multiplets by an appropriate change in the system. This aggregation is referred to as either *coagulation* or *flocculation*. The word "coagulation" is derived from the Latin verb "*coagulare*", which means "to cause to curdle". It implies that the particles are somehow driven together. Flocculation, on the other hand, derives from "*floccus*" and implies the formation of a highly porous floc, resembling a tuft of wool. Some authors use the terms coagulation and flocculation synonymously. Other authors, however, are careful to distinguish between the two terms: coagulation is the term applied when the coagulum formed is composed of particles that have approached one another closely; flocculation is reserved for a looser form of aggregation wherein the particles are held relatively far apart. To illustrate this terminology in action, we consider the aggregation of an electrostatically stabilized sol. This sol would be considered to undergo *coagulation* on the addition of an appropriate concentration of a simple electrolyte but to be *flocculated* by the addition of a suitably high molecular weight polyelectrolyte at a low concentration (a few p.p.m.). In the latter case, bridging flocculation would be expected to produce flocs with a relatively open structure.

Not all authors follow this dichotomy, originated by La Mer (1964). One reason for this is that the precise structure of the flocs formed by aggregation is rarely known or even determined. Yet the aggregation phenomenon has to be described by one or other of the terms, coagulation or flocculation.

In the following chapters, the La Mer convention will be adhered to as closely as possible. Floc structure is not usually a parameter of central importance in colloid stability theory, although it can be vital in other areas of colloid science (e.g. the rheology of aggregated systems). From the viewpoint of stability theory, the crucial distinction between the two processes, as envisaged by La Mer, is the origin of the interparticle attraction. In the one case, *viz* coagulation, the van der Waals attraction (*v.i.*) between the core particles is responsible for holding the aggregates together. In the other, the close proximity of the particles is engendered by the polymeric species. In what follows, coagulation will be employed wherever possible to specify aggregation that is induced by the van der Waals forces between the colloidal particles. The term flocculation will be reserved for particle agglomeration that is mediated by polymeric species, whether they are attached to the colloidal particles or free in solution in the dispersion medium. In some instances, it will not be possible to adhere strictly to this terminology. This will be especially the case with certain common acronyms.

1.4. The London dispersion attraction

It is a matter of common observation that naked, uncharged colloidal particles undergo very rapid coagulation. Table 1.1 displays the particle number concentrations (N_1) that give rise to various half-lives ($t_{1/2}$) as calculated using the theoretical Smoluchowski rapid coagulation rate constant ($k_2 = 4kT/3\eta$, where η = viscosity of the dispersion medium) (Smoluchowski, 1917). In water at 25 °C, to which these half-lives pertain, $t_{1/2} = 1/N_1 k_2$, where $k_2 = 6{\cdot}2 \times 10^{-15}$ dm^3 s^{-1}. In actual systems, local hydrodynamic interactions (*viz* the interactions of the "bow-waves" of the approaching particles) prevent the attainment of the hypothetical Smoluchowski rapid rate. The rate constant is approximately halved according to theoretical calculations (Derjaguin and Muller, 1967; Spielman, 1970), a result that has been confirmed experimentally (Lips *et al.*, 1971; Honig *et al.*, 1971; Lichtenbelt *et al.*, 1974). It follows that the particle number concentrations listed in Table 1.1 should be approximately doubled.

The usual range of particle number concentrations of interest in colloid science varies from, say, 1×10^{10} to 1×10^{22} dm^{-3}. It is apparent from Table 1.1 that the number concentration of particles in most practical dispersions

TABLE 1.1
Particle number concentrations for various half-lives according to Smoluchowski kinetics

$t_{1/2}$	Particle number concentration N_1/dm^{-3}
1 ns	$1{\cdot}6 \times 10^{23}$
1 μs	$1{\cdot}6 \times 10^{20}$
1 ms	$1{\cdot}6 \times 10^{17}$
1 s	$1{\cdot}6 \times 10^{14}$
1 min	$2{\cdot}7 \times 10^{12}$
1 h	$4{\cdot}5 \times 10^{10}$
1 day	$1{\cdot}9 \times 10^{9}$
1 week	$2{\cdot}7 \times 10^{8}$
1 yr	$5{\cdot}2 \times 10^{6}$

would theoretically be reduced by a factor of 2 in a time span of less than a few seconds if the particles were rendered completely unstable. The reasons for this are well-understood: colloidal particles are constantly undergoing Brownian encounters and these collisions are sticky. The particles adhere when they collide because there are operative between them fairly strong, and quite long range, van der Waals attractive forces.

The data presented in Table 1.1 can be read in another way. To obtain a measurable half-life (say, >1 s) for a colloidal dispersion of naked, uncharged particles, a number concentration of less than 10^{14} dm^{-3} would be required. This means that, when rendered unstable, dispersions of practical importance, which usually contain more than 10^{16} particles per dm^{-3}, will coagulate at an immeasurably fast rate, at least in terms of the measuring techniques developed to date (Parfitt and Peacock, 1978). Of course, some very dilute dispersions can be encountered in water treatment, for example, and their rate of coagulation will be correspondingly slower.

1.4.1. The microscopic theory of London attraction

The van der Waals attraction between gas molecules may originate from three possible sources: *permanent dipole-permanent dipole* (Keesom) forces; *permanent dipole-induced dipole* (Debye) interactions; and *transitory dipole-transitory dipole* (London) forces. This, of course, ignores the higher multipole interactions. Only the classical London dispersion forces contribute to the long-range attraction between colloidal particles. These London interactions are the self-same forces that are responsible for the liquefaction of the rare gases, such as helium and argon, at low temperatures.

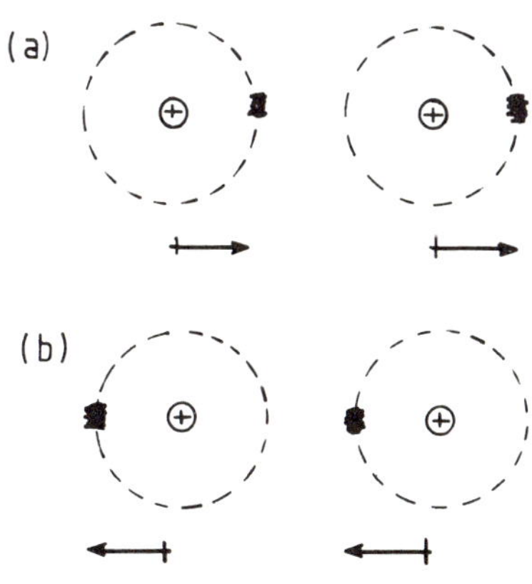

Fig. 1.1. Schematic representation of the origin of the London dispersion forces for two atoms.

The origin of the London forces for, say, two helium atoms may be depicted schematically as shown in Fig. 1.1. If two helium atoms are separated by an infinite distance, the positions of their respective electrons in their orbital motion are completely uncorrelated. Closer approach of the atoms, however, leads to correlated electronic motions. The positions of the electrons can now be considered to be such as to minimize the potential energy of the system. Thus, in Fig. 1.1a, the electrons are envisaged as being separated by their maximum possible distance. In this configuration, the transitory dipoles formed by the electrons and the positive charges associated with the nuclei are mutually attractive. This is also true for the second configuration shown in Fig. 1.1b. These instantaneous dipoles fluctuate over a period of order 0·1 fs (10^{-16} s). It is clear that summation of the transitory dipoles depicted in Figs. 1.1a and b leads to zero net permanent dipole moment. Yet in both configurations, the atoms are mutually attractive. This line of reasoning can be readily extended to encompass all orbital positions of the electrons.

The van der Waals attraction between two atoms or molecules is relatively short range, extending only over a few-tenths of a nanometre. For colloidal particles, however, each atom or molecule of one particle attracts every atom in the other particle. Typically, a colloidal particle is composed of 10^6–10^{10} atoms. The net effect of adding all of the myriad of possible atomic interactions is to generate a long range attraction between the particles that is of considerable strength.

Hamaker (1937) extended London's treatment (London, 1934) of the dispersion forces between atoms to calculate those between colloidal particles. The result for the attraction (V_A) between two spheres, each of radius a, is

$$V_A = -(A^*/6)G \tag{1.1}$$

where A^* = effective Hamaker constant and the geometrical term G is given by $G = 2/(s^2-4) + (2/s^2) + \ln([s^2-4]/s^2)$. Here the parameter $s = (H_0 + 2a)/a$, where H_0 = the minimum distance of separation between surfaces of the particles. If H_0 is small compared with the particle radius, then the first term in the expression for G is dominant, and equation (1.1) reduces to

$$V_A \sim -A^*a/12H_0 \quad (a \gg H_0) \tag{1.2}$$

The corresponding expression for the attractive potential energy per unit area for thick parallel flat plates is

$$V_A = -A^*/12\pi h^2 \tag{1.3}$$

where h = distance between the plates.

Note that in calculating A^* for sterically stabilized systems, allowance must be made not only for the intervening dispersion medium between the particles but also for the presence of any adsorbed layers (Vincent, 1973b). A^* is typically of order kT for latex particles, although it can be substantially larger for dispersions of metals or for inorganic sols (Gregory, 1970; Visser, 1972).

To gain some insight into the magnitudes of the attractive energies between colloidal particles, consider a pair of latex particles of radii 120 nm whose surfaces undergo close approach. The potential energy of attraction is of order 10 kT at a separation of 1 nm and 2 kT at a separation of 5 nm. Both would be sufficient in the absence of any repulsion to induce significant coagulation since their absolute magnitudes exceed the thermal energy available to the particles (ca 3/2 kT, for three translational degrees of freedom). We conclude that the range of significant van der Waals attraction is of order, say, 5 nm for many colloidal systems. The magnitude and range of the London attraction are decisive in determining the strategies that are feasible for imparting colloid stability. This long range character of the van der Waals attraction is severely debilitating for several conceivable modes of stabilization.

Note that the adjective "dispersion", as applied to London forces, is derived from the phenomenon of *optical dispersion*, which is a consequence of the same electronic oscillations as generate London forces.

1.4.2. The macroscopic theory of London attraction

The Hamaker approach to calculating the London attraction between colloidal particles is based upon the pairwise additivity of interatomic attractions. Moreover, it is customary to evaluate the Hamaker constant A for a substance at a single frequency, usually the electronic correlation frequency for an electron in the ground state of a hydrogen-like atom.

This approach is not without its difficulties: pairwise additivity may well be appropriate for dilute gases but, in contrast, many-body effects may become significant for condensed phases; additionally, correlations other than those propagated at ultra-violet frequencies can contribute significantly to the overall attraction.

A more satisfactory method for calculating the attraction between colloidal particles is the macroscopic continuum theory due to Lifshitz (Lifshitz, 1956; Dzyaloshinskii *et al.*, 1961) and subsequently elaborated by Ninham and coworkers (Mahanty and Ninham, 1976). This expresses the interaction in terms of the bulk dielectric properties of the two colloidal particles. The power of the Lifshitz formalism lies in its ability to encompass all many-body interactions; to deal properly with the effects of intermediate substances (here the microscopic method is quite vague); and to include contributions from all resonant electronic and molecular frequencies. Its disadvantage lies in the dramatic increase in the complexity of the calculations, although such computations are readily performed with the aid of a digital computer.

Approximate treatments have also been devised (Gregory, 1970; Visser, 1972) to provide alternative routes for calculating Hamaker constants. These alternative methods are necessarily very approximate because they improperly separate the geometric and the electromagnetic components of the attraction. As Parsegian and Ninham (1971) have pointed out, these two components are inextricably interwoven. The Hamaker constant is not really a constant, in general, but a function of the distance between the particles and the temperature (Evans and Napper, 1973c; Everett and Stageman, 1978a).

1.5. Strategies for imparting colloid stability

1.5.1. Requirements for stabilization

Given that there exist strong, long range attractive forces between colloidal particles, it follows that in order to impart colloid stability, it is necessary to provide a long range repulsion between the particles. This repulsion must be at least as strong as, and comparable in range to, the attractive interaction.

1.5.2. Current stabilization methods

At the present time, the number of options that are available for stabilizing colloidal dispersions is severely limited. Indeed, there currently exist just two general ways by which colloid stability can be imparted: *electrostatic stabilization* and *polymeric stabilization*. So-called *solvation* forces are excluded from this short list because they cannot be sufficiently well controlled as yet to be of any practical value. Indeed, Overbeek (1982) has expressed doubts as to whether such forces will ever be of any practical significance in the preparation of stable dispersions because of their relatively short range character.

1.5.2.1. Electrostatic stabilization

Overbeek's stricture noted above focuses attention on the principal reason why the current methods of imparting colloid stability are so few in number. The paramount difficulty resides in the projection of the repulsion over distances comparable to that of the attraction (5–10 nm). One way to accomplish this is to use Coulombic repulsion. In the electrostatic stabilization of aerosols, the Coulombic repulsion between the colloidal particles is of a long range character and can impart stability. In liquid dispersion media, however, the principle of electroneutrality demands that the net charge in the dispersion medium be equal, but opposite in sign, to that of the particles. This leads to a more rapid fall-off in the potential. The counterions in the dispersion medium, however, give rise to the electrical double layers that surround the colloidal particles. It is the mutual repulsion of these double layers that provides stability in electrostatic stabilization (see Fig. 1.2).

The spatial extension ("thickness") of the electrical double layers is, however, a sensitive function of the ionic strength of the dispersion medium. The thickness of an electrical double layer is normally characterized by the

Debye length $(1/\kappa)$. For a 1 : 1 electrolyte in water at 25 °C, $(1/\kappa)/\text{nm} \sim 0{\cdot}3/c^{\frac{1}{2}}$, where c = molar concentration. Some typical values of $(1/\kappa)$ in this specific instance are displayed in Table 1.2. It is apparent that at low ionic strengths ($c \lesssim 10^{-3}$ M) the thickness of the double layer exceeds, say, 5–10 nm, which is the range of the van der Waals attraction. This accounts for the observation of electrostatic stabilization in dispersion media of low ionic strength.

Increasing the ionic strength significantly diminishes the thickness of the double layer. At ionic strengths greater than, say, 10^{-1} M, the thickness of the double layer is less than 1 nm. Electrostatic repulsion is then usually of insufficient range (and, as it eventuates, of insufficient magnitude) to outweigh the van der Waals attraction. This explains why most electrostatically stabilized dispersions coagulate when the ionic strength of the dispersion medium is increased sufficiently.

The potential energy diagram

The total potential energy of interaction as a function of the distance of separation of two electrostatically stabilized colloidal particles is shown schematically in Fig. 1.3. At large distances of separation, the total

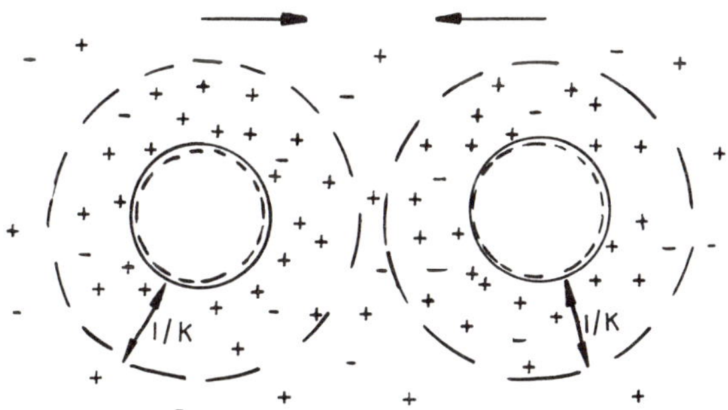

Fig. 1.2. Diagrammatic representation of the origin of electrostatic stabilization for negatively charged particles.

TABLE 1.2
The effect of the concentration of a 1 : 1 electrolyte on the thickness of the electrical double layer

Concentration/M	$(1/\kappa)$/nm
10^{-5}	100
10^{-4}	30
10^{-3}	10
10^{-2}	3
10^{-1}	1

interactional energy is obtained from the super-position of the van der Waals attraction and the electrostatic double layer repulsion. On closer approach, *Born repulsion* (which is essentially a consequence of the operation of the Pauli exclusion principle) is also operative.

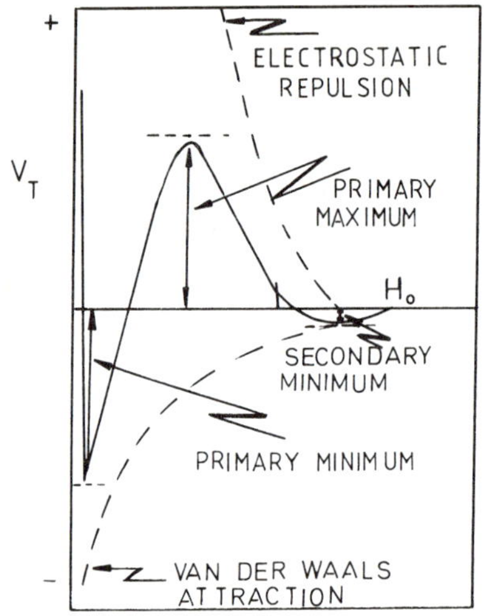

Fig. 1.3. Schematic representation of the total potential energy *vs* distance of separation for a pair of electrostatically stabilized particles. Also shown are the component electrostatic repulsion and van der Waals attraction.

One maximum and two minima are frequently identified in the total potential energy diagram. The height of the *primary maximum* (see Fig. 1.3), if sufficiently large ($\gtrsim 25\,kT$), will ensure that the rate of coagulation of the particles is so slow that the system displays long-term stability. Such stability, arising for kinetic reasons, corresponds to thermodynamic metastability.

The minimum that occurs in the potential energy diagram on very close approach is termed the *primary minimum* whereas that sometimes apparent at larger separations is called the *secondary minimum*. Both are, of course, a consequence of the van der Waals attraction. If the depth of the primary minimum is sufficiently large, it can give rise to coagulation, always provided that the primary maximum can be surmounted at a kinetically significant rate. Secondary minimum coagulation requires no such caveat. In this case, however, the particles are separated by relatively large distances after coagulation.

1.5.2.2. Polymeric stabilization

The polymer concept

The concept of polymer molecules, which nowadays is accepted without question, was championed in the early part of this century by Hermann Staudinger. For this, he was awarded the 1953 Nobel Prize for Chemistry, having demonstrated experimentally the existence of large molecules. Up until the late 1920s, the existence of *macromolecules* (a term coined by Staudinger in 1923) was almost universally denied. The occurrence in solution of high molecular weight species had long been recognized for natural polymers, such as rubber and proteins. The conventional wisdom, however, was that these were micellar in character: high molecular weight aggregates were believed to be formed from low molecular weight molecules, which were often assumed to be cyclic and unsaturated in character. According to the micellar theory, the aggregated molecules were physically associated, perhaps by Thiele's partial valencies, rather than being chemically (i.e. covalently) bonded as Staudinger maintained.

Although the idea of polymer molecules was not entirely new, having had limited currency from 1860 onwards, chemists were most reluctant to accept the concept of giant molecules. It took Staudinger over a decade to adduce the experimental evidence that allowed the concept of large molecules to become part of the orthodoxy of the chemical sciences (Napper, 1981).

A *polymer* molecule is defined as a molecule of relatively high molecular weight consisting of regularly repeating units, or chemically similar units, connected by primary covalent bonds. The word "polymer" was coined by the Swedish chemist Berzelius in 1833. It is derived from two Greek roots, "poly" which means "many" and "meros" which implies "parts".

Copolymers

If a polymer molecule is composed of more than the minimum number of repeat units (one, in the case of an addition polymer), it is termed a *copolymer*. Copolymers may be of a *statistical (random)*, *block* or *graft* type:

```
 ... AABBABABBABABB ...            statistical
 ... AAAAAAABBBBBBB ...            block
... AAAAAAAAAAAAAA ...             graft ("comb")
       B              B
       B              B
       B              B
       B              B
       B              B
       B              B
       .              .
       .              .
       .              .
```

The molecular weight distribution

Polymer molecules differ from minimolecules in that usually they display a distribution of molecular weights rather than a single value (see Fig. 1.4). Thus they are said to be *polydisperse* rather than *monodisperse*. The distribution of molecular weights is often characterized by two of its lower moments: the number average molecular weight $\langle M_n \rangle$ and the weight average molecular weight $\langle M_w \rangle$. These are defined by

$$\langle M_n \rangle = \sum_i n_i M_i / \sum_i n_i \tag{1.4}$$

$$\langle M_w \rangle = \sum_i w_i M_i / \sum_i w_i$$

$$= \sum_i n_i M_i^2 / \sum_i n_i M_i \tag{1.5}$$

where n_i = number of molecules of molecular weight M_i and w_i = the weight of these n_i molecules. We see that $\langle M_n \rangle$ is clearly the first moment of the distribution function (i.e. the exponent of M in the numerator of equation (1.4) is unity) whereas $\langle M_w \rangle$ is the second moment.

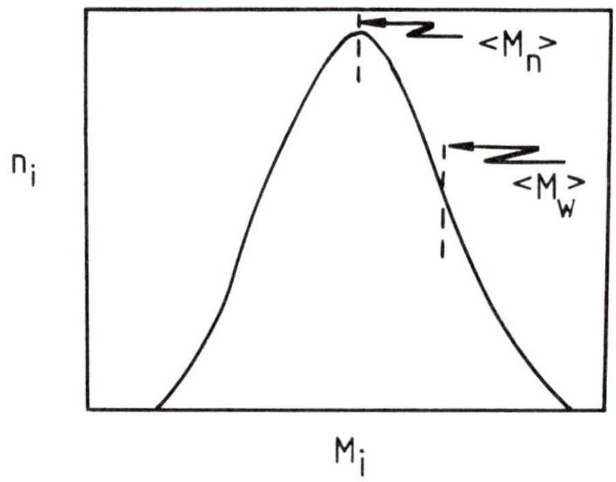

Fig. 1.4. Diagrammatic representation of the molecular weight distribution of a typical sample of polymer; n_i is the number of polymer molecules of molecular weight M_i and $\langle M_n \rangle$ and $\langle M_w \rangle$ are the number and weight average molecular weights respectively.

One polydispersity index which measures the width of the distribution function is the ratio $\langle M_w \rangle / \langle M_n \rangle$. For many addition polymers, this polydispersity index lies in the range 1·5–2·0 but other polymers, especially those prepared by Ziegler-Natta catalysts, may have a polydispersity index that is an order of magnitude greater. The polydispersity index is 1·0 for a monodisperse polymer. Anionic polymerizations can lead to polydispersity indices close to unity.

Types of polymeric stabilization

The spatial extensions of polymer molecules of even modest molecular weights are usually comparable to, or greater than, the range of the London attraction between colloidal particles. This can be illustrated as follows. The r.m.s. end-to-end distance of linear polymers can be considered crudely to be a measure of their "diameter". For many common carbon backbone polymers, this is given roughly by $\langle r^2 \rangle^{\frac{1}{2}}/\text{nm} \sim 0{\cdot}06\ \text{M}^{\frac{1}{2}}$, where M = molecular weight. Some typical chain dimensions, calculated using this formula, for polymers of differing molecular weights are presented in Table 1.3. These values are merely illustrative orders of magnitudes. The precise dimensions of polymer chains vary quite widely from one type of polymer to another at the same molecular weight, even in θ-solvents. It is none the less apparent that for high polymers, by which is meant chains with molecular weights greater than 10 000, the chain dimensions are comparable to, or in excess of, the range of the van der Waals attraction. Hence polymer molecules are of the right dimensions to be used to impart colloid stability, provided that they can generate repulsion. One major advantage of macromolecules is that their dimensions are relatively insensitive to electrolyte concentration, unlike electrical double layers.

TABLE 1.3
Typical dimensions of polymer chains of different molecular weights

M	$\langle r^2 \rangle^{\frac{1}{2}}$/nm
1 000	2
10 000	6
100 000	20
1 000 000	60

There are at present two different mechanisms whereby polymer chains can impart colloid stability: *steric stabilization* and *depletion stabilization.*

(*A*) Steric stabilization. Steric stabilization of colloidal particles is imparted by macromolecules that are attached (e.g. by grafting or by physical adsorption) to the surfaces of the particles. This is represented schematically in Fig. 1.5. Over recent years there have been an impressive number of reviews of steric stabilization to which the reader is referred (Ottewill, 1969; Napper, 1970c; Medalia, 1971; Napper and Hunter, 1972; Ottewill, 1973; Vincent, 1973a; Vincent, 1973c, 1974; Barrett, 1975; Napper and Hunter, 1975; Napper, 1977; Parfitt and Peacock, 1978; Ottewill, 1980; Tadros, 1980; Laible and Hamann, 1980; Sato and Ruch, 1980; Vincent and Whittington, 1981; Tadros, 1982). These reviews show clearly the rapid expansion in our understanding of steric stabilization in recent years. It must be acknowledged, however, that this understanding still lags far behind that of electrostatic stabilization.

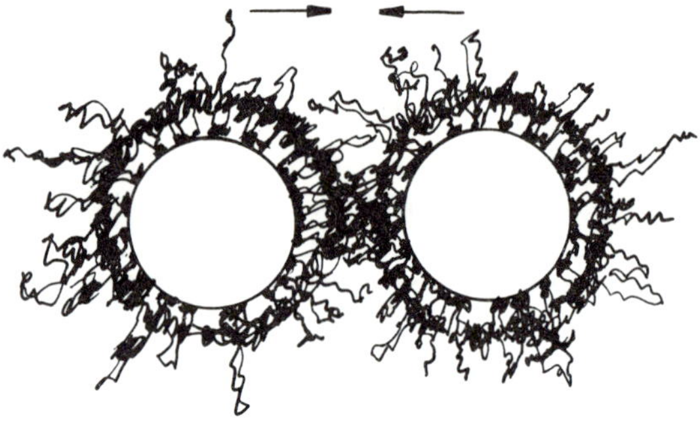

Fig. 1.5. Schematic representation of steric stabilization (not to scale).

(*B*) Depletion stabilization. This differs from steric stabilization in that stability is imparted not by attached polymer but rather by macromolecules that are free in solution (see Fig. 1.6). Investigations of this phenomenon, which was first studied experimentally in 1975 (Li-in-on *et al.*, 1975) and theoretically in 1980 (Feigin and Napper, 1980b), are still in their infancy.

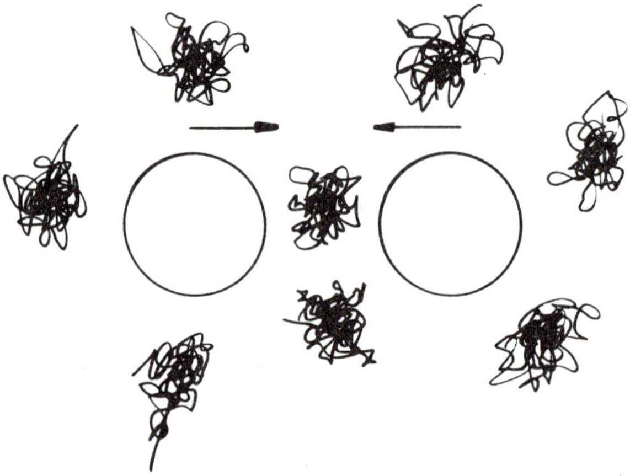

Fig. 1.6. Schematic representation of depletion stabilization (not to scale).

1.5.3. Combinations of stabilization methods

1.5.3.1. Electrosteric stabilization

It is possible to have combinations of electrostatic and steric stabilization, which has been termed *electrosteric* stabilization. The electrostatic component may originate from a net charge on the particle surface (see Fig. 1.7a) and/or charges associated with the polymer attached to the surface (i.e. through an attached *polyelectrolyte*) (see Fig. 1.7b). Electrosteric stabilization is common in biological systems.

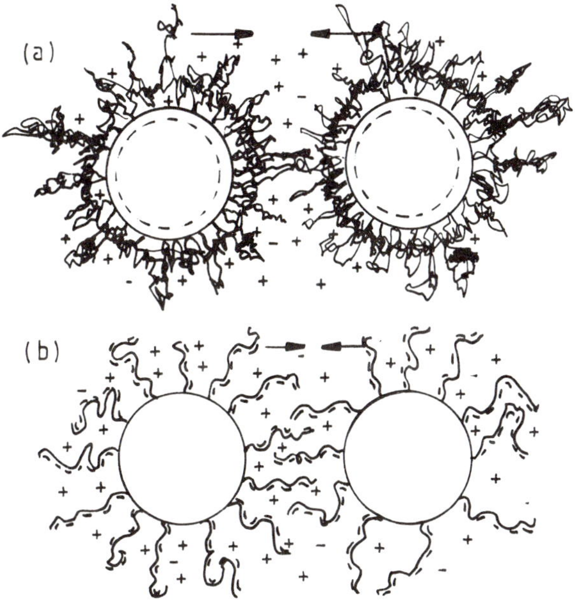

Fig. 1.7. Diagrammatic representation of electrosteric stabilization: (a) charged particles with nonionic polymers; (b) polyelectrolytes attached to uncharged particles (not to scale).

1.5.3.2. Other combinations

In addition to electrosteric stabilization, it is possible to have combinations of depletion stabilization with both steric and/or electrostatic stabilization. The combination of depletion and steric stabilization is quite common at high concentrations of free polymer in the dispersion medium.

1.5.4. Future stabilization methods

The strategies that have been devised in the past to stabilize colloidal particles are only necessary because of the attractive London interactions between the elements of matter composing the particles. It follows that the most general procedure for stabilizing colloidal particles ought to involve the "switching-off" of the van der Waals attraction. In principle this is readily achieved: all that would be required is for the electrons in one particle to be prevented from semaphoring their orbital positions to the electrons in the other. If this were accomplished, the electronic correlations that generate dispersion forces would disappear. In one qualitative sense, this idea is somewhat reminiscent of the mechanism whereby London dispersion forces are weakened by retardation. At large distances (≥ 30 nm), the two atoms are so far apart that by the time the virtual photon from one atom has polarized the other, the electronic configuration of the first atom will have changed significantly, thus weakening any correlation.

Although advances in our understanding of the origins of fundamental forces are proceeding apace, achievement of the facility whereby dispersion forces can be switched on and off at will remains very distant. In the interim, research into methods of stabilization other than those currently available, should pay handsome dividends.

1.5.5. Summary of the effects of polymer on colloid stability

Some of the effects that polymers produce on colloidal dispersions are summarized in Table 1.4. This Table illustrates dramatically the enormous complexity of the effects that polymer chains can exert on colloidal dispersions. Some 13 different phenomena are already well documented and this list is likely to be expanded in the near future as additional diverse phenomena are discovered.

The chart presented in Table 1.4 provides a road map to the structure and organization of this book. The effects of attached polymer will be discussed first, followed by consideration of the influence of free polymer. Note that an alternative schematic representation of the effects of polymeric chains on colloid stability is presented in Chapter 17, Section 17.7.

TABLE 1.4
The effects of polymers on colloidal dispersions

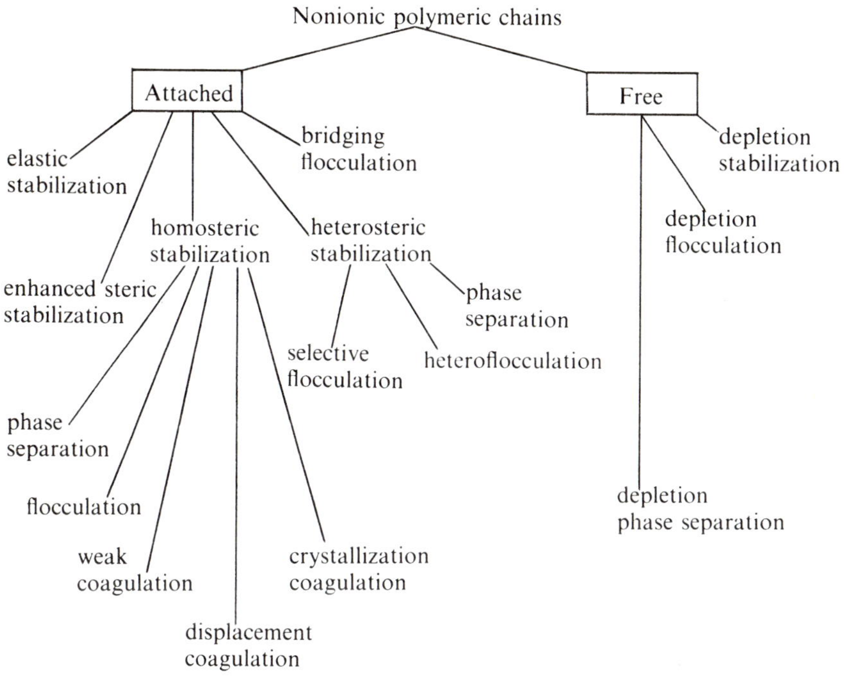

Chapter 2

Stabilization by Attached Polymer: Steric Stabilization

2.1. Early technological history

The phenomenon of steric stabilization has been exploited by man for at least five millennia. One of its earliest uses appears to have been in the preparation of ink in Ancient Egypt and China around 2500 BC (Anon, 1968a; Anon, 1968b). Ancient Egyptian ink, which revolutionized the recording of information, was extraordinarily ingenious, yet of the utmost simplicity. Particles of carbon black, which were generated by combustion, were mixed with a solution of a naturally occurring steric stabilizer. Typical of the naturally occurring steric stabilizers used were casein (from milk), albumin (from egg white) and gum arabic (a flexible polysaccharide obtained from the *Acacia* tree). The resulting carbon black dispersion was placed in a mould, shaped roughly like a modern-day pencil, and dried. When the ink was required, the stick of carbon black was simply dipped into water whereupon the particles rapidly redispersed, producing what in modern parlance would be termed "instant" ink. It was this ink that was used to make written records on papyrus, a form of early paper.

Note that the spontaneous redispersion of dried particles is a characteristic feature of most sterically stabilized systems, in marked contrast to the behaviour of many (though certainly not all) electrostatically stabilized systems. Note, too, that fresco paints were also produced by dispersing suitable pigments in water. This technology presumably had as one of its antecedents the preparation by Palaeolithic man of crude "paints" for cave paintings, an activity that chronologically preceded the preparation of ink in Egypt and China by many millennia. These "paints", which were often based on iron oxide, were, however, commonly prepared without a binder.

2.2. Practical applications

2.2.1. Technological applications

Since ancient times the phenomenon of steric stabilization has been exploited continuously in a variety of technological applications and products. These nowadays include

(i) paints, inks and glues;
(ii) photography and photocopying;
(iii) emulsions, whether they be (a) pharmaceutical, (b) food, (c) agricultural, (d) monomeric;
(iv) polymer dispersions, including latices and suspensions;
(v) detergents and other laundry products;
(vi) dry cleaning;
(vii) oils and lubricants in general;
(viii) dispersions of carbon black fillers in tyres; and
(ix) medical prostheses that have been rendered biocompatible.

2.2.2. Biological role

The widespread importance of steric stabilization in biological systems has still to be unravelled. That steric stabilization is likely to be involved in stabilizing biological particles, such as cells, stems from two recurring features of biological systems: the first is that the ionic strength of the fluid bathing such dispersions is often relatively high, thus rendering electrostatic stabilization less effective; the second is that water soluble polymers, especially polyelectrolytes, abound in nature.

One example of a biological system where steric stabilization is undoubtedly influential in controlling stability is in the occurrence of so-called casein "micelles" in milk (Payens, 1979; Darling, 1982). Payens (1979) has shown that steric stabilization must be operative in order to explain the physical properties of casein micelles. Each micelle is an aggregate of smaller particles ("submicelles") that contain various amounts of α-, β- and κ-casein. Most of the κ-casein, which is a hydrophilic, negatively charged macropeptide, resides on the surface of the micelle. Indeed, the total surface area of the micelles has been found to be directly proportional to the κ-casein concentration (Darling and Van Hooydonk, 1981).

The κ-casein provides a layer of hydrophilic polymer at the surface of the micelle that contributes to stability. Removal of the κ-casein by enzymatic hydrolysis with chymosin (rennin) results in the micelles eventually undergoing coagulation. This is the principle of cheese making.

Another pointer to the importance of steric stabilization comes from the temperature dependence of the stability of the micelles. The isoelectric point of casein occurs at pH 4·6. Darling and Dickson (1979) have shown that at temperatures below 5 °C, the casein micelles are stable at this pH. Hence a net surface charge is not essential for stability. Casein micelles are, however, usually negatively charged and electrostatic interactions must provide some contribution to their stability under normal conditions. This is an example of naturally occurring electrosteric stabilization.

2.3. The advantages of steric stabilization

Steric stabilization is widely exploited industrially and biologically because it can offer several distinct advantages over electrostatic stabilization. These will be detailed in the sections that follow.

2.3.1. Relative insensitivity to the presence of electrolytes

It was mentioned above that for 1 : 1 electrolytes, the concentration limit for the stability of an electrostatically stabilized dispersion is an ionic strength in the dispersion medium of order 10^{-1} M. This concentration is not very high: by way of specific examples, we recall that the effective ionic strength of many biological systems is comparable to this (e.g. an isotonic solution of potassium chloride is ca 0·16 M and sea water has an ionic strength of order 0·6 M).

The reason that electrostatically stabilized systems are so sensitive to electrolyte concentration has been discussed in Chapter 1, Section 1.5.2.1. Briefly, the addition of electrolyte compresses the spatial extension of the double layers to such an extent that their electrostatic repulsion is of insufficient magnitude and range to compete with the van der Waals attraction. Coagulation thereupon ensues. The dimensions of polymer chains display no such dramatic sensitivity (except near the coil-globule transition region) and sterically stabilized dispersions are relatively insensitive to the presence of electrolyte. Of course, electrolytes at high concentrations can often flocculate sterically stabilized dispersions. This, as we shall see, is almost always not the result of a *dimensional* collapse of the polymer but rather the consequence of the polymer segments undergoing a transformation from being mutually repulsive to being mutually attractive (i.e. a *thermodynamic* collapse).

2.3.2. Equal efficacy in both aqueous and nonaqueous dispersion media

A wealth of practical experience points to the conclusion that electrostatic stabilization is less effective in nonaqueous dispersion media than it is in aqueous media (Overbeek, 1982). This is primarily a consequence of the low dielectric constant (< 10) of most nonaqueous media. In contradistinction, steric stabilization is just as effective a mode of stabilization in nonaqueous media as it is in aqueous media. For this reason, steric stabilization is usually the preferred method of stabilization in nonaqueous dispersion media.

2.3.3. Equal efficacy at both high and low solids content

One advantageous feature of steric stabilization for practical applications is that it is usually equally effective at both high and low volume fractions of the dispersed phase. This would not be true, for example, for electrostatic stabilization in nonaqueous media (Albers and Overbeek, 1959; Feat and Levine, 1976). In that case, the spatial extension of the double layers is so great, due to the low dielectric constant of the dispersion medium, that the mere act of preparing high solids dispersions forces the particles close together. The net result is that little repulsion remains between their double layers and the particles coagulate.

In aqueous dispersion media, the preparation of high solids dispersions that are stabilized electrostatically is often hindered by the gel-like nature of the product. This is a consequence of the interactions between the double layers surrounding each particle. No comparable increase is evident in the viscosity of sterically stabilized dispersions at high solids content. Of course, in certain applications, such as in paints, the rheological properties exhibited by electrostatically stabilized dispersions at high solids may be a decided advantage rather than a drawback.

2.3.4. Reversibility of flocculation

If a polystyrene latex that is stabilized *solely* by an electrostatic mechanism is coagulated by the addition of electrolyte, that coagulation is usually irreversible to subsequent dilution. In contrast, sterically stabilized dispersions can usually be flocculated by the addition to the dispersion medium of a nonsolvent for the stabilizing moieties; mere dilution of the concentration of the nonsolvent to a suitably low value is often sufficient to induce the particles to redisperse spontaneously.

This difference in behaviour between electrostatically and sterically stabilized dispersions probably springs from the fact that whereas sterically stabilized particles may be thermodynamically stable, dispersions that are electrostatically stabilized are only thermodynamically metastable. In the latter case, the coagulated state is the lowest energy state and, once achieved, it is only reversed if there is an input of work into the system. It must be remembered, however, that there exist many electrostatically stabilized dispersions that undergo reversible coagulation (e.g. clays, Carey Lea silver sols). Indeed, Overbeek (1977) has argued that repeptization is the rule rather than the exception. It appears to be associated with redispersion from a shallow minimum (which may be significantly greater than zero) in the potential energy curve, rather than from a deep primary minimum.

One important consequence of the thermodynamic stability of sterically stabilized dispersions is that they redisperse spontaneously after drying. Mention has already been made of the "instant" ink of the Ancient Egyptians. More recently, Everett and Stageman (1978a) have demonstrated the spontaneous redispersion in *n*-alkanes of freeze dried poly(methyl methacrylate) particles, coated by poly(dimethylsiloxane).

Sterically stabilized dispersions also often display good freeze-thaw stability, which can be a desirable attribute in some practical applications (e.g. paint systems).

2.4. The early studies of steric stabilization

2.4.1. Aqueous dispersions

2.4.1.1. Biopolymers

The honour of conducting the first properly-designed scientific investigation into steric stabilization appears to belong to Michael Faraday (1857; see also Hatschek, 1925). Faraday generated gold sols with a "beautiful ruby" appearance by reducing gold chloride with phosphorus. He observed that the addition of sodium chloride to the sol resulted in a red to blue colour change, which he clearly recognized as being a consequence of the aggregation of the gold particles. He further found that the red to blue colour transformation could be induced by heating.

Faraday also experimented with the effects of gelatin on the red to blue colour change that was stimulated by the addition of common salt. He reported

> "If a trace of phosphorus in sulphide of carbon be added to the solution of gold in a dilute state, and *some salt be added to the warm jelly*, and the latter be then mixed gradually and with agitation with the gold solution, a ruby jelly is generally produced."

Here Faraday is clearly announcing the discovery of the phenomenon that subsequently was termed *protective action*. This is the ability of soluble macromolecules to protect a sol against coagulation by an electrolyte.

The production in these experiments of a ruby jelly, apparently uncoagulated even in the presence of considerable amounts of salt, led Faraday to postulate that the particles were associated with envelopes of "that animal substance". This clearly recognizes, albeit in a primitive way, the importance of the adsorption process in conferring protection.

It was fully forty years after Faraday's seminal experiments that Lottermoser and von Meyer (1897) also recognized that the presence of proteins could shield silver sols from coagulation by the addition of electrolyte. Their observations were greatly expanded several years later by Zsigmondy (Zsigmondy, 1901; Zsigmondy and Schulz, 1902), who reverted to studying red gold sols. Such sols are especially suitable for this type of experiment because, as mentioned in connection with Faraday's experiments, coagulation is signalled by a vivid red to blue colour change. Zsigmondy determined the *gold number* of various biopolymeric stabilizers: this he defined as the amount (expressed in milligrams) of polymer which when added to a sample of gold sol (10·0 cm^3) just prevents its flocculation on the addition of 1·0 cm^3 of a 10% sodium chloride solution. Subsequently, Wolfgang Ostwald (1919) extended this notion to the use of a dispersion of an organic substance (congorubin) instead of a gold sol whilst changing the coagulating electrolyte to potassium chloride. Some typical gold and rubin numbers are shown in Table 2.1.

Clearly, the smaller is the gold or rubin number, the more effective is the biopolymer at shielding the dispersion from coagulation by electrolytes. To describe this shielding phenomenon, Zsigmondy coined the term *protective colloid* ("Schutzkolloide"), which functions by protective action. From the results recorded in Table 2.1, it can be inferred that sodium caseinate is considerably more effective in its protective action than is, say, dextrin or potato starch. Note that both of these latter polymers are *nonionic* in character.

The protective action of biopolymers was recognized as being a consequence of the adsorption of the macromolecules on the surfaces of the particles. The biopolymers thus envelop the particles, transforming their surfaces from being hydrophobic to being hydrophilic. Hydrophilic colloids, as these biopolymers were commonly termed, were known to be intrinsically less sensitive to the presence of electrolytes than are hydrophobic colloids. This was attributed simply to the hydration of the protective agents, a theory

TABLE 2.1
Gold and rubin numbers for various biopolymers

Biopolymer	Gold number ($mg/10\ cm^3$)	Rubin number ($g/10^2\ cm^3$)
gelatin	0·005–0·01	0·025
sodium caseinate	0·01	0·004
albumin	0·1–0·2	0·008
gum arabic	0·15–0·25	—
dextrin	6–20	—
potato starch	$\sim$25	$\sim$0·2

that was proposed as early as 1900 by Spring and which has its adherents even today.

The protective action of a polymer was found to increase monotonically with increasing concentration of the macromolecule, as expected from the foregoing concepts. Electrophoretic measurements demonstrated conclusively that the protective action did not depend upon a strong electrostatic charge being imparted to the colloidal particles (Freundlich, 1926). Freundlich (1926) conjectured that the strength and toughness of the protective envelopes, together with the manner in which they are anchored to the colloidal particles, played a major role in determining the protective action. Furthermore, he surmised that there was a connection between the rigidity of the protective agent and its efficacy as a protective agent.

2.4.1.2. Synthetic polymers

By the end of the 1940s, two classical principles had been firmly established for stabilizing hydrophobic sols in water. First, stability could be imparted by the adsorption of ionic species on to the surfaces of the particles, thus generating electrostatic stabilization. Second, adsorption of relatively rigid, nonionic macromolecules (e.g. natural or synthetic polysaccharides) on to the surfaces of the particles could also impart stability because the particles were then kept apart by these rigid, hydrophilic "distance keepers". For some protective agents, such as globular proteins, both principles might be expected to apply.

Following the adsorption studies of Heller and Tanaka (1951), Heller and Pugh (1954) proposed that *flexible* macromolecules could also be useful in imparting colloid stability. The wisdom conferred by hindsight enables us to assert that in one sense, this proposal was not novel: we now know that most biopolymers in solution are not absolutely rigid structures and that the protective action demonstrated in some of Zsigmondy's experiments (e.g. by dextrin or potato starch) was a consequence *inter alia* of the flexibility of the macromolecular structures. Heller and Pugh, however, were correct in their desire to differentiate between the actions of biopolymers and those of synthetic polymers because many naturally occurring polymers, especially proteins and polynucleotides, exhibit secondary, tertiary, and even quaternary structures in their native state in solution. Synthetic polymers, with some notable exceptions, are normally devoid of these intricacies of molecular architecture.

Heller and Pugh (1954) used a slight modification of Zsigmondy's gold number experiment to measure the effect that four samples of poly(oxyethylene) had on the stability of a Faraday gold sol when potassium chloride was added. Their results are shown in Table 2.2. Listed therein are the values of the critical concentration of potassium chloride (c^*_{KCl}), required to induce the red→blue transformation in the gold sol in the presence of the polymer at a given concentration ($1{\cdot}0$ g dm^{-3}). This concentration of polymer

TABLE 2.2
The protective action of poly(oxyethylene)

	No polymer	Molecular weight of polymer 600	4 000	6 000	9 000
$10^3\ c^*_{KCl}$/M	12·5±2·5	18·0±2·5	22±5	47±5	480±30

was apparently chosen because it was near the plateau region of the adsorption isotherm.

Two important conclusions emerge from the data of Heller and Pugh. First, there is no doubt that poly(oxyethylene) can protect the gold sol from coagulation by electrolyte. Second, the protective action is enhanced by increasing the molecular weight of the polymer.

2.4.2. Nonaqueous dispersions

To this point, our discussion of the protective action of nonionic stabilizing molecules has been concerned exclusively with hydrosols. Realization that stabilization could similarly be imparted to colloidal particles in nonaqueous media was somewhat later in coming. This comment disregards, of course, the pragmatic technological application of the phenomenon in such products as paints, etc.

Rehbinder *et al.* (1930) were among the first to show that long-chain carboxylic acids could generate stable dispersions in benzene. Subsequently, Verwey and de Boer (1938) gave an impressively modern account of the stabilization of several powders, such as iron, by oleic acid in nonaqueous dispersion media (e.g. carbon tetrachloride, benzene and *n*-hexane). In hindsight we can see that their preparation of dispersions containing a high (80%) solids content is convincing evidence that a protective stabilization mechanism was operative in their experiments. Electrostatic stabilization is ineffectual at high solids content in dispersion media of low dielectric constant, as was mentioned previously, because the double layers are too highly extended. The very act of preparing a high solids dispersion then overcomes most of the interparticle electrostatic repulsion, resulting in coagulation in the absence of another stabilizing mechanism.

Verwey and de Boer surmised that the particles were surrounded by an oriented layer of oleic acid molecules. These had their polar carboxylic acid groups adsorbed at the surfaces of the particles and their nonpolar tails oriented toward the nonaqueous dispersion medium. Verwey and de Boer even represented the particles (see Fig. 2.1) by a schematic diagram that would be immediately recognizable today as depicting steric stabilization.

The stabilization of dispersions by nonionic molecules seems to have been put to one side by the Dutch school of colloid scientists until van der Waarden (1950; 1951) revived interest in it. He showed that carbon black particles could be stabilized in hydrocarbon media by aromatic hydrocarbons provided that they were substituted with aliphatic chains of sufficient length. Aromatic molecules *per se* did not impart stability, neither did aromatic molecules substituted with very-short alkyl chains. The efficacy of the alkyl tail in imparting stability increased with increasing chain length (this was studied only up to C_{16}). The presence of two alkyl tails instead of only one also improved the effectiveness of stabilization. van der Waarden clearly recognized that the aromatic part of the molecule interacted with the polar groups on the surface of the carbon black, thus anchoring the alkyl tail to the surface.

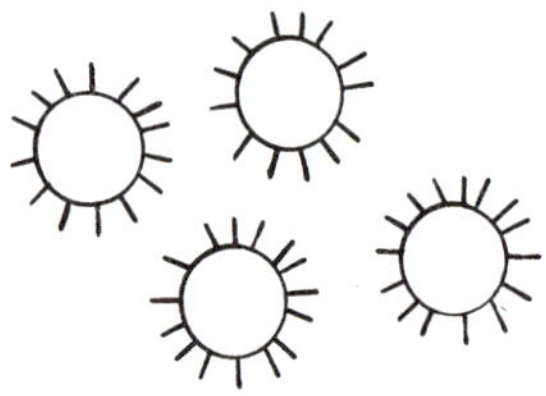

Fig. 2.1. Diagrammatic representation of the stabilization of colloidal particles by oleic acid in nonaqueous media as depicted by Verwey and de Boer (1938).

Koelmans and Overbeek (1954) later came to the conclusion that adsorbed nonionic molecules cannot impart stability to coarse particles unless the length of the stabilizer is comparable to the particle size. This theoretical conclusion, which now must be severely modified, was based on an assumed value for the Hamaker constant ($\sim 25\,kT$) that is far too large, perhaps by as much as an order of magnitude, for colloidal particles composed of organic material. These authors argued that quite modest electrical charges were sufficient to stabilize even coarse particles in dispersion media of low dielectric constant. They recognized, however, in light of van der Waarden's experiments, that very small colloidal particles could be stabilized by nonionic molecules.

2.5. The misnomer "steric stabilization"

Heller and Pugh (1954) correctly recognized that the stability imparted to colloidal particles by naturally occurring polymers often contains both electrostatic and polymeric components. Nonionic macromolecules, such as poly(oxyethylene), can only impart stability by virtue of their polymeric character. To distinguish this type of protection by nonionic polymers from protection in general, Heller and Pugh coined the phrase *steric protection*. This was later transformed into *steric stabilization* (Heller, 1966).

We note parenthetically that Koelmans and Overbeek (1954) used the term "steric hindrance" stabilization contemporaneously with Heller and Pugh to refer to stabilization in nonaqueous dispersion media by *n*-alkyl tails.

In the foregoing terminology, classical protective action is, in general, a combination of electrostatic plus steric stabilization. Heller and Pugh (1956) subsequently showed that the polyelectrolyte, sodium poly(*p*-styrene sulphonate), which would be expected to function in a likewise manner, was a very effective protective agent for iron oxide (α-FeOOH) sols, generating stability greatly in excess of that achieved by nonionic macromolecules. This is another example of electrosteric stabilization.

The choice of the term "steric" by Heller and Pugh (1954) was singularly unfortunate. The word "steric" has definite connotations in organic chemistry: "steric" interactions refer to the electron-electron (and nucleus-nucleus) repulsions that occur between nonbonded, and usually bulky, substituents that exist in close proximity to one another in a molecule. They can be considered to be covered by the general term *Born repulsion*. As such, they are very-short range in character (extending only a few tenths of a nanometre) and can be traced back to their primordial source, the operation of the Pauli exclusion principle.

The steric effects to be discussed in this book have little in common with the steric interactions of organic chemistry. They are not short range in character; nor do they derive, at least directly, from the operation of the Pauli exclusion principle. They have, in fact, a general thermodynamic origin. In this respect, the term "steric stabilization" is an unfortunate misnomer. The use of this terminology is now widespread. The history of science points to the inescapable conclusion that once a term has achieved common usage, it is already too late to rectify any such aberration in terminology. Continued use of the expression "steric stabilization" will do no serious harm provided that its disjunction from the steric effects of classical organic chemistry is constantly borne in mind.

Scheutjens and Fleer (1982) have recently suggested that the term *adsorption stabilization* replace steric stabilization. Adoption of this recommendation would be ill-advised because the stabilizing moieties that impart stability do not have to be adsorbed at the surface of the particle, merely attached (e.g. terminally). Finally, we remark parenthetically that it may prove possible in the future to impart colloid stability by a genuine steric mechanism (i.e. by the direct operation of the Pauli exclusion principle). Indeed, biopolymers with a rigid structure (e.g. helices) when attached to colloidal particles may conceivably function in this way. Additionally, certain lipid films may well exhibit some facets of bona fide steric stabilization.

2.6. The best steric stabilizers

The best steric stabilizers are found empirically to be amphipathic block or graft copolymers. One of the comonomers should generate a homopolymer that is nominally insoluble in the dispersion medium whereas the other should polymerize to form a polymer that is soluble in the dispersion medium.

The idea that the best steric stabilizers are amphipathic in character has long been recognized by manufacturers of commercial nonionic surfactants. These are classified by the empirical HLB (an acronym for *hydrophilic-lipophilic balance*) scheme which endeavours to scale the relative solubilities of the two contrasting components in aqueous and nonpolar dispersion media.

Figure 2.2 portrays the reason why amphipathic molecules are such effective steric stabilizers. It would be expected that the polymer that was nominally insoluble in the dispersion medium would attach itself to the particle. This attachment to a colloidal particle could occur by physical adsorption on to a preformed colloidal particle, or by physical and/or chemical adsorption on to (or even incorporation into) a growing particle. Whatever the mechanism of attachment, the nominally insoluble polymer serves to anchor the soluble moieties to the colloidal particles. Such polymer is accordingly referred to as the *anchor polymer*. The role of the soluble polymer is to impart steric stabilization and, for this reason, such chains are termed the *stabilizing moieties*.

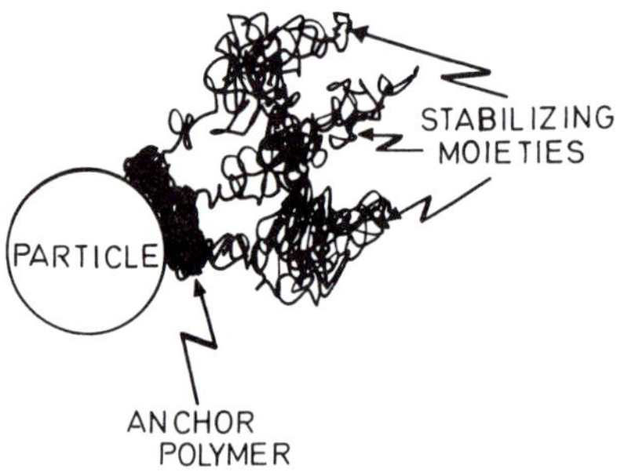

Fig. 2.2. Schematic representation of the steric stabilization of a colloidal particle by an amphipathic block copolymer (only some of the stabilizing chains are shown).

The effectiveness of amphipathic polymer molecules in imparting steric stabilization can be understood if a second identical particle is imagined to approach the one portrayed in Fig. 2.2. The stabilizing moieties must be mutually repulsive if the polymer is to impart stability. In these circumstances, the Brownian collision stresses the stabilizing molecules, which endeavour to escape from the stress zone. This escape can be effected either by desorption from the particle surface or by lateral movement over the particle surface away from the stress zone. Effective anchoring prevents the desorption of the stabilizing chains. To circumvent lateral movement of the chains away from the stress zone, it is necessary to ensure that the surfaces of the particles are fully coated. Complete surface coverage means that a stabilizing chain that is endeavouring to escape laterally is prevented from so doing by repulsion from its neighbours.

Virtually any polymer that is appropriately soluble in the dispersion medium is effective as a stabilizing moiety and any polymer that is insoluble in the dispersion medium is effective as an anchor polymer. Some typical stabilizing moieties and anchor polymers for aqueous and nonaqueous dispersion media are shown in Table 2.3.

Random copolymers are usually less effective as steric stabilizers than block or graft copolymers, although it should not be overlooked that under certain conditions they are able to confer enhanced steric stabilization (see Chapter 9).

Homopolymers are also able to impart steric stabilization, although often the stability so produced is but a pale shadow of that conferred by

TABLE 2.3
Typical stabilizing moieties and anchor polymers for sterically stabilized dispersions

Aqueous dispersions	
Anchor polymer	Stabilizing moieties
polystyrene	poly(oxyethylene)
poly(vinyl acetate)	poly(vinyl alcohol)
poly(methyl methacrylate)	poly(acrylic acid)
poly(acrylonitrile)	poly(methacrylic acid)
poly(dimethylsiloxane)	poly(acrylamide)
poly(vinyl chloride)	poly(vinyl pyrrolidone)
poly(ethylene)	poly(ethylene imine)
poly(propylene)	poly(vinyl methyl ether)
poly(lauryl methacrylate)	poly(4-vinylpyridine)
Nonaqueous dispersions	
Anchor polymer	Stabilizing moieties
poly(acrylonitrile)	polystyrene
poly(oxyethylene)	poly(lauryl methacrylate)
poly(ethylene)	poly(12-hydroxystearic acid)
poly(propylene)	poly(dimethylsiloxane)
poly(vinyl chloride)	poly(isobutylene)
poly(methyl methacrylate)	*cis*-1:4-poly(isoprene)
poly(acrylamide)	poly(vinyl acetate)
	poly(methyl methacrylate)
	poly(vinyl methyl ether)

amphipathic copolymers. Anchoring of a homopolymer, however, can be effective if the surfaces of the colloidal particles contain sites with which the homopolymer can interact specifically. It might be argued in support of the case for homopolymeric steric stabilizers that one of the most effective, and industrially most important, steric stabilizers for aqueous dispersions is poly(vinyl alcohol). It should be remembered, however, that this polymer is commonly prepared industrially by alcoholysis of poly(vinyl acetate) and is usually only partially hydrolysed (of order 88% or greater). It is therefore a copolymer of poly(vinyl acetate) and poly(vinyl alcohol), often with considerable block copolymer character.

Note that it is possible to eliminate the need for the anchor polymer if the stabilizing moieties can be covalently attached to the surface of the colloid.

2.7. The preparation of sterically stabilized dispersions

The preparation of nonaqueous sterically stabilized polymer latices was pioneered by Osmond and coworkers at ICI Paints Division. Their work is summarized in a comprehensive monograph (Barrett, 1975) that provides actual recipes for the successful generation of nonaqueous dispersions. Only the most general considerations need therefore be mentioned here.

Two general strategies present themselves for the preparation of sterically stabilized dispersions: the addition of a steric stabilizer to a preformed dispersion or the generation of the dispersion in the presence of stabilizer. The former method is really only satisfactory for aqueous dispersions that can be stabilized by an alternative method (e.g. electrostatically) during preparation. Latices are usually prepared by the latter of these two general methods. In that case, two options are open in the preparative step: a preformed steric stabilizer can be added to the system or a stabilizer precursor can be present which generates the whole stabilizer *in situ* as polymerization proceeds. The disadvantage of the latter method is that the rate of production of the whole stabilizer, which is sometimes difficult to control, must be adequately geared to the rate of generation of the latex particles.

An important recent development in the preparation of sterically stabilized dispersions is the grafting technique disclosed by Vincent and coworkers (Bridger *et al.*, 1979; Bridger and Vincent, 1980; Clarke and Vincent, 1981a) for silica particles. Basically, the silica particles are heated *in vacuo* at 700 °C to form activated siloxane bridges which can be reacted with suitably-capped stabilizing moieties, or ring-opened by reaction with polymeric carbanions. This chemical grafting obviates the need for an anchor polymer but can suffer from the disadvantage that the linking group (e.g. the carbamate of a urethane linkage) may undergo hydrolysis. Moreover, silica particles can undergo surface dissolution in aqueous dispersion media that can lead to desorption of the stabilizer on standing, resulting in coagulation.

Chapter 3

Polymer Solution Thermodynamics

3.1. Introduction

Any discussion of the role played by polymers, whether free or attached, in stabilizing colloidal dispersions must inevitably be couched in terms of polymer solution thermodynamics. The earliest, and still the best known, theory of polymer solution thermodynamics was that set forth by Flory (1941; 1942) and Huggins (1941; 1942a,b). Despite its serious inadequacies, which will be detailed later, the Flory-Huggins theory has had such a profound impact on polymer science that Manfred Gordon has properly referred to it as a "paradigm of polymer science". It embodies concepts that remain central to the theory of polymer solution thermodynamics whilst ignoring certain features that are vital to the explanation of the entire spectrum of observed phenomena.

The free volume theory, which was developed subsequently by Flory and coworkers, is much more elaborate in its innate construction. It attempts to rectify the omissions of the Flory-Huggins theory whilst retaining certain important elements therefrom. Whereas the Flory-Huggins theory has found wide application in the development of theories of polymeric stabilization, the free volume theory has not yet proved to be quite so productive, although it remains crucial to the understanding of the effects of temperature. Both theories will therefore be outlined in this chapter.

3.2. The Flory-Huggins theory

The Flory-Huggins theory aims at predicting the Gibbs free energy of mixing of pure polymer with pure solvent. This it does by separate calculation of the entropy of mixing (a combinatorial term) and the enthalpy of mixing. The two terms are then combined in the usual manner:

$$\Delta G^M = \Delta H^M - T\Delta S^M \tag{3.1}$$

where the superscript M denotes mixing.

3.2.1. The combinatorial entropy of mixing

3.2.1.1. The lattice approach

The Flory-Huggins theory as conceived by Flory is a lattice theory, although, as we shall see, this happens not to be absolutely necessary. Both the polymer segments and the solvent are placed on a grid in 3-D space. The lattice sites are all assumed to be of the same size, which we will take to be that of the solvent molecules. The polymer chains, taken to be monodisperse, are all assumed to occupy x contiguous lattice sites. Note that x is not in general equal to the degree of polymerization. The configurational entropy can be calculated from the number of different ways by which this placement can be accomplished, since the solvent molecules simply fill in one possible way the cells unoccupied by the polymer segments (see Fig. 3.1).

Consider the mixing of n_1 solvent molecules and n_2 polymer molecules, the subscripts 1 and 2 denoting solvent and polymer respectively. The total number of lattice sites n_0 is given by

$$n_0 = n_1 + xn_2 \tag{3.2}$$

Suppose i molecules have been added to the lattice. We consider the number of different ways that the $(i+1)$th molecule can be sited on the lattice. The first segment can be placed on any one of $(n_0 - xi)$ unoccupied sites. The next segment, which is connected to the first, can be placed in $z(1-f_i)$ lattice sites, where z is the coordination number of the lattice and f_i is the probability of a site being filled when at least one adjacent cell is empty. The third segment can be located in $(z-1)(1-f_i)$ different ways, since the second segment necessarily

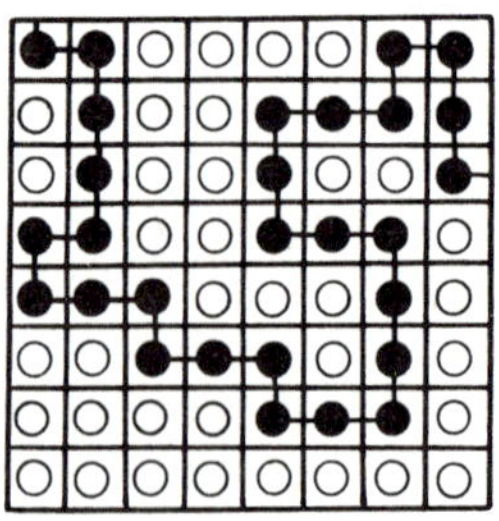

Fig. 3.1. Placement of polymer segments and solvent molecules on the lattice sites in accordance with the precepts of the Flory-Huggins theory.

occupies one of the cells contiguous to that containing the third segment. All subsequent segments have the same probability as the third, end effects having now been eliminated.

The total number of ways by which the $(i+1)$th molecule can be placed on the lattice is thus

$$\begin{aligned} A_{i+1} &= (n_0 - xi)z(1-f_i)(z-1)^{x-2}(1-f_i)^{x-2} \\ &\simeq (n_0 - xi)(z-1)^{x-1}(1-f_i)^{x-1} \end{aligned} \tag{3.3}$$

To find f_i, it is necessary to assume that the solution is *moderately concentrated* so that f_i can be replaced by its average value $(n_0 - xi)/n_0$. Under this assumption,

$$\begin{aligned} A_{i+1} &= (n_0 - xi)^x (z-1)^{x-1}/n_0^{x-1} \\ &\sim \frac{(n_0 - x[i-1])!}{(n_0 - xi)!}\left(\frac{z-1}{n_0}\right)^{x-1} \end{aligned} \tag{3.4}$$

If all of the polymer molecules were distinguishable, the total number of ways Ω_f in which the polymer molecules could be mixed with the solvent would be

$$\Omega_f = \prod_{i=1}^{n_2} A_i \tag{3.5}$$

where Π denotes the product. This overestimates Ω_f by $n_2!$ since the n_2 polymer chains are in reality indistinguishable. Therefore

$$\begin{aligned} \Omega_f &= \left(\prod_{i=1}^{n_2} A_i\right)/n_2! \\ &= \frac{n_0!}{n_1!n_2!}\left(\frac{z-1}{n_0}\right)^{n_2(x-1)} \end{aligned} \tag{3.6}$$

The first term $n_0!/n_1!n_2!$ is the expression that is obtained on mixing n_1 and n_2 equal sized minimolecules; indeed, Ω_f reduces to this expression if $x=1$. The components of the second term, which allow for the connectivity of the segments, can be interpreted as follows: $(z-1)^{n_2(x-1)}$ is the total number of internal configurations of sets of n_2 chains well separated from one another; the factor $n_0^{n_2(x-1)}$ gives rise to the drastic reduction in Ω_f due to the restrictions on the placement of the segments on a lattice which allows only single occupancy of each cell.

If we consider the mixing of pure crystal-like, perfectly ordered polymer with pure solvent, the entropy change involved is given by

$$\Delta S = k\ln\Omega_f = k\ln\left\{\frac{n_0!}{n_1!n_2!}\left(\frac{z-1}{n_0}\right)^{n_2(x-1)}\right\} \tag{3.7}$$

Applying the Stirling approximation ($\ln N! \simeq N \ln N - N$), we obtain

$$\Delta S = -k\left\{ n_1 \ln\left(\frac{n_1}{n_1 + xn_2}\right) + n_2 \ln\left(\frac{n_2}{n_1 + xn_2}\right) - n_2(x-1)\ln([z-1]/e)\right\} \tag{3.8}$$

The process of mixing can be considered to take place in two distinct steps: first, disorientation of the polymer segments in the absence of solvent; second, mixing of solvent with the disoriented polymer. The entropy associated with the disorientation (ΔS^{DO}) is obtained by setting $n_1 = 0$ in equation (3.8). This eliminates the mixing step whence

$$\Delta S^{DO} = k\{n_2 \ln x + n_2(x-1)\ln([z-1]/e)\} \tag{3.9}$$

The entropy of mixing disoriented (amorphous) polymer with solvent is found from

$$\Delta S^{M} = \Delta S - \Delta S^{DO}$$
$$= -k\left\{ n_1 \ln\left(\frac{n_1}{n_1 + xn_2}\right) + n_2 \ln\left(\frac{xn_2}{n_1 + xn_2}\right)\right\} \tag{3.10}$$

Now the volume fraction of polymer is given by

$$v_2 = \frac{xn_2}{n_1 + xn_2} \tag{3.11}$$

and that of solvent by

$$v_1 = \frac{n_1}{n_1 + xn_2} \tag{3.12}$$

These relationships permit the entropy of mixing to be expressed in the especially simple and revealing form:

$$\Delta S^{M} = -k\{n_1 \ln v_1 + n_2 \ln v_2\} \tag{3.13}$$

Note that the Flory-Huggins expression for the entropy of mixing of polymer and solvent corresponds to *volume fraction* statistics. This should be compared with the analogous *mole fraction* statistics that are exhibited by ideal minimolecules in the Bragg-Williams approximation. Mole fraction statistics are inappropriate to polymer-solvent systems because the disparity in molecular weights means that the mole fraction of solvent is always close to unity, except at extremely high polymer volume fractions.

Note that the only entropy change taken into account above is that associated with combinatorial effects. Any specific interactions between the polymer and solvent are ignored.

3.2.1.2. Free volume approach to the combinatorial entropy

The combinatorial entropy of mixing can be more readily derived by a free volume approach which renders the assumptions inherent in the Flory-Huggins theory more transparently obvious. Anticipating what is to be presented in Section 3.3, *vis-à-vis* the equation-of-state theory, we present a brief account of this alternative derivation.

The free volume of a liquid can be crudely considered to represent the volume of a system not occupied by the geometrical volumes of the constituent molecules. It is therefore space accessible to the centres of mass of the liquid molecules. If the free volume is some fraction v_f of the total (external) volume of a liquid, then the total free volume of the solvent is just $n_1 V_1 v_f$, where V_1 = measured external molecular volume ($V_1 = \bar{V}_1/N_A$, where $\bar{V}_1$ = partial molar volume). Suppose the (unphysical) assumption is made that the same free volume fraction v_f applies to the polymer in the amorphous state as to the solvent. This implies that the accessible free volume for the polymer is $n_2 V_2 v_f$. It is next assumed that on mixing, the free volume is conserved, i.e. no volume change accompanies mixing. The total accessible free volume after mixing is thus $(n_1 V_1 + n_2 V_2) v_f$.

The thermodynamic probability W in the Boltzmann equation $S = k \ln W$ may be taken to be proportional to the accessible free volume. Thus the entropy change of the solvent on mixing is simply $n_1 k \ln \{(n_1 V_1 + n_2 V_2)/n_1 V_1\} = -n_1 k \ln v_1$, with an analogous expression ($-n_2 k \ln v_2$) for the polymer. Summation of these two terms leads immediately to equation (3.13).

The preceding derivation makes clear, in the context of the free volume approach, what assumptions are necessary to derive the Flory-Huggins combinatorial entropy of mixing. Specifically, these are that the solvent and polymer possess identical free volume fractions and that the total free volume is conserved on mixing.

3.2.2. The enthalpy of mixing

The enthalpy of mixing is introduced into the Flory-Huggins theory in the spirit of the classical van Laar approach. The mixing can be considered as a quasichemical reaction between solvent contacts and segment contacts:

$$1{-}1 + 2{-}2 \rightarrow 2(1{-}2)$$

The change in internal energy for the formation of one set of 1—2 contacts is simply

$$\Delta u_{12} = u_{12} - (\tfrac{1}{2})(u_{11} + u_{22}) \tag{3.14}$$

where u_{ij} = internal energy of a contact between i and j. Suppose that p_{12} such sets are formed. The overall change in internal energy on mixing is

$$\Delta U^M = p_{12}\,\Delta u_{12} \tag{3.15}$$

In the Flory-Huggins theory, it is assumed that all volume changes that occur on mixing can be ignored so that

$$\begin{aligned}\Delta H^M &= \Delta U^M + \Delta(PV) \simeq \Delta U^M \\ &= p_{12}\,\Delta u_{12}\end{aligned} \tag{3.16}$$

To find ΔH^M, it is necessary to evaluate p_{12} in a solution of given composition. Each polymer segment has two connected segmental neighbours, leaving $(z-2)$ adjacent lattice sites that could be filled by solvent or polymer. The maximum number of segment-solvent contacts per polymer molecule is thus $(z-2)x+2 \simeq xy$, the 2 being added here to account for the two ends. Of course, in a polymer solution, some of these contacts must be with other polymer segments. The probability of a given cell being occupied by solvent is just v_1, so that for each polymer molecule the number of segment-solvent contacts is just $v_1 xz$. Therefore, for n_2 molecules, $p_{12} = n_2 v_1 xz$. Accordingly

$$\Delta H^M = n_2 v_1 xz \Delta u_{12} \tag{3.17}$$

The lattice parameters x and z are eliminated as follows. First, equation (3.11) for v_2 shows that $xn_2v_1 = n_1v_2$. Second, a dimensionless interaction parameter χ_1 is introduced such that

$$\chi_1 = z\Delta u_{12}/kT \tag{3.18}$$

The subscript 1, referring to the solvent, is used here because $\chi_1 kT$ is just the difference in energy of a solvent molecule immersed in pure polymer compared with that in pure solvent. Note that χ_1 is a pivotal parameter in Flory-Huggins theory. With these relationships, the enthalpy of mixing can be expressed as

$$\Delta H^M = n_1 v_2 \chi_1 kT \tag{3.19}$$

3.2.3. The free energy of mixing

The free energy of mixing can be found by combining the enthalpy and entropy of mixing according to equations (3.1), (3.13) and (3.19):

$$\Delta G^M = kT\{\underbrace{n_1 \ln v_1 + n_2 \ln v_2}_{\text{Combinatorial}} + \underbrace{n_1 v_2 \chi_1}_{\text{Contact dissimilarity}}\} \tag{3.20}$$

This is the celebrated Flory-Huggins expression for the free energy of mixing of solvent with amorphous polymer. It is composed of the two combinatorial entropy terms and a contact dissimilarity (enthalpy) term. As χ_1 is normally positive (i.e. mixing is endothermic) for nonaqueous systems, the contact dissimilarity terms opposes mixing. Both of the combinatorial entropy terms promote mixing but the number of polymer molecules n_2 is comparatively small because of the high molecular weight of the polymer. The dominant term promoting mixing is therefore $n_1 \ln v_1$.

The Flory-Huggins equation is interesting on several scores. First, it contains no lattice parameters. Second, the dominant term is $n_1 \ln v_1$, which physically means that the mixing of the polymer and solvent occurs because of the additional space available to the solvent when the domains of the polymer molecules become accessible to the solvent. The detailed structure of the polymer is irrelevant, rods being just as effective in providing space for the solvent molecules as polymer coils. This point is stressed because it is sometimes claimed that polymer molecules dissolve as a result of their increase in configurational entropy in solution. This is contrary to the precepts of the Flory-Huggins theory.

3.2.4. The chemical potentials

Differentiation of the Flory-Huggins expression for the free energy of mixing yields the chemical potentials of the solvent and the polymer segments. Since the Gibbs free energy of mixing is given by

$$\Delta G^M = G_{12} - G_1 - G_2 \tag{3.21}$$

where G_{12} = the free energy in the mixed state and G_i = the free energy of pure i, it follows that

$$\begin{aligned}\frac{\partial(\Delta G^M)}{\partial N_1} &= \left(\frac{\partial G_{12}}{\partial N_1}\right) - \left(\frac{\partial G_1}{\partial N_1}\right) \\ &= (\mu_1 - \mu_1^{\ominus})\end{aligned} \tag{3.22}$$

where N_1 = number of moles of solvent. Evaluation of $\partial(\Delta G^M/\partial N_1)$ yields:

$$\mu_1 - \mu_1^{\ominus} = RT\{\ln v_1 + (1 - 1/x)v_2 + \chi_1 v_2^2\} \tag{3.23}$$

Notice that the parameter x appears here as a reciprocal and, since x is large for high polymers, its precise value is irrelevant. The first two terms in equation (3.23) are of opposite signs and all terms may be important.

An analogous procedure for the polymer molecules leads to

$$\mu_2 - \mu_2^{\ominus} = N_A \frac{\partial(\Delta G^M)}{\partial n_2} = RT\{\ln v_2 - (x-1)(1-v_2) + x\chi_1(1-v_2)^2\} \tag{3.24}$$

The difficulty with this expression for the polymer chemical potential is that, since x is large, the dominant term is $(x-1)(1-v_2)$. To evaluate this quantity, it is necessary to specify x, which is essentially a lattice parameter. This requirement renders the expression for the chemical potential of the polymer somewhat less useful than that for the solvent, although x may be approximated by $x=(\bar{V}_2/\bar{V}_1)$, where $\bar{V}_2$ = molar volume of the polymer.

3.2.5. The osmotic pressure

In an osmotic pressure experiment, the additional pressure Π exerted on the polymer solution results in the solvent chemical potential being rendered equal to that of the pure solvent at ambient pressure p:

$$\mu_1^{\ominus} = \mu_1 + \int_p^{p+\Pi} \left(\frac{\partial \mu_1}{\partial p}\right) \mathrm{d}p$$

$$= \mu_1 + \int_p^{p+\Pi} \bar{V}_1 \, \mathrm{d}p$$

$$= \mu_1 + \Pi \bar{V}_1$$

so that

$$\Pi = -(\mu_1 - \mu_1^{\ominus})/\bar{V}_1$$

$$= -(RT/\bar{V}_1)\{\ln(1-v_2) + (1-1/x)v_2 + \chi_1 v_2^2\} \tag{3.25}$$

If it is possible to assume that v_2 is small, which is somewhat in conflict with the assumptions adopted above, then

$$\Pi = (RT/\bar{V}_1)\{v_2/x + (\tfrac{1}{2}-\chi_1)v_2^2 + \ldots\} \tag{3.26}$$

where the logarithmic expansion was truncated after the squared term. Now x can be set equal to $\bar{V}_2/\bar{V}_1$ so that $x\bar{V}_1 = \bar{V}_2$. Moreover, $v_2 = c_2\bar{v}_2$, where c_2 = polymer concentration and $\bar{v}_2$ = partial specific volume, i.e. $\bar{v}_2 = \bar{V}_2/M_2$, where M_2 = molecular weight of the polymer. These relationships result in $v_2/x\bar{V}_1 = c_2/M_2$. Therefore

$$\Pi/c_2 = RT\{1/M_2 + (\bar{v}_2^2/\bar{V}_1)(\tfrac{1}{2}-\chi_1)c_2 + \ldots\} \tag{3.27}$$

The first virial term in this power series is the classical van't Hoff infinite dilution expression for the osmotic pressure. It is associated with the movement of the centres of mass of the solute molecule over the available volume. The second term, which turns out not to be entirely correct for a real polymer system, is the two-body interaction term. According to this theory, the second virial coefficient B_2 in the expansion

$$\Pi/c_2 = RT\{B_1 + B_2c_2 + B_3c_2^2 + \ldots\} \tag{3.28}$$

must be such that

$$B_2 = (\bar{v}_2^2/\bar{V}_1)(\tfrac{1}{2} - \chi_1) \tag{3.29}$$

Experiments show that whereas B_2 is predicted by the Flory-Huggins theory to be independent of molecular weight, measured values of B_2 decrease slowly with increasing molecular weight of the polymer. Much more sophisticated theories (e.g. perturbation theories), which lie outside our current scope, are required to account for this behaviour.

3.2.6. The dilution parameters

The interaction parameter χ_1 as introduced above is merely a contact dissimilarity internal energy (or, if volume effects are ignored, enthalpy). The form of the final equation, however, would be unaltered if the interaction parameter were a *contact dissimilarity free energy*, i.e. if χ_1 contained both enthalpic and entropic components.

According to equation (3.27), the excess chemical potential involved in two-body interactions is just

$$(\mu_1 - \mu_1^{\ominus})^E = -\Pi^E \bar{V}_1 \simeq -(\tfrac{1}{2} - \chi_1)v_2^2 RT \tag{3.30}$$

Flory (1953) has defined enthalpy (κ_1) and entropy (ψ_1) of dilution parameters whereby

$$\Delta\bar{H}_1 = RT\,\kappa_1\, v_2^2 \tag{3.31}$$

$$\Delta\bar{S}_1 = R\,\psi_1\, v_2^2 \tag{3.32}$$

Here $\Delta\bar{H}_1$ and $\Delta\bar{S}_1$ are the relative partial molar heats of dilution such that $\Delta\bar{G}_1 = \Delta\bar{H}_1 - T\Delta\bar{S}_1 = (\mu_1 - \mu_1^{\ominus})^E$. The use of a squared term in these definitions is simply recognition that the excess functions must involve two-body (and higher) interactions. Equations (3.31) and (3.32) when coupled with equation (3.30) yield

$$(\tfrac{1}{2} - \chi_1) = \psi_1 - \kappa_1 \tag{3.33}$$

This important result allows the interaction free energy to be related to its enthalpic and entropic components.

3.2.7. The theta-point

Examination of equation (3.29) shows that the second virial coefficient B_2

becomes zero when $\chi_1 = \frac{1}{2}$. Of course, higher virial coefficients than the second need not vanish under these conditions, although the third virial coefficient, whilst not vanishing, is relatively small. The point at which the second virial coefficient vanishes is termed (and we shall use it as the definition) the *theta* (*θ*)*-point*. It has also been called the *Flory point* since Flory (1949) was the first to recognize its character, which was placed on a formal basis by Flory and Krigbaum (1950). Usually, in a polymer-solvent system, the second virial coefficient becomes equal to zero at some temperature, called the *θ-temperature*. As a θ-solvent is a slightly poor solvent, the θ-condition can also be achieved by the addition of nonsolvent. The θ-temperature is formally introduced into the Flory-Huggins theory via equation (3.33):

$$\tfrac{1}{2} - \chi_1 = \psi_1 - \kappa_1 = \psi_1(1 - \theta/T) \qquad (3.34)$$

where $\theta = (\kappa_1/\psi_1)T$. Clearly, if $T = \theta$, then $\chi_1 = \frac{1}{2}$, which is the ideal value for χ_1.

3.2.7.1. Analogy with the Boyle point

Flory (1953) has pointed out a far-reaching analogy between the θ-point for polymer molecules in solution and the Boyle point for nonideal gas molecules. This is useful in understanding the nature of the θ-point, which is crucial to much that is to follow on steric stabilization.

Noninteracting point molecules, if they existed, would obey the ideal equation of state:

$$p\bar{V} = RT \qquad (3.35)$$

where $\bar{V}$ = molar volume. Of course, real molecules occupy finite volumes in space and manifest intermolecular van der Waals attraction. These two effects, the first of which is entropic in character and the second of which is enthalpic, were incorporated by van der Waals into a revised equation of state *viz*:

$$(p + a/\bar{V}^2)(\bar{V} - b) = RT \qquad (3.36)$$

where a, b are constants for a particular gas. The a term provides allowance for the attraction whilst the b term adjusts for excluded volume effects. Notice that the two terms are of opposite signs.

The compressibility factor z, defined as $p\bar{V}/RT$, is clearly unity for an ideal gas. Deviations from unity are a measure of the nonideality of a gas: negative deviations arise if the intermolecular attraction is dominant whereas positive deviations appear if the finite volume effects are paramount. By way of specific example, we present in Fig. 3.2 the compressibility factor of methane as a function of pressure at several temperatures. At low temperatures, the attractive deviations are dominant and negative deviations from ideality result. At high temperatures, the excluded volume repulsion is paramount and positive deviations from ideality ensue. It is apparent that by adjusting the temperature, it is possible to so balance the two opposing effects that compensation ensues over a wide range of pressures (e.g. up to several hundred atmospheres). The temperature at which a nonideal gas behaves ideally up to quite high pressures is called the *Boyle temperature* (since Boyle's law is obeyed at this point). Of course, ultimately the serendipitous cancelling of the two opposing effects breaks down and deviations from ideality appear at high pressures, even at the Boyle point. These correspond to the intrusion of higher virial coefficient terms.

The solution analogue of the compressibility factor of a gas is the reduced osmotic pressure (Π/c_2). This quantity is shown schematically in Fig. 3.3 for polymer molecules under different solvency conditions. In a poor solvent for the polymer, negative deviations from ideality are apparent. This can be envisaged as arising because the polymer molecules are in dynamic association under such solvency conditions. Since osmotic pressure is a colligative

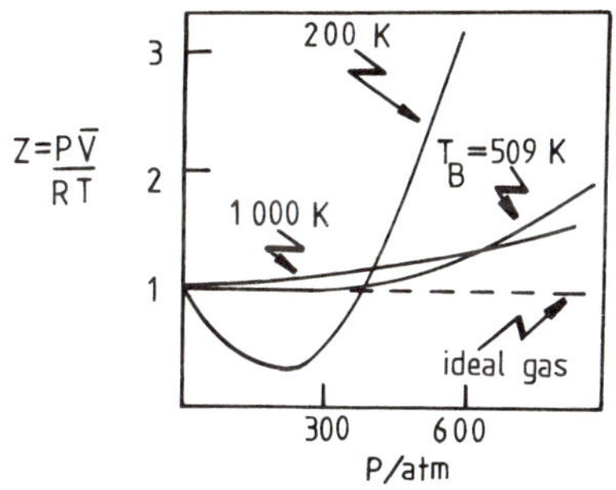

Fig. 3.2. The dependence of the compressibility factor z upon pressure for methane at various temperatures including the Boyle temperature T_B.

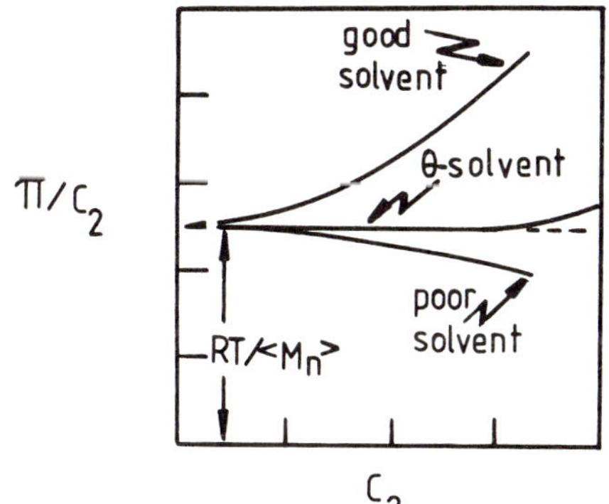

Fig. 3.3. Schematic representation of the concentration dependence of the reduced osmotic pressure of polymer solutions under different solvency conditions.

property, any reduction in the effective number of discrete molecules reduces the osmotic pressure below the ideal van't Hoff limit.

By way of contrast, the segments are mutually repulsive in a good solvent since, by definition, contacts with solvent molecules are enthalpically favoured. This tends to cause the polymer chains to swell, a process that is counteracted by the loss in configurational entropy as the chains expand. Nonetheless, the polymer molecules are mutually repulsive so that the volume available in the polymer solution is effectively reduced below the nominal volume. This causes the effective polymer concentration to be greater than that expected for an ideal system and results in positive deviations from ideality.

It is scarcely surprising in light of these positive and negative deviations from ideality, that it is possible to adjust the solvency so that the polymer molecules obey the ideal van't Hoff equation of state, i.e. the second virial coefficient is zero. This is the θ-point. Under these solvency conditions, the macromolecules, which may, for example, be several millions in molecular weight, behave like ideal minimolecules (i.e. molecules comparable in size and interactions to the molecules of solvent). Of course, obedience to ideal behaviour is evident only up to a concentration of several per cent. Beyond this, the higher virial coefficients involving the higher order many-body interactions manifest themselves and deviations from the ideal van't Hoff limit appear.

The θ-point is important on several scores. First, at the θ-point, the polymer chains are able to telescope one another with impunity insofar as free energy changes are concerned. Second, the θ-point demarcates the region wherein the segmental interactions switch over from being repulsive to being attractive. This, as we shall see, is vital to the proper understanding of the flocculation behaviour of certain sterically stabilized dispersions. Third, the polymer molecules adopt their unperturbed dimensions at the θ-point (see Section 4.3.5) since the polymer segments behave as if they were volumeless lines in space.

There are several other comments that need to be made about the θ-point. The θ-point represents the theoretical limit of solubility of a polymer of infinite molecular weight. de Gennes (1975) has also proposed that the θ-point is a tri-critical point that separates a first order change in a worse than θ-solvent (the coil-to-globule transition) and a second order change in a better than θ-solvent (coil expansion).

It has been argued that the θ-point is really a θ-region extending, for example, over a finite temperature range (Domb, 1974; Daoud *et al.*, 1975). Whilst this may well be true, there now exists a significant body of experimental evidence to suggest that this range must be quite small ($< 1\ ^\circ C$). If this is correct, the distinction between the θ-temperature and θ-range is unimportant to what follows in this book.

Finally, we note that if $\chi_1 < \frac{1}{2}$, the solvent is a better than θ-solvent and the second virial coefficient B_2 in equation (3.29) is positive. Conversely, a value of $\chi_1 > \frac{1}{2}$ implies a solvency worse than a θ-solvent. It follows that as χ_1 *decreases* in numerical value so the quality of the solvency *improves*.

3.2.7.2. Determination of the theta-point

It was shown above that at the θ-point the second virial coefficient vanishes. This allows the θ-point to be located by light scattering or osmotic pressure measurements. It can also be found by the rather time-consuming determination of the critical solution temperature extrapolated to infinite molecular weight.

Quicker, but less soundly based, methods have been developed by Elias (1959; 1961), Cornet and van Ballegooijen (1966) and Talamini and Vidotto (1967). These allow the rapid determination of either the θ-temperature or θ-composition. In these methods, the temperature of phase separation, for example, over a range of polymer concentrations ($10^{-5} \leq v_2 \leq 10^{-2}$) is measured for a given polymer sample. The polymer may even be polydisperse in molecular weight. A plot of the reciprocal temperature *vs* the logarithm of the volume fraction of polymer is usually linear (see Fig. 3.4). Extrapolation to $\log v_2 = 0$ yields the value of $1/\theta$. This method is, of course, inapplicable to polymers that crystallize on phase separation. Note that Elias (1977) has reviewed in an extremely comprehensive manner the rapid methods of determining the θ-point.

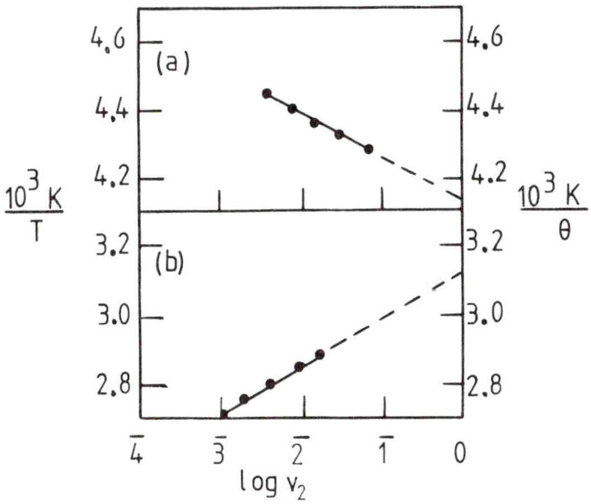

Fig. 3.4. Experimental plots of the reciprocal phase separation temperature *vs* logarithm of the volume fraction of polymer showing the extrapolation required to determine the θ-temperature: (a) poly(12-hydroxystearic acid) in *n*-heptane; (b) poly(oxyethylene) (as Carbowax 6000) in 0·39 M magnesium sulphate solution (after Napper, 1968b; 1969b).

The theoretical basis of the rapid methods

If a polymer solution undergoes phase separation into two liquid phases then the solution cannot be too far removed from the θ-point. It follows that the Flory-Huggins theory, which was derived for moderately concentrated solutions, can be applied to both of the polymer solution phases, one of which is dilute and the other concentrated. The chemical potentials of both the polymer and the solvent must be equal in the two co-existing phases:

$$\mu_1^c = \mu_1^d \tag{3.37}$$

$$\mu_2^c = \mu_2^d \tag{3.38}$$

where the superscripts c and d denote the concentrated and dilute phases respectively. Insertion of equation (3.23) into equation (3.37) for the solvent chemical potentials leads to (Napper, 1968a; 1969a)

$$\ln(1 - v_2^c) - \ln(1 - v_2^d) + (v_2^c - v_2^d) + \chi_1[(v_2^c)^2 - (v_2^d)^2] = 0 \tag{3.39}$$

where we have set $\chi_1^c = \chi_1^d = \chi_1$ (i.e. the interaction parameter is concentration independent). The assumptions $v_2^c >> v_2^d \leq 0.01$ are now invoked to give

$$\ln(1 - v_2^c) + v_2^c + \chi_1(v_2^c)^2 = 0 \tag{3.40}$$

Since $v_2^c < 1$, we may expand the logarithm as a power series so that

$$(v_2^c)^3/3 - (v_2^c)^2(\chi_1 - \tfrac{1}{2}) = 0 \tag{3.41}$$

where terms higher than third order have been neglected. The nontrivial solution of this cubic equation is just

$$v_2^c = 3(\chi_1 - \tfrac{1}{2}) \tag{3.42}$$

On substituting the Flory-Huggins expressions for chemical potential into equation (3.38), we find that

$$\ln v_2^d = \ln v_2^c + x\, v_2^c + x\chi_1\, v_2^c(v_2^c - 2) \tag{3.43}$$

Equation (3.42) can be used to eliminate v_2^c from this equation whereupon

$$\log v_2^d = \log[3(\chi_1 - \tfrac{1}{2})] + 3{\cdot}908x(\chi_1 - \tfrac{1}{2})^2(\chi_1 - 2/3) \tag{3.44}$$

and x = number of solvent-sized segments per polymer molecule. This equation relates the interaction parameter to the logarithm of the volume fraction of the polymer solution at the point of incipient phase separation. It does not seem possible to make χ_1 the subject of this equation but χ_1 is readily evaluated numerically once v_2^d and x are specified. Plots of χ_1 *vs* log v_2^d are shown in Fig. 3.5. It is apparent that in the dilute solution range $10^{-5} \leq v_2^d \leq 10^{-2}$, there exists a linear relationship between χ_1 and log v_2^d, as is

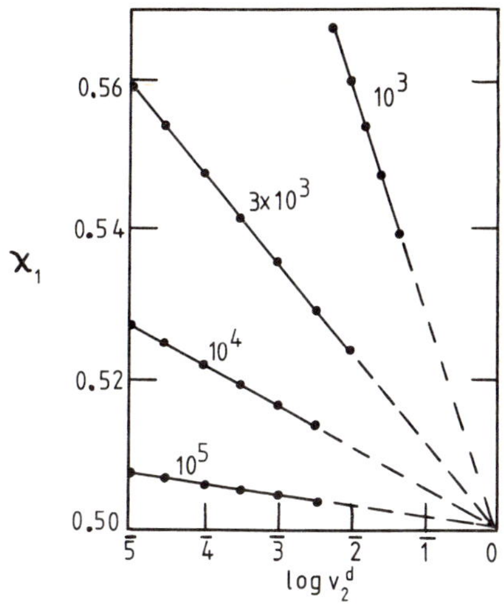

Fig. 3.5. Theoretical plots of the critical value of the interaction parameter for phase separation *vs* logarithm of the volume fraction of polymer showing the extrapolation to the θ-point ($\chi_1 = \frac{1}{2}$) at $v_2{}^d = 1{\cdot}00$ for different values of x (after Napper, 1969a).

observed experimentally. Inspection of the linear plots displayed in Fig. 3.5 shows that extrapolation of the plot of χ_1 *vs* log $v_2{}^d$ to $v_2{}^d = 1$ yields $\chi_1 = \frac{1}{2}$ (i.e. the θ-point). Of course, according to equation (3.34), χ_1 is linearly related to $1/T$ so that linear plots of $1/T$ *vs* log $v_2{}^d$ extrapolate to $1/\theta$ at $v_2{}^d = 1$.

The linearity of the plots in Fig. 3.5 can be derived by an alternative procedure. Another rendition of equation (3.44) is

$$\log v_2{}^d = \log [3(\chi_1 - \tfrac{1}{2})] + 9x(\chi_1 - \tfrac{1}{2})\, \Lambda/2{\cdot}303 \tag{3.45}$$

where

$$\Lambda = \chi_1{}^2 - (7/6)\chi_1 + 1/3 \tag{3.46}$$

and the slopes of the plots of χ_1 *vs* log $v_2{}^d$ are inversely related to Λ. On differentiating Λ with respect to χ_1 we obtain

$$\mathrm{d}\Lambda/\mathrm{d}\chi_1 = 2\chi_1 - 7/6 \sim 0$$

since $0{\cdot}5 < \chi_1 < 0{\cdot}6$. Thus Λ is essentially independent of small variations in χ_1 in the phase separation regime. Inspection of equation (3.45) shows that χ_1 must assume the value $\frac{1}{2}$ if $v_2{}^d = 1$, i.e. extrapolation leads to the θ-point. The linearity of the plots obtained in the rapid methods for determining the θ-point is accordingly explained and their extrapolation to the θ-point given a firm thermodynamic basis.

3.2.8. Failure of the Flory-Huggins theory

The rudimentary approach adopted by the Flory-Huggins theory incorporates several of the important features that are indispensable to the explanation of the properties of polymer solutions. The theory allows for the connectivity of the polymer segments in a way that is substantially correct. It also allows for the contact dissimilarity free energy, but only in a pragmatic fashion. The value of χ_1 is not predicted by the Flory-Huggins theory but must be evaluated by experimental determination (e.g. by measurement of the solvent activity $a_1 \simeq v_1 e^{v_2(1+\chi_1 v_2)}$). According to the Flory-Huggins scheme, the value of χ_1 so determined should be independent of the concentration of polymer (at least in non-dilute solutions).

Measured values of the interaction parameter for a wide range of polymer-solvent systems show two important departures from the expectations of the Flory-Huggins scheme. The first is that the actual values of the interaction parameter are usually strongly concentration dependent. More surprising, however, is the finding that for many systems the interaction free energies, as embodied by χ_1, prove to be predominantly entropic in origin, the enthalpic contributions being relatively small. Yet χ_1 was originally introduced solely as an enthalpic term. These two unexpected results could perhaps be glossed over were it not for a third devastating observation: most nonaqueous polymer solutions undergo phase separation on heating. This usually occurs near to the critical point of the solvent. Each of these three difficulties will now be discussed in turn.

3.2.8.1. The concentration dependence of the interaction parameter

Illustrative values of the interaction parameter, measured as a function of the volume fraction of polymer for several nonaqueous systems, are shown in Fig. 3.6. These results clearly indicate that the interaction parameter can be a strong function of the volume fraction of polymer. Usually χ_1 increases with the volume fraction of polymer (e.g. for polyisobutylene in benzene) but this behaviour is by no means universal (e.g. χ_1 for polystyrene in toluene decreases with increasing polymer concentration).

Mathematical representation

Orofino and Flory (1957) have suggested that the concentration dependence of the interaction parameter can be expressed by the perturbation expansion

$$\chi = \chi_1 + \chi_2 v_2 + \chi_3 v_2^2 + \ldots \tag{3.47}$$

The problem is to determine the values of the higher order interaction parameters χ_2, χ_3, etc. Some guide to these values can be obtained in the following manner.

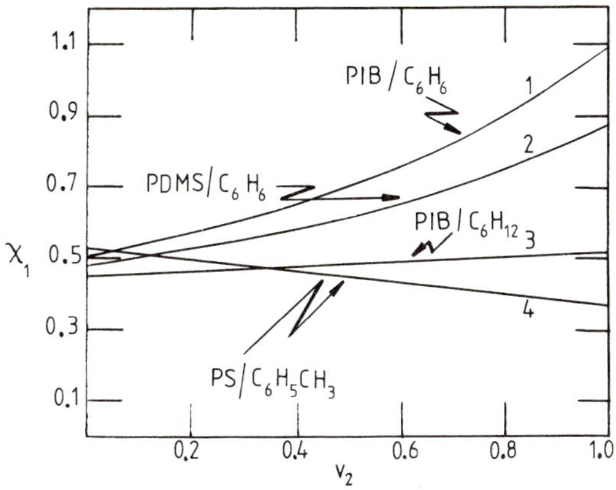

Fig. 3.6. Experimental plots of the concentration dependence of the interaction parameter for (1) poly(isobutylene) in benzene, (2) poly(dimethylsiloxane) in benzene, (3) poly(isobutylene) in cylcohexane, and (4) polystyrene in toluene.

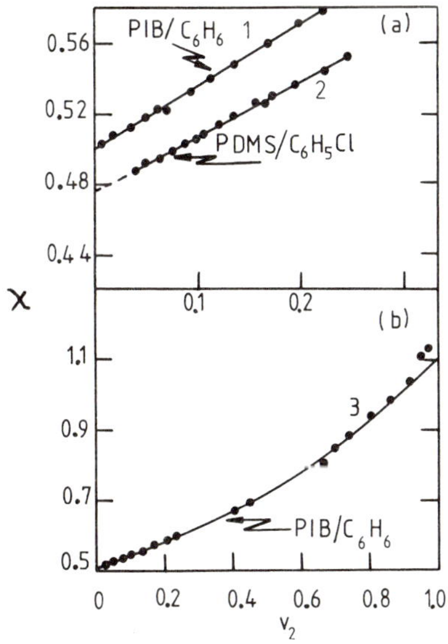

Fig. 3.7. Plots of the concentration dependence of the interaction parameter at 298 K: (a) comparison of the experimental points with equation (3.48) for (1) poly(isobutylene) in benzene and (2) poly(dimethylsiloxane) in chlorobenzene; (b) comparison of experimental points with equation (3.49) for poly(isobutylene) in benzene over the entire concentration range (after Evans and Napper, 1977).

It has been shown (Evans and Napper, 1977) that the interaction between polymer chains, which display a concentration dependent interaction parameter, vanishes when $\chi_i = 1/(i+1)$ for all positive integral values of i. This relationship suggests that near to the θ-point $\chi_1 = \frac{1}{2}$, $\chi_2 = 1/3 = (2/3)\chi_1$ and $\chi_3 = 1/4 = (\frac{1}{2})\chi_1$. In these circumstances, we might guess that the simplest volume fraction dependence might be

$$\chi = \chi_1(1 + 2v_2/3) \tag{3.48}$$

and that at the next level of complexity, the relationship would be

$$\chi = \chi_1(1 + 2v_2/3 + v_2^2/2) \tag{3.49}$$

Figure 3.7 displays a comparison of these empirical mathematical forms against some experimental data. It is found that for several polymer/solvent systems (e.g. poly(dimethylsiloxane) in chlorobenzene), the simplest equation is adequate up to volume fractions $v_2 \sim 0{\cdot}25$. To span a larger volume fraction range, it seems necessary to use the more complex equation (3.49). The data for poly(isobutylene) in benzene at 298 K obey this simple equation quite accurately over almost the entire volume fraction range.

These relationships provide convenient mathematical expressions for the concentration dependence of the interaction parameter that can be used to incorporate this dependence into the theory for steric stabilization.

The UNIFAC group contribution method

A semi-empirical method for calculating the concentration dependence of the interaction parameter has been developed by Oishi and Prausnitz (1978) and Gottlieb and Herskowitz (1981) based upon the UNIFAC group contribution method (Fredenslund *et al.*, 1977). This method regards a solution as a mixture of functional groups whose solution properties can be characterized by just four parameters. The first two quantities are the reduced Bondi group volume and group surface area, which can be calculated from the van der Waals radii and the covalent bond lengths. The third parameter, which must be determined experimentally, is an interaction quantity. Once the interaction term has been measured for two groups, its value remains unchanged for all systems in which they occur together. The fourth parameter, which accounts for the free volume effect, is the number of external degrees of freedom per solvent molecule $3c_1$. Oishi and Prausnitz (1978) have suggested that $c_1 = 1{\cdot}1$ for many solvents. The results calculated by the UNIFAC group contribution method appear to be in good agreement with the results of experiments for many systems.

3.2.8.2. The entropic component of the interaction parameter

If χkT is presumed to be an exchange free energy, then it can be written in terms of its enthalpic and entropic components:

$$\chi = \chi_H + \chi_S \tag{3.50}$$

where $\chi = \Delta G/kT$, $\chi_H = \Delta H/kT$ and $\chi_S = -\Delta S/k$. Note that

$$\lim_{v_2 \to 0} \chi_H = \kappa_1 \text{ and } \lim_{v_2 \to 0} \chi_S = (\tfrac{1}{2} - \psi_1)$$

which establishes the connection between χ_H and χ_S and the Flory-Huggins parameter. It follows from equation (3.50) that

$$\frac{\partial(\chi)}{\partial T} = \frac{\partial(\Delta G/kT)}{\partial T} = \frac{-\Delta H}{kT^2} = \frac{-\chi_H}{T} \tag{3.51}$$

and

$$\frac{\partial(T\chi)}{\partial T} = \frac{\partial(T\Delta G/kT)}{T} = \frac{\partial(\Delta G/k)}{\partial T} = \frac{-\Delta S}{k} = \chi_S \tag{3.52}$$

These relationships are valid irrespective of any dependence of χ_H and χ_S upon T. Consequently, it is possible from measurements of the temperature dependence of χ to resolve it into its enthalpic and entropic contributions.

Some results for poly(methyl methacrylate) in a range of different solvents at 20 °C are presented in Table 3.1. These show that the exchange free energy per solvent molecule usually lies in the range (0·3–0·5)kT. One remarkable feature of the data in Table 3.1 is the relatively small values of χ_H, which can be positive or negative (i.e. endothermic or exothermic). An even more surprising result is that the overwhelming contribution to χ comes from the entropic component χ_S. This shows that even at room temperature there is a strong *decrease* in entropy, which is not properly accounted for by the Flory-Huggins combinatorial expression, that disfavours the mixing of poly(methyl methac-

TABLE 3.1
The enthalpic and entropic components of the interaction parameter for poly(methyl methacrylate) in various solvents at 20 °C (from Casassa, 1970)

Solvent	χ	χ_H	χ_S
chloroform	0·365	−0·075	0·44
benzene	0·429	−0·017	0·45
tetrahydrofuran	0·446	0·026	0·42
acetone	0·481	0·028	0·45
toluene	0·452	0·028	0·42

rylate) with organic solvents. This is a general result for most polymers in nonaqueous media.

3.2.8.3. Phase separation on heating

According to the precepts of the Flory-Huggins theory, the mixing of a polymer with a solvent is governed by the combinatorial entropy (ΔS^{COMB}), which is positive. As a result, its contribution to the mixing free energy ($-T\Delta S^{COMB}$) increases in absolute value, albeit relatively slowly, on raising the temperature above the ambient value. Polymer solutions are thus predicted to become more stable on heating. Conversely, cooling should lead to the situation (assuming that the freezing of the solvent does not intervene) where phase separation occurs.

The latter prediction appears to be well substantiated by experiment. On the other hand, it appears, as first shown by Freeman and Rowlinson (1960), that most, if not all, polymer solutions phase separate not only on cooling but also on heating. This constitutes a serious contravention of the predictions of the Flory-Huggins theory. In what follows, it will become apparent that the Flory-Huggins theory correctly identifies the driving force behind the mixing of polymer and solvent but overlooks a major factor that disfavours mixing.

The binary phase diagram for typical polymer-solvent systems is shown schematically in Fig. 3.8. Nonpolar polymer solutions usually display an upper and a lower critical solution temperature. In the limit of infinite polymer molecular weight, these would correspond to an upper (θ_U) and lower (θ_L) theta-temperature. This can be readily seen as follows.

For a polymer of infinite molecular weight, n_2 in equation (3.20) is zero so that the chemical potential of the solvent swelling the polymer is given by

$$\mu_1 - \mu_1^{\ominus} = RT[\ln(1 - v_2) + v_2 + \chi_1 v_2^2] \tag{3.53}$$

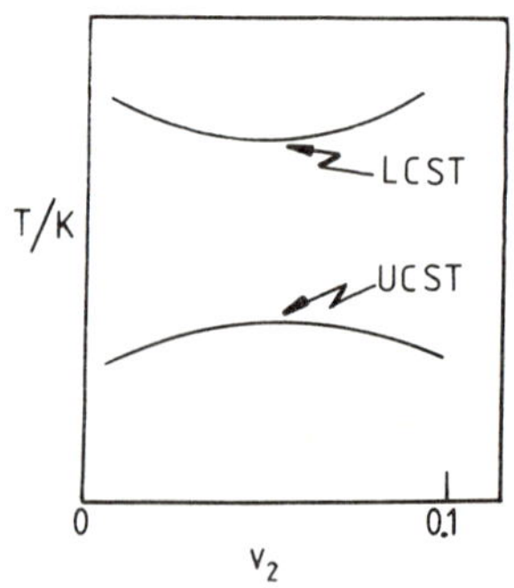

Fig. 3.8. Schematic representation of a typical phase diagram for polymers dissolved in nonaqueous media.

The requirements for phase separation are simply

$$(\partial\mu_1/\partial v_2)_{T,P}=0 \tag{3.54}$$

$$(\partial^2\mu_1/\partial v_2{}^2)_{T,P}=0 \tag{3.55}$$

These criteria result in the simultaneous equations

$$1/(1-v_2)-1=2\chi_1 v_2 \tag{3.56}$$

$$1/(1-v_2)^2=2\chi_1 \tag{3.57}$$

Solution of these equations for v_2 leads to only one acceptable value for χ_1. This is $\chi_1=\frac{1}{2}$, a value corresponding to the attainment of theta-conditions (whether θ_U or θ_L).

3.2.8.4. The defects in the Flory-Huggins theory: An explanation

The derivation of the Flory-Huggins theory was predicated on the assumption that the volume changes that occur on mixing are negligible and, indeed, the volume changes are often quite small ($\sim 1\%$). It turns out, however, that what is important is not so much the overall volume but the *free volume.*

The clue to the understanding of the phenomenon of phase separation on heating resides in the recognition that it usually takes place as the critical point of the solvent is approached. Near to the critical point, the liquid becomes more gas like, with a large free volume (this perhaps exaggerates the effect but it seems justified for the purposes at hand). There are no covalent bonding constraints on the separation of the solvent molecules on heating, the only constraints arising from the van der Waals attractions. The same cannot be said for the segments of the polymer molecules: their concatenation via covalent bonds severely restricts the permissible increase in free volume of the polymer that can take place on heating. Hence on heating a polymer, its increase in free volume is significantly less than that for a low molecular weight solvent. This diminution is reflected in the large difference in the coefficients of thermal expansion ($\alpha_v=(1/v)(\partial v/\partial T)_p$) for high polymers and minimolecules. By way of specific example, we cite the value for benzene $\alpha_v=1{\cdot}22\times10^{-3}\,K^{-1}$ which is almost double that for natural rubber ($\alpha_v=0{\cdot}65\times10^{-3}\,K^{-1}$) at room temperature.

The introduction of this free volume difference leads to the following explanation for phase separation on heating. Near to the critical point, the solvent is gas like, with a large free volume. When placed in contact with a polymer, the solvent "condenses" on to the segments. This ordering provides a negative contribution to the entropy of the system. If this decrease in entropy is sufficiently large, mixing will be thermodynamically disfavoured. Thus, on heating near to the critical point of the solvent, phase separation allows most of the solvent molecules to become more gas like. Their increase in entropy promotes phase separation. The corresponding enthalpy requirement is, like the latent heat of vaporization, positive and opposes phase separation.

The contribution to the interaction parameter from free volume dissimilarity persists down to room temperature and below. It accounts for the dominant positive values of χ_S noted above (see Table 3.1). It further explains the observed increase in χ with increasing volume fraction of polymer found with some systems: the extent of "condensation" of the solvent is related to the amount of polymer present. The greater is the amount of polymer present, the more will the solvent molecules "condense" onto the polymer segments, thus increasing the numerical magnitude of χ_S. If χ does not increase with the polymer volume fraction, as not infrequently happens (e.g. polystyrene in toluene), then specific interactions must provide compensatory effects. Such specific interactions appear at present to be beyond the grasp of any fundamental theoretical analysis.

The free volume dissimilarity provides one of the important conceptual features that is missing from the Flory-Huggins theory. It rationalizes (i) the observed phase separation on heating; (ii) the strong entropic contribution to χ that opposes mixing; and (iii) the observed increase in χ with volume fraction of polymer in certain systems. Qualitatively, we can write for the mixing of polymer and solvent at room temperature:

$$\Delta V^M = \Delta V^{\mathrm{CON}} + \Delta V^{\mathrm{VOL}} \tag{3.58}$$

where the superscripts CON and VOL denote *contact energy dissimilarity* and *free volume dissimilarity* respectively. The first term is usually positive for nonaqueous systems whereas the second term is negative. The small net volume change observed for ΔV^M is a consequence of the mutual cancelling of the two effects. This shows that free volume effects can still be operative, even if ΔV^M is zero (i.e. mixing obeys the Flory-Huggins assumption).

In a likewise fashion, we can write

$$\Delta H^M = \Delta H^{\mathrm{CON}} + \Delta H^{\mathrm{VOL}} \tag{3.59}$$

The first term is usually positive whereas the second term, as noted above, is negative. Again, the mutual cancellation of the two effects on summation accounts for the unexpectedly small values of χ_H observed in most nonaqueous systems (see Table 3.1).

Finally, we note that

$$\Delta S^M = \Delta S^{\mathrm{COMB}} + \Delta S^{\mathrm{VOL}} + \Delta S^{\mathrm{STRUCT}} \tag{3.60}$$

where the superscripts COMB and STRUCT denote *combinatorial* and *structural* contributions respectively. The first term is positive, the second term is negative and the third term may be positive or negative but would normally be expected to be negative. This explains why χ_S is often found to be positive.

3.3. Flory's equation-of-state theory

Flory (1965; 1970) has developed an elaborate *equation-of-state* theory that endeavours to incorporate the free volume contribution into the various thermodynamic functions. Because of its inherent complexity, only the barest outline of the theory will be adumbrated at this point. Patterson (1968) has developed a separate, but in many ways similar, *corresponding states* theory. Both theories are in many respects only satisfactory at the semi-quantitative level for nonpolar polymer molecules in apolar solvents.

3.3.1. Equation-of-state for noninteracting disconnected particles in 1-D

Flory's theory starts with consideration of noninteracting, disconnected particles in 1-D. The results obtained are then generalized to 3-D and the interactions and connectivity introduced in a subtle, yet elegant, fashion.

Consider the n unconnected particles shown in Fig. 3.9. Let the length of the system be L and l^* be the core particle length (diameter). The average space in

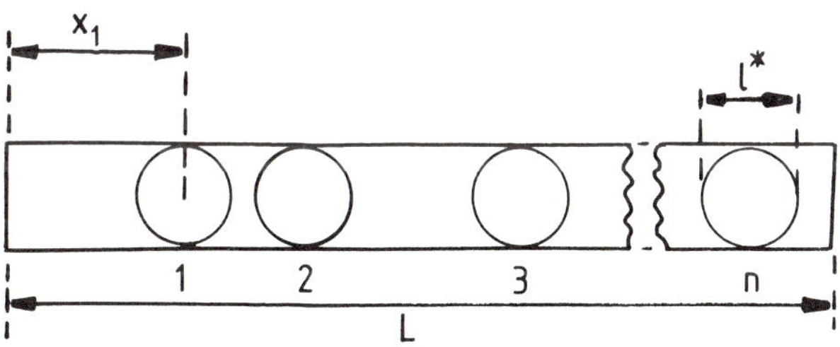

Fig. 3.9. Schematic representation in 1-D of unconnected particles with free volume.

1-D available per particle is simply $l = L/n$ ($> l^*$). The totality of configurational space available to the system Ω in the x-direction is

$$\Omega = \int_{(2n-1)l^*/2}^{(L-l^*/2)} \cdots \int_{\frac{3}{2}l^*}^{(x_3-l)} \int_{l^*/2}^{(x_2-l)} \mathrm{d}x_1 \mathrm{d}x_2 \ldots \mathrm{d}x_n$$

$$= (L - nl^*)^n / n! \qquad (3.51)$$

The reader can easily check the validity of the integration for small values of n. The inverse of Sterling's approximation for ln $n!$ gives, after some elementary algebra, a second rendition of Ω:

$$\Omega = [(l - l^*)e]^n \qquad (3.62)$$

3.3.2. Generalization to 3-D

The expression for Ω is readily generalized to 3-D as follows. The hard-core length l^* is related to the hard-core volume v^* by $l^* = (gv^*)^{1/3}$, where g = geometrical factor. A similar relationship also holds for the available space: $l = (gv)^{1/3}$. It follows that $(l - l^*) = g^{1/3}(v^{1/3} - v^{*1/3})$. This leads in 3-D to

$$\Omega = [(l - l^*)^3 e^3]^n$$
$$= [g(v^{1/3} - v^{*1/3})^3 e^3]^n \quad (3.63)$$

The term e^{3n} is the communal entropy according to this free volume approach.

If we define the *reduced volume* $\tilde{v} = v/v^*$ then

$$\Omega = [gv^*(\tilde{v}^{1/3} - 1)^3 e^3]^n \quad (3.64)$$

3.3.3. Inclusion of contact dissimilarity

To allow for interactions, the mean intermolecular energy E_o is introduced as follows:

$$Z = \Omega \exp\{-E_o/kT\} \quad (3.65)$$

where

$$\exp\{-E_o/kT\} = (1/\Omega)\int \ldots \int \exp\{-E(x)/kT\} \mathrm{d}x_1 \ldots \mathrm{d}x_n \quad (3.66)$$

The configuration integral on the right hand side of equation (3.66) ensures that E_o is the average energy of the entire system. It follows that the partition function for the system is given by

$$Z = (gv^*)^n (\tilde{v}^{1/3} - 1)^{3n} e^{3n} \exp\{-E_o/kT\} \quad (3.67)$$

Note that Flory chose, somewhat arbitrarily, to employ a van der Waals interactional energy. This sets the attraction of molecules in a dense fluid as being inversely related to the particle volume ($E_o \propto 1/r^3$).

3.3.4. Incorporation of connectivity

The partition function derived above applies to gas molecules that are devoid of connectivity. To allow for the connectivity, Flory considered each polymer molecule to be composed of r isometric segments, each of which possesses 3c external degrees of freedom. Note that if $c = 1$, the segments must be disconnected since each segment would have 3 translational degrees of freedom. This means that the range of c is specified by $1/r < c < 1$. The lower bound indicated here is set by the fact that each free polymer molecule has 3cr degrees of freedom, which must be at least as large as 3. This approach is intuitive rather than rigorous and the numerical values of c should not be taken too seriously.

3.3.5. The reduced equation-of-state

Once the partition function of a system is known, the thermodynamic functions are readily found. The total partition function of the system can be written in this case as

$$Z = Z_{\text{COMB}} \underbrace{(gv^*)^{nrc}(\tilde{v}^{1/3} - 1)^{3nrc}}_{\text{Equation-of-state}} \underbrace{\exp\{-E_o/kT\}}_{\text{Contact dissimilarity}} \tag{3.68}$$

where Z_{COMB} is the partition function relevant to the evaluation of the combinatorial term, including the communal entropy.

To determine the pressure in the system, we use the relationship

$$p = -(\partial A/\partial V) = kT\,\partial(\ln Z)/\partial V \tag{3.69}$$

On evaluation of this differential, the following equation-of-state is obtained:

$$\tilde{p}\tilde{v}/\tilde{T} = \tilde{v}^{1/3}/(\tilde{v}^{1/3} - 1) - (1/\tilde{v}\tilde{T}) \tag{3.70}$$

where $\tilde{T} = T/T^*$ and $\tilde{p} = p/p^*$ and

$$p^*v^* = crkT^* = -E_o\tilde{v}/nr \tag{3.71}$$

It so happens that $\tilde{p} \sim 0$ under normal experimental conditions which means that the equation-of-state defined by equation (3.70) can be simplified to

$$\tilde{T} = (\tilde{v}^{1/3} - 1)/\tilde{v}^{4/3} \tag{3.72}$$

Note that these equations-of-states are quite general, applying to both pure macromolecules and pure minimolecules alike.

3.3.6. Evaluation of the reduced variables

The reduced variables $\tilde{v}$, $\tilde{p}$ and $\tilde{T}$ can be evaluated from measurements of the coefficient of thermal expansion α_v and compressibility κ_v. These are defined by

$$\alpha_v = (1/v)(\partial v/\partial T)_p = (\partial \ln v/\partial T)_p \tag{3.73}$$

$$\kappa_v = -(1/v)(\partial v/\partial p)_T = -(\partial \ln v/\partial p)_T \tag{3.74}$$

It can be shown that

$$\tilde{v}^{1/3} = 1 + \alpha_v T/3(1 + \alpha_v T) \tag{3.75}$$

so that once α_v is measured, $\tilde{v}$ is readily calculated. This allows the value of v^* to be determined since $v^* = v/\tilde{v} = 1/\rho\tilde{v}$, where ρ = density of the substance.

Once $\tilde{v}$ is known, $\tilde{T}$, and perforce T^*, can be calculated from the equation-of-state with $\tilde{p} \sim 0$. The value of p^* is calculated from $p^* = (\alpha_v/\kappa_v)T\tilde{v}^2$ and this leads to $\tilde{p}$. If required, c can be calculated from $p^*v^* = ckT^*$ but its actual value, as noted previously, is of little physical significance.

3.3.7. Application to mixtures

The preceding theories apply to pure species. Flory applied these concepts to mixtures with two principal assumptions: (i) that the core volumes are additive; (ii) that the intermolecular energy is proportional to the intermolecular surface area of contact. These assumptions render reasonable the further assumption that the equation-of-state also holds for the mixtures, provided that proper allowance is made in calculating the reduced variables of the mixture. The prescriptions required to achieve this end have been worked out by Flory and, in light of their complexity, they will not be reproduced here. Suffice to state that the reduced residual chemical potential, which is the equation-of-state analogue of the interaction parameter, is given by

$$\chi^R = (\mu_1 - \mu_1^{\ominus})^R / RT\phi_2^{\,2} \tag{3.76}$$

where

$$(\mu_1 - \mu_1^{\ominus})^R = p_1^* V_1^* \{3\tilde{T}_1 \ln \left[\frac{\tilde{v}_1^{\,1/3} - 1}{\tilde{v} - 1}\right] + (1/\tilde{v}_1 - 1/\tilde{v}) + V_1^* X_{12} \Theta_2^{\,2}/\tilde{v}\} \tag{3.77}$$

Here V_1^* $(= M_1/\rho\tilde{v}_1)$ = hard-core molar volume of the solvent (denoted by the subscript 1), X_{12} = contact dissimilarity exchange parameter, Θ_2 = fraction of sites occupied by polymer segments and the unsubscripted variables refer to the mixture. The exact equivalence of χ^R with χ above is lacking only because of the use of the segment fraction ϕ_2 rather than the volume fraction.

3.3.8. Comparison with experiment

The parameter X_{12} in equation (3.77) is the sole adjustable parameter of the free volume theory when comparison is effected with experiment. The theory predicts that χ^R increases with the segment fraction ϕ_2, as found experimentally for some systems (see Fig. 3.10). Values of X_{12} are always found to be positive and fall into two well-defined groups depending upon the chemical constitution of the polymer and the solvent. For aliphatic-aliphatic interactions, $X_{12} \leq 10\,\mathrm{J\,cm^{-3}}$ whereas for aliphatic-aromatic interactions, $X_{12} \sim 40\,\mathrm{J\,cm^{-3}}$. These relative magnitudes seem intuitively reasonable.

Although some systems (e.g. poly(isobutylene) in benzene) display good

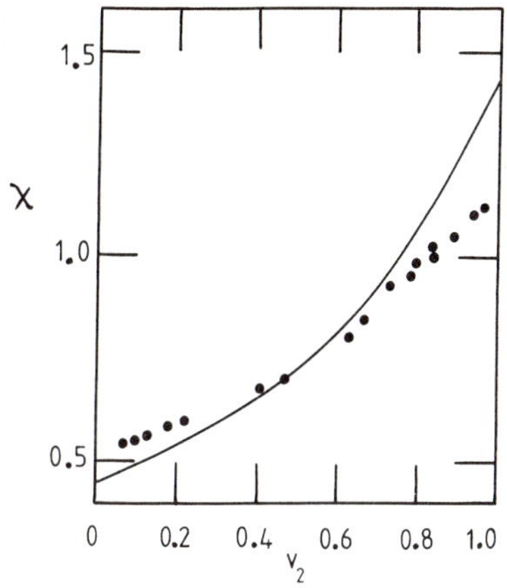

Fig. 3.10. Comparison of the predictions of the equation-of-state theory (full line) with the results (●) of experiment for the concentration dependence of the interaction parameter of poly(isobutylene) in benzene at 25 °C (after Eichinger and Flory, 1968).

agreement in an absolute sense between the predictions of free volume theory and experiment, other systems (e.g. natural rubber in benzene) require a second adjustable parameter, in addition to X_{12}, to obtain concordance between theory and experiment. Moreover, the free volume theory does not appear to be able to comprehend those systems for which the interaction parameter decreases with increasing volume fraction of polymer. One prosaic example of this is polystyrene dissolved in toluene (Lewis and Johnson, 1969; Gaeckle *et al.*, 1973), for which the heat of dilution is unquestionably negative. It would appear that in these systems, specific structural interactions are responsible for the observed concentration dependence. In the example cited, the specific interactions responsible for the negative heats of dilution might be envisaged to be the preferential orientation of the anisotropic toluene molecules in the neighbourhood of the polystyrene chains. This might optimize the attractive interactions between the benzene rings present in both the solvent and polymer molecules. Any specific ordering would lead to losses of configurational freedom, the energy associated with this loss of freedom manifesting itself as an exotherm. Almost by definition, specific interactions are beyond the scope of the currently available theoretical apparatus.

3.3.9. Lattice fluid theory

Flory's equation-of-state theory is quite complex, requiring as it does the manipulation of many interrelated parameters. Sanchez (Sanchez and Larcombe, 1978; Sanchez, 1980) has therefore developed a somewhat simpler theory based on a lattice fluid or Ising model. This approach also proceeds via an equation-of-state and certain mixing rules. To some extent, the theory might be described as a "Flory-Huggins theory with holes". The chains are distributed over a lattice but not all of the lattice sites are occupied. The lattice fluid theory will not be described in greater detail here because its predictions add relatively little that is new to our understanding of polymer solution thermodynamics. One exception to this conclusion relates to polymer solutions that display negative heats of mixing. According to the lattice fluid theory, the negative heat of mixing arises from the "condensation" of the expanded liquid which results in more intermolecular interactions at lower potential energy. This is accompanied by a corresponding loss in entropy.

3.4. Aqueous polymer solutions

The foregoing theories provide a semi-quantitative account of some, although certainly not all, solutions of nonpolar polymers dissolved in apolar solvents. They are not, however, applicable to aqueous polymer solutions, except in a purely pragmatic fashion. The Flory-Huggins expressions can be used to determine values for χ_1, κ_1 and ψ_1 from experimental data. The values so obtained can subsequently be used to calculate the magnitude of the steric repulsion, for example.

The major difficulty associated with any theoretical description of aqueous polymer solutions is the strong specific interactions (e.g. H-bonding or the weaker dipole-dipole interactions) that can occur between the water molecules and the segments of the polymer chain. This difficulty is compounded by the extreme uncertainties still surrounding the structure of liquid water itself. A further problem that can attend measurements performed on aqueous polymer solutions is the difficulty of ensuring that complete dissolution of the polymer takes place and, even if it does, that subsequent aggregation of the dissolved species does not occur.

The enthalpy and entropy of dilution parameters for polymers (e.g. poly(oxyethylene), poly(vinyl alcohol)) in aqueous solutions are frequently found experimentally to be negative. This implies that mixing is exothermic, which contrasts with the endothermic mixing usually observed at room temperature for nonpolar systems. Negative values for the dilution parameters connote the occurrence of phase separation on heating. This often occurs at temperatures well below the critical temperature of water, suggesting that factors other than free volume effects (e.g. dehydration) could be responsible. The phase diagrams for aqueous polymer solutions can even display closed loops of nonmiscibility. This means that phase separation occurs on heating the polymer solution but that complete solution occurs on heating to higher temperatures (see Fig. 3.11) (Burchard, 1983).

Closed loop coexistence curves are well documented for binary solutions of

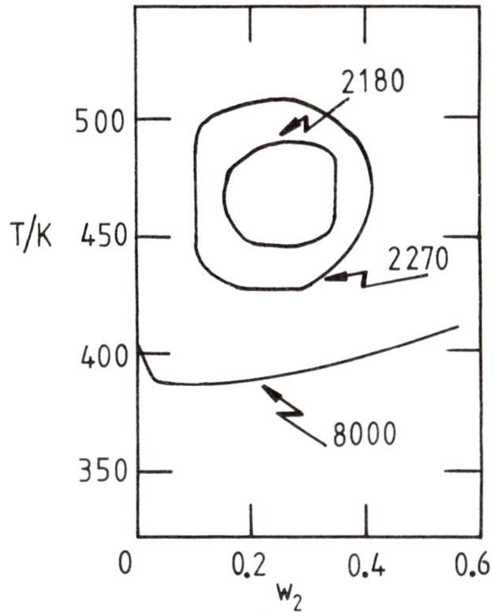

Fig. 3.11. The phase diagrams of poly(oxyethylene) of differing molecular weights in water (after Saeki *et al.*, 1976).

small molecules, the classic example being the nicotine + water system. Hirschfelder *et al.* (1937) were perhaps the first to suggest that this curious behaviour was a consequence of highly directional, short-ranged interactions which require quite specific orientations of the molecules (such as the linearity demanded by hydrogen bonds). At low temperatures, the decrease in free energy due to these specific interactions gives rise to miscibility. As the temperature rises, however, entropy effects favour the more random orientation of the molecules. This tends to destroy the directional bonds, resulting in an unfavourable free energy of solution and thus in phase separation. At still higher temperatures, however, the combinatorial entropy effects win out, resulting in an upper critical solution temperature.

Attempts have been made by Barker and Fock (1953) and by Wheeler (1975) to quantify these ideas using a lattice or a decorated lattice model. In their model, there were two components for which it was assumed that the only nonzero interactions occurred between unlike molecules. These interactions could be either attractive or repulsive. Each molecule was taken to have both a special contact point and a number of ordinary contact points. The condition for attraction was that the special contact point of one type of molecule must be directed towards a neighbouring molecule of the other type. This model was found to predict closed loop coexistence.

Presumably the foregoing concepts may be applied not only to small molecules but also to macromolecules that display closed loop coexistence. It should be remembered, however, that such molecules might be expected on free volume considerations to display a second upper critical solution temperature on heating to near the critical point of water. Symmetry considerations would then suggest the existence of a second lower critical solution temperature at temperatures below the melting point of ice. Such a hypothetical lower critical temperature was first proposed for quite different reasons by Molyneux (1975). He estimated it to occur at $-15\,^{\circ}C$ for poly(oxyethylene) in water on the basis of the decreasing solvency that water displays for that polymer on cooling below $45\,^{\circ}C$. A qualitatively similar conclusion was reached independently by Chew and Couper (1976), who estimated the corresponding θ-temperature to be $5\,^{\circ}C$.

The interaction parameter for some aqueous systems increases with increasing polymer concentration and may even pass through a maximum (see Fig. 3.12) (Burchard, 1983). Molyneux (1975) has reviewed in great detail the solution thermodynamics of seven of the most commonly encountered water soluble synthetic polymers: poly(vinyl alcohol), poly(oxyethylene), poly(acrylamide), poly(acrylic acid), poly(methacrylic acid), poly(methacrylamide) and poly(vinyl pyrrolidone). In the case of poly(oxyethylene), for example, calorimetric measurements of the heats of mixing of the polymer with water point to a maximum exotherm at a 2:1 molar ratio of water to oxyethylene units. This suggests that each oxyethylene unit may hydrogen bond cooperatively with pairs of water molecules, presumably through the two lone-pair hybrid orbitals of the oxygen atom:

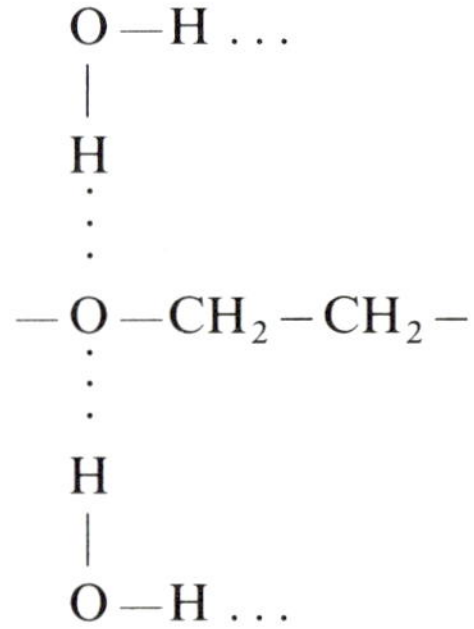

The calorimetric measurements, however, suggest that in polymer-rich systems the introduction of the first few water molecules gives rise to a much reduced exotherm, perhaps because the water molecules now hydrogen bond to two separate oxyethylene units:

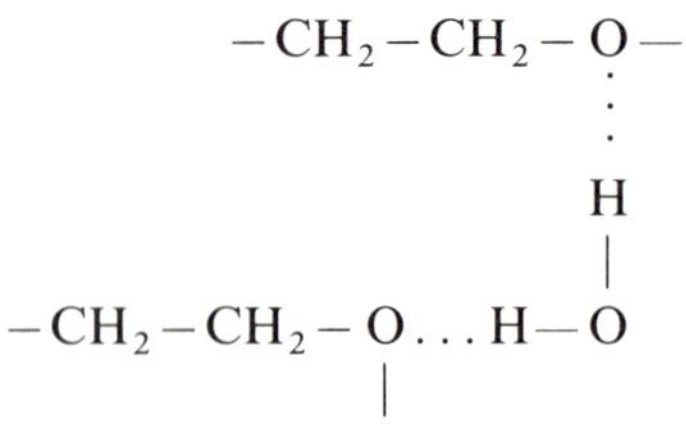

As a result, the two hydrogen bonds that are broken in removing each water

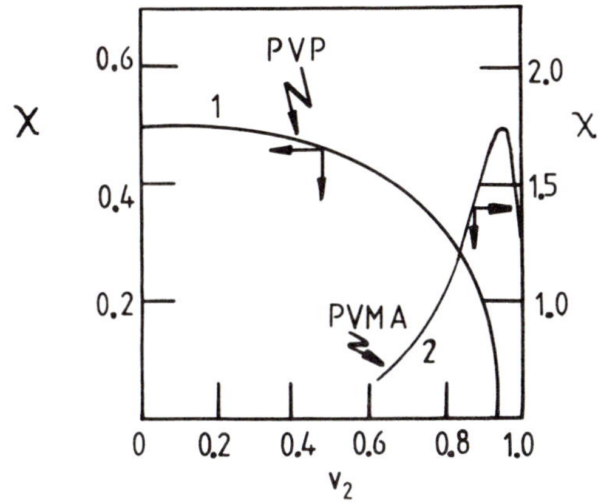

Fig. 3.12. The experimental concentration dependence of the interaction parameter for aqueous solutions of (1) poly(vinyl pyrrolidone), PVP and (2) poly(vinyl methoxyacetal), PVMA.

molecule from the bulk water structure would be replaced by two slightly stronger hydrogen bonds bridging the separate oxyethylene units. The foregoing picture is broadly in agreement with the observation by Malcolm and Rowlinson (1957) that the excess (i.e. noncombinatorial) entropy of dilution of poly(oxyethylene) in water is negative and increases strongly with increasing polymer concentration.

The above mentioned concepts imply that at high dilutions, the exothermic interactions between the polymer and the water molecules promote mixing. As the polymer concentration increases, however, this contribution to mixing declines. Further, the decrease in entropy due to structuring increases in absolute magnitude. Both factors would reduce the solvency of the water for the polymer and lead to an increase in the magnitude of the interaction parameter as the polymer concentration increases. Note that, in one crude sense, hydrogen bonding can take the place of the free volume interactions that occur near to the critical point in nonaqueous polymer solutions. This may possibly explain why some aqueous polymer solutions exhibit a lower critical solution temperature so far removed from the critical point of water. The latter inference, however, is mere conjecture at this point in time.

Finally, we note that not all aqueous polymer solutions exhibit the foregoing behaviour. Thus, for example, poly(acrylic acid) and poly(acrylamide) at low pH display positive values for the enthalpy and entropy of dilution parameters (Silberberg *et al.*, 1957). Day and Robb (1981) have confirmed calorimetrically that the heat of dilution of poly(acrylamide) in water is endothermic, although, in contrast, they found that the substituted polyacrylamides poly(N-methyl acrylamide) and poly(N,N-dimethyl acrylamide) displayed negative heats of dilution. It was proposed that the endothermic character of the poly(acrylamide) solutions was a consequence of the energy required to separate the associated amide dipoles.

Chapter 4

The Conformation of Polymer Molecules

4.1. Introduction

The stabilizing effects of polymer molecules are critically dependent upon their spatial extension and, thus, upon their conformation. It is therefore necessary to preface our detailed consideration of the stabilizing effects of polymer molecules by a discussion of their conformational properties. The number of conformations that are accessible to the normal carbon backbone polymer is of order 2^n, where n is the number of bonds in the polymer. Clearly, even for quite low molecular weights (e.g. 10 000), the number of individual conformations available to each polymer molecule is of order 2^{100} ($\sim 10^{30}$), which is astronomically large. Consideration of individual conformations of polymers is clearly precluded. Statistical procedures have therefore been devised for describing polymer conformations.

4.2. The dimensions of a polymer chain

4.2.1. rms end-to-end length

Synthetic macromolecules do not in general exhibit relatively fixed conformations as do some biopolymers. Rather, each polymer molecule in solution is bombarded by the solvent molecules and, as a result of the rotations about bonds, scans through the huge range of accessible conformations. The chains assume neither their fully extended nor their fully compressed conformations since these are both highly ordered microstates. Instead, they adopt on average an intermediate conformation, usually referred to as a *random coil* (see Fig. 4.1). In actual fact, the chains are not truly random because the valency angles are fixed and there are severe restrictions to free rotation about all bonds. The conformation might better be termed a *pseudo-random coil.*

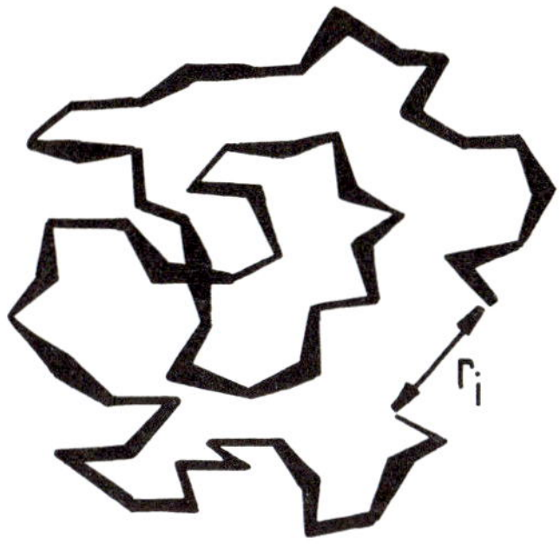

Fig. 4.1. Schematic representation of the 3-D random coil conformation of a linear polymer showing the end-to-end distance $\mathbf{r}_i$.

One parameter that is commonly used to specify the size of a linear polymer molecule is the *root-mean-square (rms) end-to-end length*

$$\langle r^2 \rangle^{\frac{1}{2}} = \left(\frac{\sum_i n_i r_i^2}{\sum_i n_i} \right)^{\frac{1}{2}} = \left(\frac{\int_0^\infty r^2 W(r)\, dr}{\int_0^\infty W(r)\, dr} \right)^{\frac{1}{2}} \tag{4.1}$$

where n_i = number of conformations with the end-to-end length r_i and $W(r)$ = distribution function for the end-to-end distance. A crude physical visualization of $\langle r^2 \rangle^{\frac{1}{2}}$ is as the mean "diameter" of the polymer chain.

4.2.2. rms radius of gyration

The end-to-end distance is an ambiguous quantity for a branched chain. In such cases, the *rms radius of gyration* R_g is used as a measure of the polymer size

$$R_g = \langle s^2 \rangle^{\frac{1}{2}} = \left(\frac{\sum_i n_i s_i^2}{\sum_i n_i} \right)^{\frac{1}{2}} \tag{4.2}$$

In this definition

$$s_i = \left(\frac{\sum_j m_{ij}\, q_{ij}^2}{\sum_j m_{ij}} \right)^{\frac{1}{2}} \tag{4.3}$$

which is the radius of gyration of the ith conformation. Here m_{ij} = the mass of the jth element in the ith conformation and q_{ij} = distance of the jth element from the centre of mass of the ith conformation. Note that this definition of the radius of gyration is rather different from that for rotors in classical physics in that no axis of rotation is specified, all distances being measured relative to the centre of mass. The radius of gyration of a molecule can be measured directly by light scattering.

For linear polymers, there is a very simple relationship between $\langle r^2 \rangle^{\frac{1}{2}}$ and R_g

$$R_g = \langle r^2 \rangle^{\frac{1}{2}} / 6^{\frac{1}{2}} \tag{4.4}$$

so that R_g corresponds to the "radius" of the chains.

4.3. Models for a polymer chain

4.3.1. The random flight chain

The simplest, and most primitive, model for a polymer molecule is the *random flight chain*, also termed the *freely jointed chain*. In this model, the bonds are represented by volumeless lines in space and there are no restrictions on the valency angles or on the rotations about bonds.

The end-to-end vector $\mathbf{r}$ is simply the sum of the individual bond vectors $\mathbf{l}_k$

$$\mathbf{r} = \sum_{k=1}^{n} \mathbf{l}_k \tag{4.5}$$

The scalar product of $\mathbf{r}$ with itself gives r^2 which can then be averaged. Following this prescription, we have

$$r^2 = \mathbf{r}.\mathbf{r} = \sum_{k=1}^{n} \sum_{j=1}^{n} \mathbf{l}_k.\mathbf{l}_j \tag{4.6}$$

so that

$$\langle r^2 \rangle = \sum_{k=1}^{n} \sum_{j=1}^{n} \langle \mathbf{l}_k.\mathbf{l}_j \rangle \tag{4.7}$$

There are two different types of terms involved in those summations. First, there are terms of the type $\mathbf{l}_k.\mathbf{l}_j$ $(k \neq j)$ that equal $l^2 \cos\tau$, where τ is the angle between the two vectors. Each of these terms, on averaging, yields $l^2 \langle \cos\tau \rangle$ which must be zero, since the angle τ is random. The other terms, and there are n of them, are characterized by $k = j$ so that $\mathbf{l}_k.\mathbf{l}_k = l^2 \cos 0$, which averages to l^2. Accordingly

$$\langle r^2 \rangle_0^{\frac{1}{2}} \phi\tau = l n^{\frac{1}{2}} \tag{4.8}$$

where the subscripts $0\phi\tau$ have been appended to signify a freely jointed chain. The subscript zero denotes that the model is composed of volumeless lines in space whereas ϕ and τ denote the absence of restriction on bond rotations and valency angles respectively.

The foregoing problem of the random flight chain was first considered, using a different approach, by Lord Rayleigh in 1919 (Rayleigh, 1919). The problem, however, had been first posed formally in 2-D by the statistician Karl Pearson in a letter to *Nature* some years earlier (Pearson, 1905). In reply, Rayleigh (1905) pointed out that the problem of a random walk in 2-D is identical with that of a number of isoperiodic vibrations of unit amplitude whose phases are distributed at random. This question Rayleigh had treated many years earlier (Rayleigh, 1880) in connection with the theory of sound.

Note that equation (4.8) predicts that the dimensions of a freely jointed chain vary as $M^{\frac{1}{2}}$, where M = molecular weight. For real chains, this prediction is strictly only correct under certain specific conditions: in θ-solvents, in the melt and, perhaps, in the so-called concentrated regime (see Section 4.6).

4.3.1.1. The equivalent freely jointed chain

It will be seen below that the conformation of a particular bond in a polymer chain is determined *inter alia* by the conformation of the preceding and subsequent bonds in the chain. These bond correlations expand the dimensions of the chain to greater than those predicted by the freely jointed chain. Nonetheless, it is permissible in the limit of high molecular weight to represent the polymer conformation by an equivalent freely jointed chain.

The concept underlying this equivalence is illustrated in Fig. 4.2. Whereas the conformations of bonds $(i+1)$, $(i+2)$, etc. are correlated with the conformation of bond i, that of bond $(i+m^*)$ is essentially independent of

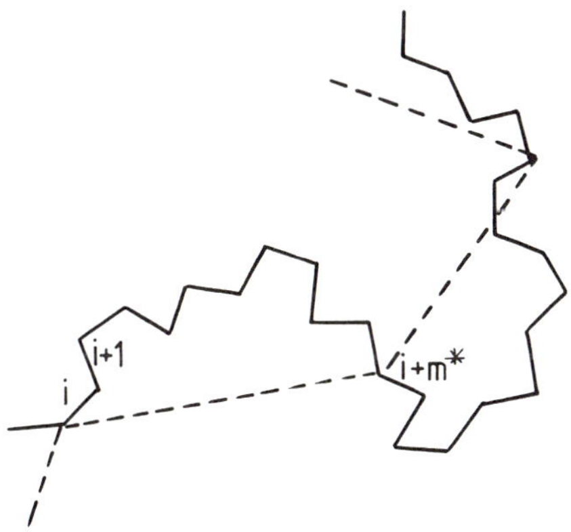

Fig. 4.2. Schematic representation of a section of a polymer chain and its representation by an equivalent freely jointed chain.

bond i because it is sufficiently far removed from that bond for it not to feel any significant influence. The intervening m^* bonds can be replaced by an equivalent single bond. Each set of m^* bonds can be similarly replaced by an equivalent bond. The directions of these equivalent bonds are quite independent of one another. Suppose there are n^* such equivalent freely jointed bonds, then clearly

$$\langle r^2 \rangle_0 = C_n l^2 n = n^* l^{*2} \tag{4.9}$$

where C_n = characteristic ratio (see below) and l^* = length of an equivalent bond. One problem with this relationship is that there are two undetermined parameters: n^* and l^*. This indeterminacy can be overcome by introducing an additional arbitrary scaling assumption. Suppose, for example, that it is assumed that the contour length of the equivalent freely jointed chain is identical with that for the real chain. The number of bonds per equivalent link $m^* = n/n^*$. Moreover, $n^* l^* = r_{\text{max}} = nl\sin(\tau/2)$, where τ = valency angle (assumed to be constant along the backbone) and r_{max} = contour length. These relationships yield

$$m^* = C_n/\sin^2(\tau/2) \tag{4.10}$$

Since C_n is characteristically of order 7 (see Table 4.1), the value of m^* for a tetrahedral valency angle (with $\sin^2(\tau/2) \sim 0{\cdot}67$) is ca 10. These 10 bonds span a distance

$$l^* = C_n l/\sin(\tau/2) \tag{4.11}$$

which equals 1·3 nm for a carbon backbone polymer.

4.3.1.2. The characteristic ratio

The result in equation (4.8) for a freely jointed chain leads to the definition of the characteristic ratio C_n of a real polymer molecule composed of n bonds. This is

$$C_n = \langle r^2 \rangle_0 / nl^2 \tag{4.12}$$

where $\langle r^2 \rangle_0$ = rms end-to-end length in a θ-solvent (the unperturbed dimensions). Values of C_∞ for polymer chains typically lie in the range 4–10, the smaller values signifying polymers with greater flexibility. Obviously, the characteristic ratio of a freely jointed chain is unity.

The characteristic ratio of a real chain can be considered crudely in terms of a time averaged spherical model (polymers adopt at any instant a shape that is more sausage like than spherical but undergo rotational random motion). Accordingly, C_n is the ratio of the average cross-section of the real chain compared with that of the corresponding random flight chain. Some values of C_∞ for common polymers are shown in Table 4.1.

TABLE 4.1
The characteristic ratio of some common polymers

polymer	C_∞
poly(oxyethylene)	3·9
poly(dimethylsiloxane)	6·4
poly(methylene)	6·9
poly(oxymethylene)	8·5
polystyrene	10·0

4.3.1.3. The span and the Hollingsworth radius

The *span* of a polymer chain in a given direction is defined as the maximum extension of the chain in that direction for a given conformation. The conformation averaged span is thus a measure of the mean diameter of the molecule. A set of spans in three orthogonal directions serves to define the dimensions of the average rectangular box occupied by a polymer chain. The typical shape of a circumscribing box for a polymer chain is significantly noncubic, with relative lengths of order 2·5 : 1·5 : 1. This is a consequence of asymmetric conformations being more numerous than symmetric ones.

The average span $\langle S_x \rangle$ of a random flight chain is given by

$$\langle S_x \rangle = (8/3\pi)^{\frac{1}{2}} l n^{\frac{1}{2}} = 0{\cdot}92 l n^{\frac{1}{2}} \tag{4.13}$$

The *Hollingsworth radius* $\langle R_{\text{Holl}} \rangle$ of a polymer chain is the radius of the smallest sphere which contains all the segments of a chain and is centred on the starting point of the chain. For a random coil chain,

$$\langle R_{\text{Holl}} \rangle = (\pi^3/24)^{\frac{1}{2}} l n^{\frac{1}{2}} = 1{\cdot}14 l n^{\frac{1}{2}} \tag{4.14}$$

Note that the value of $\langle r^2 \rangle^{\frac{1}{2}}$ for the freely jointed chain is intermediate between the average span and the Hollingsworth radius

$$\langle S_x \rangle : \langle r^2 \rangle^{\frac{1}{2}} : \langle R_{\text{Holl}} \rangle = 0{\cdot}92 : 1{\cdot}00 : 1{\cdot}14$$

4.3.2. The freely rotating chain

One model that is one notch higher in complexity from the random flight chain is the freely rotating chain. This model assigns a fixed valency angle (τ) to the chains whilst maintaining unhindered rotation about bonds. For high molecular weight polymers ($n \to \infty$), it can be shown (Flory, 1969) that

$$\langle r^2 \rangle_{0\phi}^{\frac{1}{2}} = l n^{\frac{1}{2}} \left\{ \frac{1 - \cos\tau}{1 + \cos\tau} \right\}^{\frac{1}{2}} \tag{4.15}$$

where the subscript 0ϕ denotes the freely jointed chain. A carbon backbone polymer can be considered to have a tetrahedral valency angle, ignoring the usually minor distortion therefrom, for which $\cos\tau = -1/3$. This leads to

$$\langle r^2 \rangle_{0\phi}^{\frac{1}{2}} = l(2n)^{\frac{1}{2}}$$

and a characteristic ratio $C_\infty = 2$. This value is just double that of C_n for a random flight chain.

4.3.3. Restrictions to free rotation

Up to this point it has been assumed that there are no restrictions to free rotations about bonds. This is physically unrealistic. Both macromolecules and minimolecules display bond rotational potentials that depend strongly upon the respective torsional angle. These restrictions can be incorporated into the expression for the unperturbed dimensions *viz*

$$\langle r^2 \rangle_0^{\frac{1}{2}} = l n^{\frac{1}{2}} \left\{ \frac{1 - \cos\tau}{1 + \cos\tau} \right\}^{\frac{1}{2}} \sigma_f \qquad (4.16)$$

Here $\sigma_f (= C_n^{\frac{1}{2}}(1 + \cos\tau)^{\frac{1}{2}}/(1 - \cos\tau)^{\frac{1}{2}})$ is the *steric factor*, which clearly also is a measure of the flexibility of the polymer chain. Usually $1{\cdot}5 \lesssim \sigma_f \lesssim 2{\cdot}5$ and for many carbon backbone polymers $\sigma_f \sim 2{\cdot}0$ (corresponding to $C_n \sim 8$).

4.3.3.1. Independent bond rotational potentials

In developing a more realistic model of a polymer, the simplest assumption to make is that the bond rotational potentials are independent [i.e. the conformation of bond i is independent of that of bonds $(i-1)$ and $(i+1)$]. The origin of the bond rotational potentials is illustrated schematically in Fig. 4.3 by the Newman projections for n-butane. These are *first order* effects in that they arise from the interactions between substituents separated by only one bond. The *eclipsed* conformations correspond to the peaks in the potential energy diagram whereas the *staggered* conformations, of which there are three, correspond to the minima. The lowest potential energy conformation, corresponding to a torsional angle $\phi = 0$, is that of the *trans* (t) conformation of the central bond. The minima that are symmetrically placed about the trans conformation correspond to the two equivalent *gauche* (g^+ and g^-) conformations.

It can be shown that the rms size of a polymer molecule of high molecular weight subject to *independent bond rotational potentials* is given by

$$\langle r^2 \rangle_0^{\frac{1}{2}} = l n^{\frac{1}{2}} \left(\frac{1 - \cos\tau}{1 + \cos\tau} \right)^{\frac{1}{2}} \left(\frac{1 + \langle \cos\phi \rangle}{1 - \langle \cos\phi \rangle} \right)^{\frac{1}{2}} \qquad (4.17)$$

where the angular brackets $\langle\,\rangle$ refer to the statistical mechanical average value over all torsional angles.

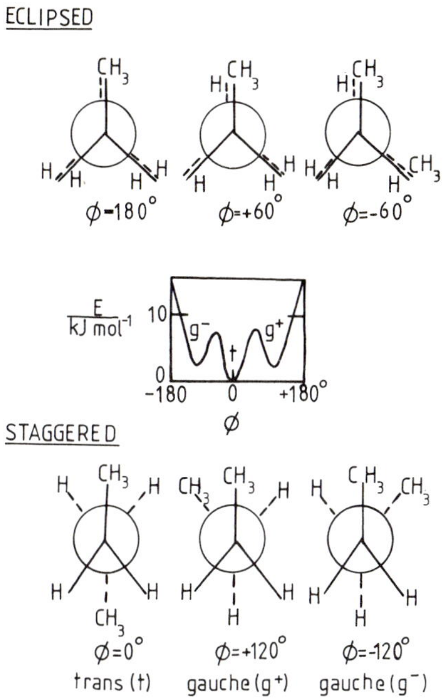

Fig. 4.3. Newman projections of the conformers of *n*-butane, together with its potential energy diagram showing the barriers to free rotation.

Comparison with experiment

The characteristic ratio of the poly(methylene) chain, which might be considered to be the simplest polymer molecule, as a function of chain length is shown in Fig. 4.4 for the simple models considered thus far. All save the simple freely jointed chain display end effects, manifest by an increase in C_n with chain length at small n, which will not be considered further. In what follows, only the asymptotic limit of the characteristic ratio for $n \to \infty$ (C_∞) will be discussed. The values of C_∞ for the freely jointed chain, the freely rotating chain and the chain with independent bond rotational potentials increase in value from 1 through 2 to ca 3·5. The latter value is, however, only ca one-half of the experimentally determined value of $C_\infty = 6{\cdot}9$ for poly(methylene). This serious discrepancy points to the fact that the bond rotational potentials are definitely not independent, i.e. the conformation of bond i depends upon the conformations of bonds $(i-1)$ and $(i+1)$.

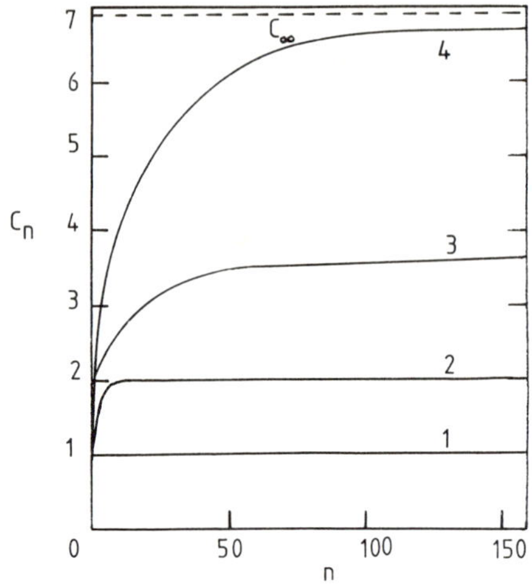

Fig. 4.4. Plots of the characteristic ratio of poly(methylene) at 140 °C as a function of the number of bonds using different theoretical models: (1) the freely jointed chain; (2) the rotating chain; (3) independent bond rotational; (4) interdependent bond rotational potentials (after Flory, 1969).

4.3.3.2. Interdependent bond rotational potentials

The origins of the *interdependence* of the bond rotational potentials in poly(methylene) can be understood in terms of the conformations of *n*-pentane (hence the phenomenon is termed the *pentane effect*).

Pentane has four bonds, of which the conformations of the central two can be specified. If both bonds are *trans* (*tt*), the pentane has minimal conformational energy (see Fig. 4.5). If only one bond is rotated to give either of the *gauche* conformations (tg^+ or tg^-), no great injury is inflicted on the terminal bonds in terms of their mutual interaction. When the second bond is rotated, however, the mutual interaction of the terminal bonds depends upon the resultant conformation: if the conformation is g^+g^+ (or g^-g^-), the terminal bonds lie on either side of the plane containing the two central bonds and there is still no cataclysmic interaction. In sharp contrast, the g^+g^- conformation (or its equivalent g^-g^+) results in the terminal bonds being in such close proximity that their steric interactions raise the energy massively (by ca 10 kJ mol^{-1}). This drastically reduces the incidence of g^+g^- conformations, which are highly compressed in terms of their spatial extension. The low probability (e.g. just a few per cent) of these compressed conformations causes the overall chains to be much more expanded than predicted by assuming independent bond rotational potentials. The large value measured for C_∞ for poly(methylene) is thus explained by these *second order* interactions (i.e. interference between groups separated by two bonds).

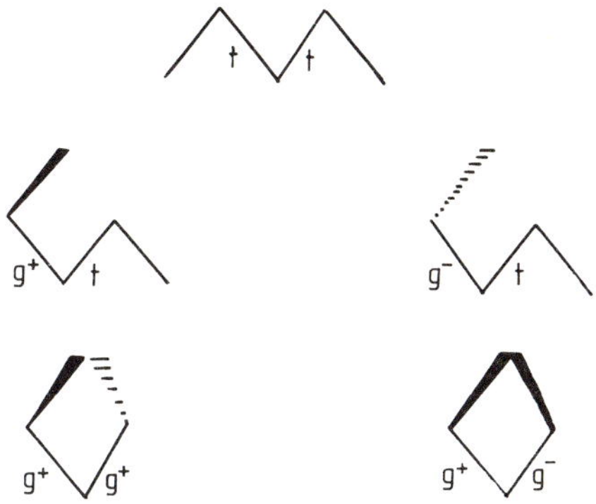

Fig. 4.5. Schematic representation of the pentane effect.

4.3.4. The rotational isomeric state theory

Flory (1969; 1971; 1974) has developed the rotational isomeric state theory for predicting the conformation of polymer molecules. This incorporates the interdependence of the bond rotational potentials. This theory is fully explained in Flory's comprehensive monograph (Flory, 1969) and only the barest details will be mentioned here.

The rotational isomeric state theory is essentially a matrix procedure that, in principle, allows all of the myriad of conformations accessible to a polymer molecule to be appropriately averaged. The end-to-end length **r** is simply the sum of the individual bond vectors $\mathbf{l}_i$

$$\mathbf{r} = \mathbf{l}_1 + \mathbf{l}_2 + \mathbf{l}_3 + \ldots + \mathbf{l}_n \tag{4.18}$$

These bond vectors are readily expressed in their own coordinate system as the column vector col (l 0 0). To find **r**, however, the vectors must all be expressed in the same coordinate system, which we will take to be that of the first bond (see Fig. 4.6). This can be accomplished by using the transformation matrix $\mathbf{T}_i$. Premultiplication by this matrix transforms a vector expressed in reference frame $(i+1)$ into its representation in reference frame i of the preceding bond:

$$\mathbf{r} = \mathbf{l}_1 + \mathbf{T}_1\mathbf{l}_2 + \mathbf{T}_1\mathbf{T}_2\mathbf{l}_3 + \mathbf{T}_1\mathbf{T}_2\mathbf{T}_3\mathbf{l}_4 + \ldots \tag{4.19}$$

It can be shown that

$$\mathbf{T}_i = \begin{pmatrix} \cos\theta_i & \sin\theta_i & 0 \\ \sin\theta_i\cos\phi_i & -\cos\theta_i\cos\phi_i & \sin\phi_i \\ \sin\theta_i\sin\phi_i & -\cos\theta_i\sin\phi_i & -\cos\phi_i \end{pmatrix}$$

where θ_i = supplement of the valency angle between bonds $(i+1)$ and i and

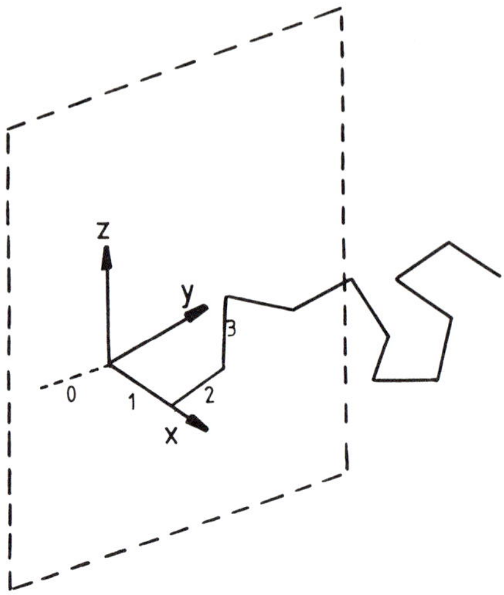

Fig. 4.6. Establishment of the rectangular co-ordinate origin on the first bond for rotational isomeric state calculations: bond 2 is *trans* to the hypothetical bond 0; the x-direction is along bond 1; the y-direction has a positive intercept on the projection of bond 0; the z-direction completes a right-handed Cartesian system (after Flory, 1969).

ϕ_i = dihedral angle between the planes defined by the respective pairs of bonds $(i-1)$, i and i, $(i+1)$.

The above expression can be written in a much more compact form by introducing generator matrices $\mathbf{A}_i$, the serial multiplication of which generates the unwieldy sum in equation (4.19). Thus

$$\mathbf{r} = \mathbf{A}_{[1}\mathbf{A}_2^{(n-2)}\mathbf{A}_{n]}, \quad n \geq 2 \tag{4.20}$$

where

$$\mathbf{A}_i = \begin{pmatrix} \mathbf{T} & \mathbf{l} \\ \mathbf{0} & \mathbf{1} \end{pmatrix}_i \tag{4.21}$$

Here i denotes the serial index of the matrix $\mathbf{T}$ and bond vector $\mathbf{l}$ within the generator matrix. The superscript $(n-2)$ in equation (4.20) denotes the serial multiplication of $(n-2)$ matrices beginning with $\mathbf{A}_2$. The initial matrix $\mathbf{A}_{[1}$ is defined by the first row of $\mathbf{A}_1$ and the final matrix $\mathbf{A}_{n]}$ is correspondingly the last column of $\mathbf{A}_n$.

The elegance of the matrix expression for $\mathbf{r}$ resides in associating each matrix with a corresponding bond so that the serial multiplication mimics the concatenation of the bonds in the chain.

Once $\mathbf{r}$ has been calculated, the moments of the distribution function can be found. For chains in free solution, Flory has developed an elegant generator matrix technique that by-passes $\mathbf{r}$ and calculates the second moment $\langle r^2 \rangle$ directly. Unfortunately, for attached polymer chains, such as terminally attached tails, the presence of an impenetrable interface breaks the symmetry of the system. This precludes the application of the direct matrix multiplication technique, so successful for chains in free solution. It is necessary in the presence of an interface to resort to the numerical Monte Carlo method for the computation of the conformational properties (Feigin and Napper, 1979; Mattice and Napper, 1981).

In the Monte Carlo method, the bonds are assigned to the different rotational states according to their conditional probabilities. These are calculated, using the eigenvalue method outlined by Flory (1969), from the corresponding statistical weight matrix:

$$
\begin{array}{cc} & \begin{array}{cccc} (i-1)/i & (t) & (g^+) & (g^-) \end{array} \\ \mathbf{U}_i = \begin{array}{c} (t) \\ (g^+) \\ (g^-) \end{array} & \begin{pmatrix} 1 & \sigma & \sigma \\ 1 & \sigma\psi & \sigma\omega \\ 1 & \sigma\omega & \sigma\psi \end{pmatrix} \end{array} \qquad (4.22)
$$

In this matrix, the convention is adopted that all statistical weights are assigned according to the premise that all succeeding bonds are *trans* (i.e. *third order* interactions are neglected). The parameters in equation (4.17) have the following meaning: σ is the Boltzmann weighting factor for a *gauche* state relative to a *trans*; $\sigma\omega$ corresponds to the statistical weight for bond i in the *gauche* state if bond $(i-1)$ is in the *gauche* state of opposite sign; the statistical weight $\sigma\psi$ is that for bond i in a *gauche* state when bond $(i-1)$ is in a *gauche* state of the same sign.

Once the conditional probability matrix has been calculated, a random number is generated by the computer (hence the name *Monte Carlo*) and its value used to assign the conformation of each bond, conditional upon the conformation of the preceding bond. The system is initialized by taking the initial bond as being in the *trans* conformation. By calculating many thousands of such conformations, the average conformational properties of the chain can be estimated with reasonable precision.

4.3.5. The unperturbed dimensions

Rotational isomeric state theory predicts the dimensions of polymer chains in θ-solvents, i.e. the unperturbed dimensions $\langle r^2\rangle_0^{\frac{1}{2}}$, where the subscript 0 denotes that the bonds are considered to be volumeless lines in space. As stated previously, these can be expressed as

$$\langle r^2\rangle_0^{\frac{1}{2}} = ln^{\frac{1}{2}}\left(\frac{1-\cos\tau}{1+\cos\tau}\right)^{\frac{1}{2}} \quad \sigma_f \qquad (4.23)$$

where σ_f = steric factor. The value of σ_f usually lies in the range 1·5–2·5, the smaller values signifying the more flexible polymers. The unperturbed dimensions are also adopted by macromolecules in the melt and the amorphous state, as well as, perhaps, in the so-called concentrated regime.

4.4. The excluded volume problem

4.4.1. The intramolecular expansion factor

The bonds of the polymer molecule have been considered thus far as volumeless lines in space. In contradistinction, the bonds of real polymer molecules occupy finite volumes in space. The repulsion between segments that are far removed from one another along the contour of the chain but spatially contiguous will cause the polymer chain to expand beyond the unperturbed dimensions. The magnitude of this dimensional expansion is measured by the intramolecular expansion factor α

$$\langle r^2 \rangle^{\frac{1}{2}} = \alpha \langle r^2 \rangle_0^{\frac{1}{2}} \tag{4.24}$$

where $\langle r^2 \rangle^{\frac{1}{2}}$ = the measured rms end-to-end distance of the chain. Obviously in a θ-solvent, the reference conformational state, the value of $\alpha = 1{\cdot}00$. Usually, α lies in the range 1·0–1·5 in good solvents. In poor solvents, values of α significantly less than 1 are difficult to achieve. The reason for this is that in worse than θ-solvents, the segments are mutually attractive ("sticky") and so *intermolecular* association usually occurs at the expense of intramolecular collapse (the *coil-to-globule* transition). This intermolecular association is manifest as phase separation. Intramolecular collapse has, however, been observed in exceedingly dilute solutions of polystyrene in cyclohexane at temperatures below the θ-temperature. Values of α as small as 0·4 were then observed (Swislow *et al.*, 1980).

4.4.2. Theories of excluded volume

4.4.2.1. Free chains

Flory (1953) has presented a celebrated theory of the excluded volume effect that relates the expansion factor to the thermodynamic properties of the polymer-solvent system. Basically what Flory calculated was the free energy of mixing of polymer segments with solvent that is associated with the expansion of the coil dimensions in a good solvent. Such expansion is opposed, however, by the loss of configurational entropy of the chain. The latter corresponds, of course, to an *elastic* contractive force. Expansion proceeds until the two opposing effects are in equilibrium. Flory's result was

$$\alpha^5 - \alpha^3 = 2C_M(\tfrac{1}{2} - \chi_1)M^{\frac{1}{2}} = 2{\cdot}60z \tag{4.25}$$

where

$$C_M=(27/2^{5/2}\pi^{3/2})(\bar{v}_2{}^2/N_A\bar{V}_1)/(\langle r^2\rangle_0/M)^{3/2}$$

and

$$z=(4/3^{3/2})C_M(\tfrac{1}{2}-\chi_1)M^{\frac{1}{2}}$$

In these equations, M=molecular weight, χ_1=interaction parameter, $\bar{v}_2$=partial specific volume of the polymer, $\bar{V}_1$=molar volume of the solvent and $\langle r^2\rangle_0$=unperturbed rms end-to-end dimensions.

Note that for polymer in the bulk, $\bar{V}_1$ is very large. This means that $C_M \sim 0$ just as it is if $\chi_1=\frac{1}{2}$. Accordingly, $\alpha=1$ in the melt, a result that is identical with that for a θ-solvent. The expansion factor α increases slowly without limit with increasing molecular weight M.

There has been a proliferation of theories of the excluded volume problem using a variety of methods to decouple the many-bodied problem. Their results can be summarized crudely by

$$\lim_{z\to\infty} \alpha^{\gamma}=a+bz \tag{4.26}$$

where $1\leq\gamma\leq5$ and a, $b\simeq$constant. The many results have been well summarized by Yamakawa (1971). Perturbation theory near to the θ-point yields the exact result

$$\alpha^2{}_{R_g}=1+(4/3)z+\ldots \tag{4.27}$$

where the subscript R_g refers to the radius of gyration. The Flory theory for α near unity yields

$$\alpha^3(\alpha^2-1)\simeq\alpha^2-1=2{\cdot}60z \tag{4.28}$$

Comparison of equations (4.27) and (4.28) shows that the Flory theory in near θ-solvents predicts a numerical coefficient (2.6) that is too large by almost a factor of 2. Stockmayer (1955; 1960) has therefore suggested that Flory's original equation should be arbitrarily adjusted so that

$$\alpha^5-\alpha^3=(4/3)z \tag{4.29}$$

in accordance with the dictates of perturbation theory. Comparison of theory with experiment shows that this latter semi-empirical expression gives excellent agreement with experiment (Yamakawa, 1971).

4.4.2.2. Attached chains

It should be pointed out that Tanaka (1977) has derived the first order perturbation theory result for the expansion factor α_t for tails in near θ-conditions. The result

$$\alpha_t{}^2=1+1{\cdot}26z+\ldots \tag{4.30}$$

can be compared with equation (4.27) for free polymers where the numerical coefficient of 1·33 has been replaced by 1·26. This slight decrease in the numerical factor arises from the greater elongation of the tails normal to the surface, which reduces the probability of segmental exclusion. The difference in the expansion factors for free chains and tails is only a few per cent, even for z values as large as 10. It can be concluded from this that the application of the free chain intramolecular expansion factor to *isolated* tails will not introduce serious error. The effects of close packing of the tails on intramolecular expansion are, as yet, incompletely understood.

It is illuminating to compare the third-power type of equation for free chains and for tails. These both follow the general formula

$$\alpha^3 - 1 = Kz \tag{4.31}$$

where $K = 2{\cdot}00$ for the free chain and $K = 1{\cdot}89$ for tails. Although tails show close concordance with free polymer chains for the expansion factor in near θ-conditions, the situation may perhaps be different for loops because of their greater segment density.

Finally, we remark that the excluded volume problem for attached chains has commonly been modelled by resorting to the self-avoiding random walk (Whittington, 1975; Mark *et al.*, 1975; Lax, 1975; Middlemiss and Whittington, 1976; Wall and Seitz, 1977; Guttman and Whittington, 1978; Guttman *et al.*, 1978; Barber *et al.*, 1978). Whilst this model may be of some value in casting light upon asymptotic behaviour, the results are not generally applicable to real polymer chains because the all-or-nothing character of the self-avoiding walk simulation violates the continuously variable nature of the excluded volume of real polymer molecules.

4.5. Attached polymer chains

The spatial extension of polymer molecules at an interface depends critically upon the mode of attachment. If the chains are attached terminally at one end then they are referred to as *tails* (or sometimes as *cilia*). Terminal attachment at both ends results in what are termed *loops*. Loopy-type adsorption, as expected for homopolymers, leads to some segments abutting the interface (these are termed *trains*), together with both *tails* and *loops* (see Fig. 4.7).

4.5.1. The rotational isomeric state method

At present the only rigorous way of treating isolated polymer molecules at an interface is by a combination of rotational isomeric state theory and Monte Carlo procedures. Some results for tails have been published by Feigin and Napper (1979) and Mattice and Napper (1981). These strictly apply only in a

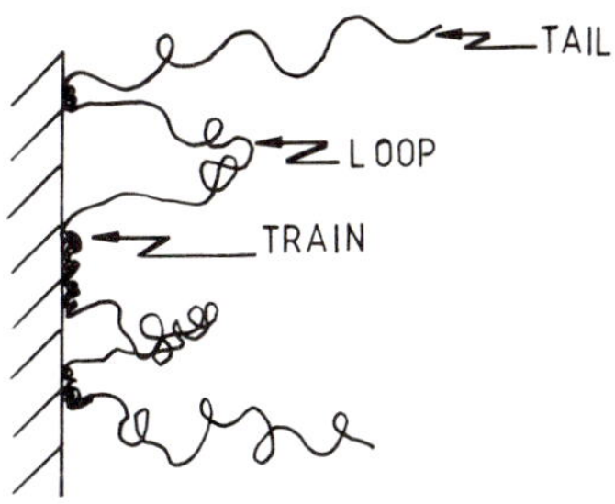

Fig. 4.7. The adsorption of a homopolymer chain at the surface of a colloidal particle showing schematically trains, loops and tails.

θ-solvent or a melt of the polymer, although it was noted in the previous section that the expansion factor for isolated chains of at least modest molecular weight is predicted theoretically to be only marginally smaller than that in free solution.

One striking feature of the conformation of isolated tails is that the spatial extension of the chain is expanded in a direction normal to the interface whilst being unaffected parallel to the interface. This is illustrated by Figs 4.8 and 4.9. These show that for poly(methylene) the mean square displacement normal to the interface for high molecular weight polymers asymptotes in value to twice that for chains in free solution. This result was first derived by DiMarzio and McCrackin (1965) for a 1-D random walk normal to an impenetrable interface. The doubling of the mean square displacement can be understood in terms of the halving of the available space normal to the interface by the presence of the impenetrable barrier. The components of the mean square displacement parallel to the interface are unaffected by its presence.

The expansion normal to the interface is reflected in the value of the

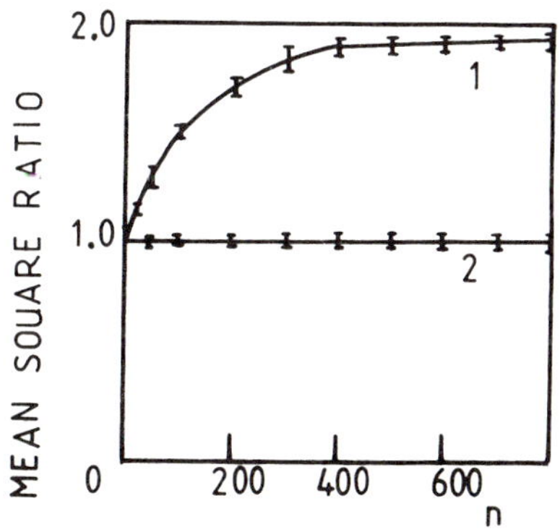

Fig. 4.8. The ratio of the mean square displacements for attached and free poly(methylene) chains: (1) normal to the interface; (2) parallel to the interface (after Feigin and Napper, 1979).

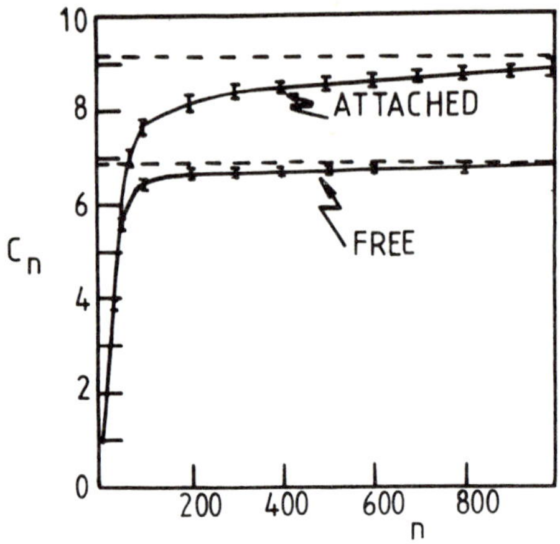

Fig. 4.9. The characteristic ratios of free and attached poly(methylene) chains at 140 °C as a function of the number of bonds in the chain. The dotted lines correspond to the respective C_∞ values (after Feigin and Napper, 1979).

characteristic ratio C_∞: this increases by the ratio 4/3 compared with the free solution value (see Fig. 4.9). This increased value would be expected because

$$\begin{aligned} C_n^{IF} &= (\langle x^2 \rangle^{IF} + \langle y^2 \rangle^{IF} + \langle z^2 \rangle^{IF})/nl^2 \\ &= (2\langle x^2 \rangle^{F} + \langle y^2 \rangle^{F} + \langle z^2 \rangle^{F})/nl^2 \\ &= (4/3)\, C_n^{F} \qquad (n \to \infty) \end{aligned}$$

for large n. In this derivation, the superscripts IF and F refer to the attached tail at the interface and the free polymer respectively.

Another striking feature of the Monte Carlo result for isolated tails is that the chain length dependent parameters converge to the limiting values only very slowly (Feigin and Napper, 1980c). The number of bonds required for comparable agreement with the limiting value is an order of magnitude larger for tails than for free chains.

A third striking feature of the results for tails is the divergence of the *persistence vector* $\mathbf{a} = \langle r \rangle$. This arises because the x-component of the persistence vector is divergent. A comparison is presented in Table 4.2 of the effects of intramolecular excluded volume with that due to an interface. It is apparent that the two different types of excluded volume exert different effects on the conformation of the polymer chain.

TABLE 4.2
Comparison of the effects of intramolecular excluded volume and that due to an interface

Parameter	Intramolecular excluded volume: Absent		Intramolecular excluded volume: Present	
	Free	Attached	Free	Attached
C_n or generalized $\tilde{C}_n(=\frac{\langle r^2\rangle}{nl^2})$	convergent	convergent	divergent	divergent
$\lvert\mathbf{a}\rvert$	convergent	divergent	convergent	divergent
$\langle x\rangle$	convergent	divergent	convergent	divergent
$\langle y\rangle$, $\langle z\rangle$	convergent	convergent	convergent	convergent

4.5.2. The Gambler's Ruin method

Finally, we draw the reader's attention to a far reaching analogy. It has been pointed out by DiMarzio and coworkers (Guttman *et al.*, 1981; 1982) that polymer chains sandwiched between parallel crystal surfaces display features closely analogous to those treated in the classical Gambler's Ruin problem of mathematical statistics. This analogy has been developed to explore the nature of the amorphous phase in semi-crystalline polymers. Its importance resides in the fact that many of the results in the Gambler's Ruin problem have already been established (Feller, 1968). It should prove possible in the future to exploit this analogy to advantage in the case of polymers attached to colloidal particles.

4.6. Scaling law theories

Scaling law theories are concerned with the *universal* properties of polymer molecules. These originate in their *global* characteristics. The detailed *local* properties are ignored. This concept is well illustrated by the radius of gyration, R_g, of a polymer chain. Flory has shown that R_g is given by

$$R_g = KN^{\gamma} \tag{4.32}$$

where K = constant, N = degree of polymerization and γ = universal exponent which is $\frac{1}{2}$ for a θ-solvent but is close to 3/5 for a good solvent. Both values of γ

are the same for all flexible polymers provided that the chains are of sufficiently high molecular weight. The value of K is not universal: it depends critically upon the detailed chemical structure of both the polymer and the solvent. Of course, the value of γ with excluded volume depends critically upon the dimensionality (d) of the problem: in fact, it may be shown that $\gamma = 3/(d+2)$ so that the excluded volume problem can be ignored for dimensionalities greater than four. The value of γ can also depend upon chain size but we shall be concerned here solely with very high molecular weight polymers.

As discussed by de Gennes (1979), scaling law theories focus attention on the exponents and eschew the determination of any prefatory constants, which are usually much more difficult to determine. Notice that the exponent γ in equation (4.32) specifies what happens to R_g when N is, say, doubled: the value of R_g will be increased by a factor of 2^γ; hence the name *scaling law*.

4.6.1. The basic scaling law

Scaling law theories are especially concerned with transitions from one regime to another. Suppose there exists a transition at some critical value x^* of the parameter x where the variable S changes its form. Then the fundamental postulate of scaling law theory is that S can be written:

$$S = S_0 f(x/x^*) \tag{4.33}$$

where the two x regimes are specified by

$$\text{(i)} \quad f(x/x^*) = 1 \qquad \text{if } x/x^* < 1$$

and

$$\text{(ii)} \quad f(x/x^*) = (x/x^*)^m \qquad \text{if } x/x^* > 1$$

Obviously, S_0 is the form adopted by S in regime (i). Such a transition is represented schematically in Fig. 4.10. Note that whilst scaling law theory can

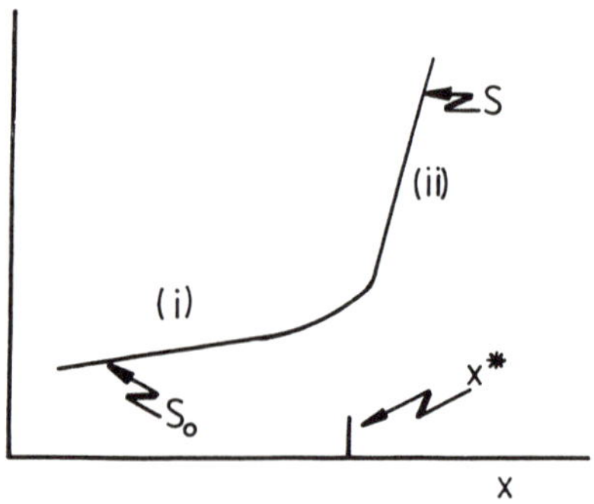

Fig. 4.10. Schematic representation of the fundamental scaling law postulate for two different regimes.

treat the two regimes far removed from the transitional region, that region itself is beyond the compass of the scaling laws. Renormalization group ideas provide some justification for the form adopted in regime (ii).

It will be apparent in what follows that all prefatory constants by and large will be ignored in most formulae, in conformity with the introductory remarks to this section. The scaling concepts as applied to polymer science are very much an outgrowth of the recognition by de Gennes (1972) and des Cloizeaux (1970) of a far-reaching analogy between the properties of ferromagnets under the action of an external field and polymer chains in more concentrated systems.

4.6.2. The dilute to semi-dilute transition

The power of scaling law theory is well exemplified by considering the transition from the dilute to the semi-dilute regime for a polymer in a good solvent. In the dilute regime, which is well-studied and reasonably well-understood, the polymer chains behave as essentially independent entities. In the semi-dilute region, the polymer chains begin to interpenetrate each other (see Fig. 4.11) but the segmental density is not large. This regime has proved difficult to handle by classical theoretical techniques but scaling law theories allow predictions to be made in this domain, some of which have already been verified experimentally.

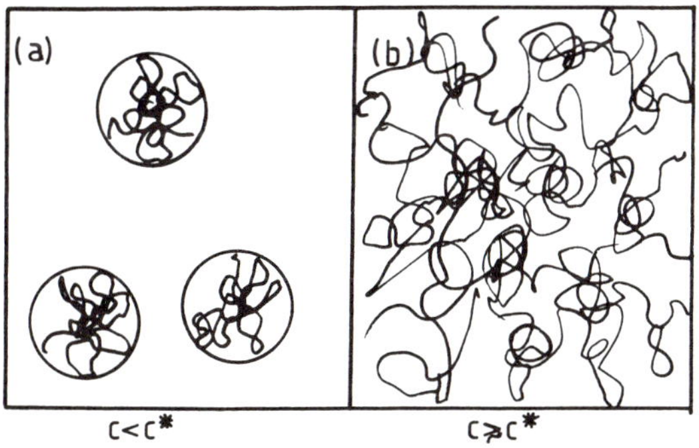

Fig. 4.11. Schematic representation of the dilute ($c < c^*$) and semi-dilute ($c \gtrsim c^*$) regimes for polymer solutions.

The threshold region at which overlap commences, signalling the transition from the dilute to the semi-dilute regime, would be expected to occur when the total volume of the chains just fills the available volume. This occurs at a polymer concentration c_2^* (whose value must be comparable to that in the polymer coil), which is given by

$$c_2^* = (\eta M/N_A)/(\eta 4\pi R_g^{\,3}/3) \tag{4.34}$$

where M = polymer molecular weight and η = number of chains. In scaling law theory this would be written as

$$c_2^* \simeq M/N_A R_g^3 \tag{4.35}$$

where the symbol $\simeq$ signifies in this context that all numerical coefficients have been ignored. As M is necessarily proportional to the number of segments, i.e. $N^{1\cdot0}$, we have from equations (4.32) and (4.35)

$$c_2^* \sim N^{1-3\gamma} \sim N^{-4/5} \tag{4.36}$$

since $\gamma = 3/5$ in a good solvent. In this equation, the symbol $\sim$ denotes that only the dependence on N has been retained ($\sim$ could have been replaced by the proportionality sign $\propto$). Note that for large N, c_2^* is small.

4.6.3. Scaling laws in the semi-dilute regime

Whereas the laws governing the dependence of many fundamental properties of polymers and polymer solutions are now well understood in the dilute regime, the same could not be said for the semi-dilute regime. Thus, for example, R_g is independent of the polymer concentration and depends upon chain length as $R_g \sim N^{3/5}$ in the dilute regime. In contrast, the concentration dependence of R_g in the semi-dilute regime is not known from classical polymer solution theories. This is one area where the scaling law procedure displays quite remarkable powers. We will illustrate this by means of three parameters, each of which is experimentally determinable: R_g, the osmotic pressure Π, and ξ. The last parameter ξ is the *correlation length* or *average mesh size*: it represents the mean distance between intersections of the trajectories of the polymer chains, as shown in Fig. 4.12. The correlation length can be readily determined from neutron scattering experiments.

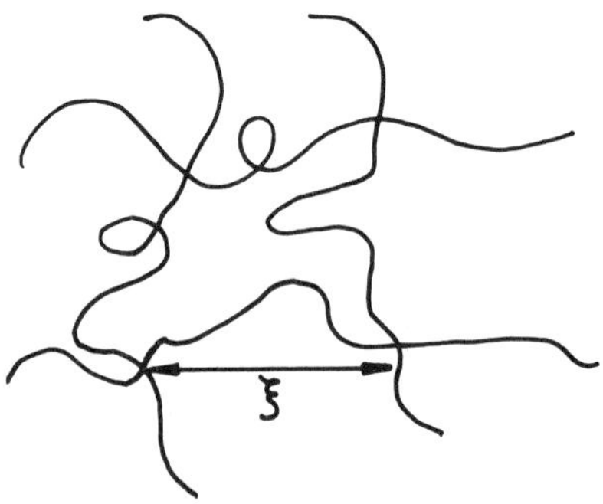

Fig. 4.12. Schematic representation of the correlation length ξ.

4.6.3.1. The radius of gyration

We have already seen that the radius of gyration in the dilute region varies as $N^{3/5}$ in a good solvent for polymers of high (infinite) molecular weight. What is required now is the concentration dependence of the radius of gyration in the semi-dilute regime.

To this end, we write the basic scaling law postulate as

$$R_g^{SD} = R_g^D f(c_2/c_2^*) = R_g^D (c_2/c_2^*)^m \tag{4.37}$$

Here the superscripts D and SD refer to the dilute and semi-dilute regimes respectively and m = universal scaling exponent. Substitution of equation (4.36) into this expression yields

$$R_g^{SD} \sim R_g^D c_2^m N^{(4/5)m} \tag{4.38}$$

In the semi-dilute range, R_g^{SD} is seen to be a function of the polymer concentration and the chain length. The scaling law theories assume that in the semi-dilute range $R_g^{SD} \sim N^{\frac{1}{2}}$. This challenging assumption, which appears to have been substantiated indirectly by the results of experiment, is surprising at first sight because it is known that this relationship is valid in the bulk phase, a point first recognized by Flory (Flory, 1953). Substitution of this relationship into equation (4.38) leads, on equating the exponents for N, to the algebraic equation:

$$1/2 = 3/5 + 4/5(m) \tag{4.39}$$

so that $m = -1/8$. Therefore

$$R_g^{SD} \sim N^{\frac{1}{2}} c_2^{-1/8} \tag{4.40}$$

This demonstrates that R_g^{SD} decreases slowly with the polymer concentration in the semi-dilute regime. Of course, in a θ-solvent, the corresponding value of γ leads to $m = 0$, i.e. $R_g \sim N^{\frac{1}{2}}$.

Harking back to the assumption introduced above that $R_g \sim N^{\frac{1}{2}}$ in the semi-dilute range, it should be noted parenthetically that the concept of *blobs* also leads to the same relationship. Blobs are regions of polymer chain of size ξ, the correlation length, inside which excluded volume effects are operative (see Fig. 4.13). Beyond the correlation length, the excluded volume correlations are negligible so that the blobs obey Gaussian statistics. Again, this implies that $R_g \sim N^{\frac{1}{2}}$.

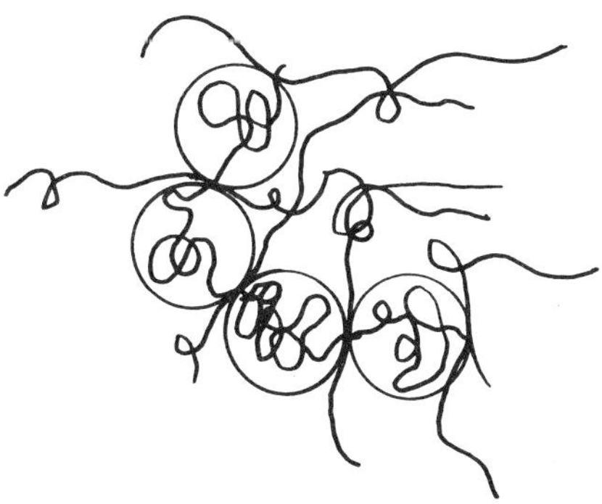

Fig. 4.13. Schematic representation of the blob concept.

4.6.3.2. The correlation length

The correlation length in the semi-dilute regime is given by

$$\xi^{SD} = \xi^D (c_2/c_2^*)^m \tag{4.41}$$

In dilute solutions ξ^D can be set equal to $R_g{}^D$. The validity of this equality can be readily seen at a polymer concentration close to c_2^*, when the coils just touch but do not interpenetrate. Recalling that $c_2^* \sim N^{-4/5}$ [see equation (4.36)], we have

$$\begin{aligned} \xi^{SD} &\sim R_g{}^D c_2{}^m N^{+(4/5)m} \\ &\sim c_2{}^m N^{(3/5+(4/5)m)} \end{aligned} \tag{4.42}$$

In the semi-dilute regime, the mesh size depends only upon the concentration of polymer, at least for polymers of high molecular weight. This seems intuitively reasonable because the mesh size should be determined solely by the total segmental concentration, being independent of chain length if that length is large in comparison with the correlation length. Thus $\xi \sim N^0 c^m$, so that the exponent equation is

$$0 = 3/5 + (4/5)m \tag{4.43}$$

leading to $m = -3/4$. Accordingly,

$$\xi^{SD} \sim c_2{}^{-3/4} \tag{4.44}$$

The corresponding analysis for a θ-solvent differs only insofar as $c_2^* \sim N^{-\frac{1}{2}}$ and $R_g{}^D \sim N^{\frac{1}{2}}$. These relationships lead to $m = -1$, i.e. $\xi_\theta^{SD} \sim c_2{}^{-1}$.

4.6.3.3. The osmotic pressure

The scaling law relationship for osmotic pressure is

$$\Pi^{SD} = \Pi^D (c_2/c_2^*)^m \tag{4.45}$$

In dilute solutions, Π^D can be approximated by the van't Hoff equation $\Pi^D \sim c_2/M$. This yields

$$\Pi^{SD} \sim c_2{}^{1+m} N^{(4/5)m}/M \sim c_2{}^{1+m}/N^{1-(4/5)m} \tag{4.46}$$

This value of Π^{SD} in the semi-dilute range must be independent of the molecular weight for high molecular weights. This is readily grasped by recognizing that the local properties (e.g. energies, entropies, etc.) are determined solely by the local segmental concentrations. They would be the same for a solution of many chains, each containing N monomers, as for a single chain ($N \to \infty$) that fills the whole vessel to the same extent. Thus $\Pi^{SD} \sim N^0 c_2{}^{1+m}$. This result, on substitution into equation (4.46), yields $m = 5/4$. Consequently,

$$\Pi^{SD} \sim c_2{}^{9/4} \tag{4.47}$$

The analogous result for a θ-solvent is derived in a similar fashion except that $c^* \sim N^{-\frac{1}{2}}$ so that $m=2$. Accordingly, $\Pi_\theta^{SD} \sim c_2^3$, which expresses the importance of three-body interactions. This is scarcely a new result. The virial expansion for the osmotic pressure.

$$\Pi = RT(c_2/M + B_2 c_2^2 + B_3 c_2^3 + \ldots) \tag{4.48}$$

reduces for a θ-solvent ($B_2=0$, by definition) to $\Pi_\theta = RTB_3\, c_2^3$ for very high molecular weight polymer. Note, however, that just as the second virial coefficient exhibits a weak ($N^{-0.2}$) dependence upon molecular weight, so might B_3 be weakly dependent upon $N^{-0.2}$.

4.6.3.4. Summary of some scaling law predictions

Some predictions of scaling law theory for polymers of high molecular weight in both good solvents and θ-solvents in the semi-dilute regime are summarized in Table 4.3. Notice that $\Pi \sim T/\xi^3$ and, further, that $\langle r^2 \rangle$ scales in the same fashion as R_g^2.

TABLE 4.3
Some scaling law predictions in the semi-dilute regime

Property	Scaling law Good solvent	θ-solvent
$c_2^* \sim$	$N^{-4/5}$	$N^{-\frac{1}{2}}$
$R_g^2 \sim$	$Nc_2^{-1/4}$	N
$\xi \sim$	$c_2^{-3/4}$	c_2^{-1}
$(\Pi/T) \sim$	$c_2^{9/4}$	c_2^3

4.6.3.5. Temperature dependence

Up to this point, the scaling results have been derived at constant temperature but for two different qualities of the solvent: a θ-solvent and a good solvent. The transformation of a θ-solvent into a good solvent can be accomplished by changing the temperature T. The natural temperature variable to express this deviation from the θ-point is the reduced temperature change $\tau = (T-\theta)/\theta$. A diagram schematically displaying the various regimes, and the transition zones between them, is shown in Fig. 4.14 (Cotton *et al.*, 1976; Richards *et al.*, 1981). There is, of course, a similar diagram associated with the lower critical solution temperature (sometimes termed the *Rowlinson temperature*). This will be a mirror-image, relative to some intermediate temperature, of the one given. Note that these constructions are sometimes loosely termed *phase diagrams*.

One interesting feature about this diagram is that it predicts a second transitional region at higher polymer concentrations between the semi-dilute regime and that relevant to the melt-like behaviour. Flory (1953) has pointed out that the behaviour of polymer chains in melts parallels that in θ-solvents so that at some higher polymer concentration, the exponents relevant to melts must prevail. Table 4.3 shows that the melt exponents are quite different from those operative in the semi-dilute regime. This implies the existence of a second transition zone at c_2^{**} ($c_2^* < c_2^{**}$). Above the polymer concentration c_2^{**}, the radius of gyration R_g is independent of the polymer concentration, just as it is in a θ-solvent.

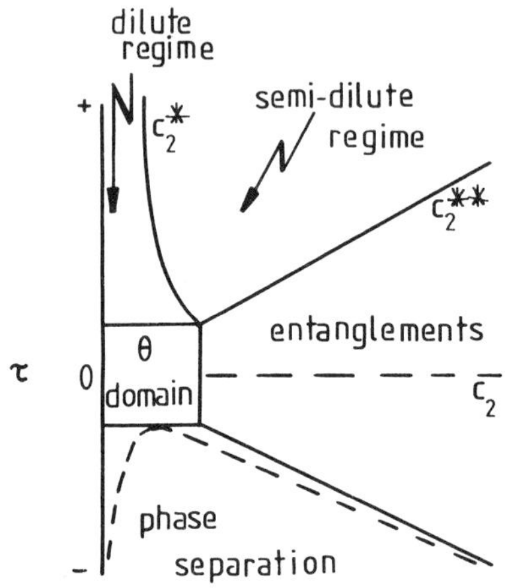

Fig. 4.14. Schematic representation of the various concentration regimes for polymer solutions under both good and bad solvency conditions (after Cotton *et al.*, 1976).

We note in passing that Edwards and Jeffers (1979) have used a mean field theory to predict an additional concentration regime not predicted by the scaling law approach. This regime is at higher concentration than the semi-dilute region and approaches the concentrated region. It has been termed the *semi-concentrated region* and some experimental evidence for its existence has been obtained for the polystyrene/cyclohexane system using small angle neutron scattering (Richards *et al.*, 1981a,b,c).

The boundaries between the various domains shown in Fig. 4.14 need not be very sharp. They represent regions wherein a crossover occurs in the scaling exponents. The value of c_2^*, for example, might be calculated to lie in the range ($3M/4\pi N_A R_g^3 \leq c_2^* \leq (M/N_A R_g^3)$. Here the lower bound assumes that the polymer molecules behave like hard spheres whereas the upper bound assumes that they occupy cubic cells. Actually, there is evidence to suggest that overlap begins at polymer concentrations below the hard sphere lower bound (Hardgraft *et al.*, 1979).

As mentioned above, Edwards has used a self-consistent mean field theory to derive similar (though not identical) results to those obtained from scaling theory (Edwards, 1966; Edwards and Jeffers, 1979). One virtue of this approach is that the relationships were obtained from a single extrapolation formula which encompasses all concentrations from an infinitely dilute solution to bulk polymer. This means that a somewhat more intuitively acceptable picture of polymer solutions emerges, one where there is a gradual change from one concentration regime to another.

Note that according to scaling theory the θ-point is a tricritical point for a polymer of high (infinite) molecular weight. This implies that there is a first order transition in the worse than θ-solvent domain (the coil-to-globule transition) and a second order transition in the better than θ-solvent domain (the conformational expansion manifest therein). Scaling theory also predicts a θ-solvent region rather than a θ-point *per se*. The experimental evidence of Pritchard and Caroline (1981) suggests, however, that this temperature region must be extremely narrow, less than, say, 1 K.

4.6.3.6. Experimental tests of the scaling law predictions

Experimental tests of the scaling law predictions presented in Table 4.3 have been reported in the now classic paper of Daoud *et al.* (1975). The experimental data were obtained as a function of concentration primarily by small angle neutron scattering (SANS) on protonated or deuterated polystyrene in the semi-dilute regime. The chains overlapped strongly in this regime, although the good solvent used, carbon disulphide, was still at a relatively low volume fraction. A comparison is presented in Table 4.4 between the predictions of theory for the concentration dependence of R_g, ξ and Π and the results of these experiments in the semi-dilute regime. It is apparent that the measured exponents are in excellent accord with the scaling

TABLE 4.4
Comparison of experimental exponents in the semi-dilute regime with the theoretical predictions

Parameter	Experiment	Scaling law Theory: Scaling law	Scaling law Theory: Mean field
$R_g \sim$	$c_2^{-(0\cdot25\pm0\cdot02)}$	$c_2^{-\frac{1}{4}}$	c_2^{0}
$\xi \sim$	$c_2^{-(0\cdot72\pm0\cdot06)}$	$c_2^{\frac{3}{4}}$	$c_2^{-\frac{1}{2}}$
$\Pi \sim$	$c_2^{(2\cdot25\pm0\cdot04)}$	$c_2^{\frac{9}{4}}$	c_2^{2}

law predictions, which are somewhat different from those of the mean field theory of classical polymer science.

The consonance noted above between the predictions of scaling law theories and the results of experiments suggests that the assumptions underlying the theories are correct. Mattice (1981), however, has criticized the basic assumptions of the blob concept because they are not verified by rotational isomeric state calculations that incorporate the excluded volume term, admittedly in a somewhat empirical fashion. A major difficulty envisaged by Mattice in the simple application of the blob concept lies in its failure to account for the consequences of the repulsive interaction of the subchain with atoms elsewhere in the main chain.

4.6.4. Scaling law theory applied to attached polymer chains

4.6.4.1. Segment density profiles

To show the applicability of scaling law theory to colloid stability problems, we present here some results, first published by de Gennes (1980), on the concentration profiles of the segments of chains attached terminally to flat plates in a good solvent. Two cases were distinguished: low surface coverage (case I in Fig. 4.15) and high surface coverage (case II in Fig. 4.15).

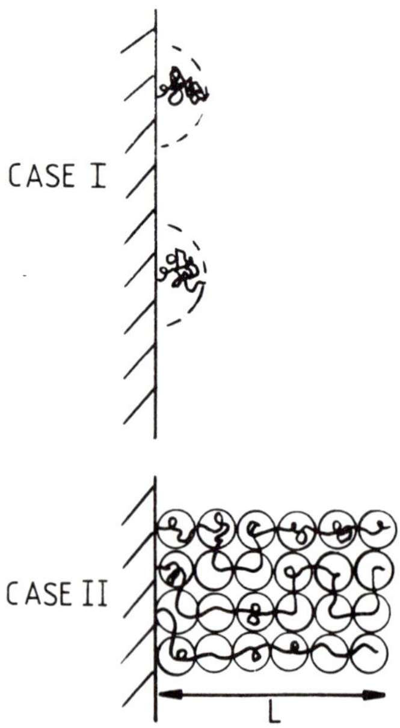

Fig. 4.15. Schematic representation of tails: case I, low surface coverage; case II, high surface coverage (after de Gennes, 1980).

Low surface coverage

In the lower limit of surface coverage, de Gennes adopted the oversimplified view that each chain, composed of N segments each of length l, can be represented by a hemisphere that is centred on the point of attachment of the terminus of the chain to the surface. Its radius is the Flory radius of a coil in a good solvent, i.e. $R_g{}^D \sim N^{3/5}$. For the chains to be well spaced and not touching, their total projected area on the surface must be less than the total available surface area. For ν chains per unit area, this means that $\nu 4\pi R_g{}^2/l^2 < 1$ (or $\nu < N^{-6/5}$) where division by l^2 normalizes the area occupied.

Consider now the segment density distribution function in the x-direction normal to the impenetrable interface. This we shall denote by $\rho(x)$. At its upper limit (which occurs at $x \sim R_g$), the segment density $\rho(x)$ can be approximated by the product of the average segment concentration inside the coils multiplied by the fraction of the wall area occupied by coils. The multiplicand is roughly $(N/R_g{}^3)$ whereas the multiplier is $(\nu R_2{}^2/l^2)$. Thus

$$\rho(x = R_g) \simeq \nu N/R_g{}^D \sim \nu N^{2/5} \tag{4.49}$$

At its lower limit (at $x \sim l$), the value of $\rho(x)$ should be approximately equal to ν. Interpolation between these two extremes by a power law gives

$$\rho(x) = \nu(x/l)^m \tag{4.50}$$

At $x = \mathrm{R}_g$, we therefore have

$$\nu(R_g{}^D/l)^m \sim \nu(N^{(3/5)m}) = \nu N^{2/5} \tag{4.51}$$

from which we conclude that $m = 2/3$. It follows that

$$\rho(x) \sim (x/l)^{2/3} \tag{4.52}$$

which is represented schematically in Fig. 4.16a. This is a very different dependence from that predicted by scaling laws for unattached chains

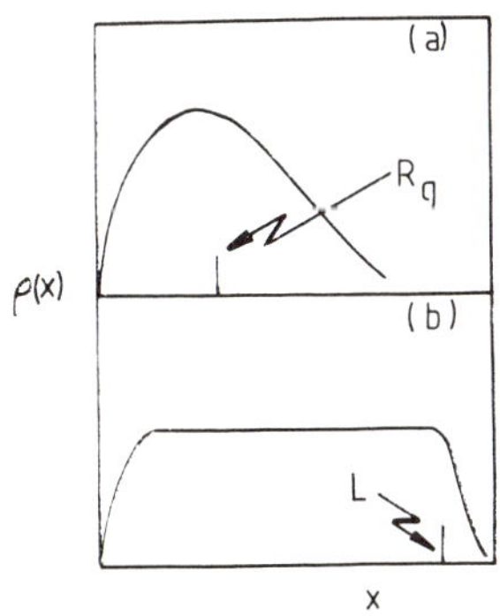

Fig. 4.16. Schematic representation of the segment density distribution functions for tails: (a) low surface coverage; (b) high surface coverage (after de Gennes, 1980).

($m = 5/3$). Note, too, that for a θ-solvent, the analogous derivation yields $m = 1{\cdot}0$ for an attached chain, a dependence in good agreement with the results of rotational isomeric state calculations (Feigin, 1980). The fall-off in $\rho(x)$ beyond $x \sim R_g$ is quite rapid.

High surface coverage

High surface coverage is undoubtedly the case which is of greater practical importance in polymeric stabilization. It occurs when the average distance D between the points of attachment is significantly less than $R_g{}^D$ (i.e. if $\nu > N^{-6/5}$, as suggested above). Since by definition $\nu D^2 = l^2$, we see that $D = l\nu^{-\frac{1}{2}}$.

A grafted chain can be considered to be composed of blobs of linear size D, each containing g_D segments. On the microscopic scale, the excluded volume interactions are operative so that D is related to g_D by $D \sim l g_D{}^{3/5}$ in a good solvent. Thus if $D \ll R_g{}^D$ ($\sim N^{3/5}$), it follows that $g_D \ll N$. In the region occupied by the attached chains, the blobs act as hard spheres and fill the space densely. Thus the average polymer segment density in each blob is given by

$$\rho \simeq l^3 g_D / D^3 \tag{4.53}$$

Substituting $\nu^{-\frac{1}{2}} = D/l$ and $g_D = (D/l)^{5/3}$ yields

$$\rho \simeq \nu^{2/3} \tag{4.54}$$

The thickness L of the attached layer follows immediately. The segment density is the number of segments in one chain divided by the average volume occupied by each chain (i.e. $N/LD^2 = \nu N/Ll^2$). This can be substituted into the lhs of equation (4.54) to yield

$$L \simeq N l \nu^{1/3} \tag{4.55}$$

showing that L is directly proportional to the molecular weight. This can only be true if the polymer chain is in a stretched, as distinct from a random coil [which requires $L \sim R_g(\sim N^{3/5})$], type of conformation. This conclusion is confirmed by writing the thickness as

$$L \simeq (N/g_D) D \tag{4.56}$$

Equation (4.56) shows that the (N/g_D) blobs, each of size D, composing each chain are extended normal to the interface.

In this stretched regime, the shape of the curve of the segment density *vs* distance from the surface is essentially flat, except for two adjustment regions at each end of the chains (see Fig. 4.16b). In the region within each blob ($l < x < D$), the equation $\rho(x) \sim (x/l)^{2/3}$, which was derived in the preceding section, is still applicable. For $x > D$, $\rho \sim \nu^{2/3}$ which is constant.

In summary, it can be seen that, at both low and high coverage, depletion layers exist near the particle surface for a terminally attached chain (tail). Two conformations exist in the extreme cases examined: one relatively unstretched and the other stretched. Of course, in real systems, all intermediate degrees of stretching are possible but the precise details of the transitional zone lie outside the ambit of scaling theory at present.

4.6.5. Polymers confined to thin slabs

Lax *et al.* (1981) have applied scaling law theory to a free polymer behaving as a self-avoiding walk in a thin slab of thickness h. This limits the span of the polymer chains leading to distortions in the direction parallel to the slab plane. The rms end-to-end distance of a self avoiding walk of isolated chains is given by

$$\langle r^2 \rangle^{\frac{1}{2}} \sim N^{\gamma} \tag{4.57}$$

where $\gamma = 3/5$ in 3-D and $\gamma = 3/4$ in 2-D. Suppose $\langle r^2{}_y \rangle$ is the component of $\langle r^2 \rangle$ parallel to the constraint; this component would be expected to scale in the same way as $\langle r^2 \rangle$, except that it refers to 2-D.

The general scaling law hypothesis suggests that

$$\langle r^2{}_y \rangle \sim \langle r^2 \rangle (h/\langle r^2 \rangle)^{-m} \tag{4.58}$$

a negative exponent being chosen because $\langle r^2{}_y \rangle$ increases with decreasing h. This leads to the exponent equation

$$3/2 = 6/5(1+m) \tag{4.59}$$

so that $m = 1/4$. This means that $\langle r^2{}_y \rangle \sim h^{-1/4} N^{3/2}$.

Chapter 5

Thermodynamically Limited Steric Stabilization: The Phenomenology of Incipient Flocculation

5.1. Introduction

Two strategies have been exploited to study experimentally the stability of sterically stabilized dispersions:

(i) the dispersions are taken to the brink of instability and the factors that induce flocculation determined;
(ii) measurements are made of the properties of robustly stable dispersions.

The first strategy, studies of incipient instability, provides insights into the thermodynamic factors that control stability. Further, they show how optimum stability can be imparted and allow the bounds of stabilization (e.g. the upper and lower limits to temperature stability) to be determined. Necessarily, however, incipient instability studies examine dispersions in what might be described as their death throes. This could well change the very properties that we wish to study in order to understand how stability is imparted. It is therefore important to complement incipient instability studies with measurements on robustly stable dispersions. In this and the two subsequent chapters, the phenomenology of incipient flocculation will be discussed. The presentation of the properties of robustly stable dispersions will be deferred until Chapter 13.

5.2. General strategy for inducing flocculation

5.2.1. The reduction in the solvent quality of the dispersion medium

Many sterically stabilized dispersions can be induced to flocculate simply by decreasing the solvency of the dispersion medium for the stabilizing moieties. This may be achieved for some systems by changing the temperature and/or pressure. Other dispersions, however, resolutely defy all such efforts to induce flocculation merely by adjusting the ambient conditions. This is especially evident with certain aqueous sterically stabilized dispersions. For example poly(vinyl acetate) latices stabilized by poly(oxyethylene) in pure water are stable at 100 °C. In these instances, the addition of a substance that reduces the solvency of the dispersion medium for the stabilizing moieties usually permits flocculation to be observed. In aqueous systems, for example, the addition of electrolytes will commonly reduce the solvent power that water displays for the polymeric stabilizing moieties. In nonaqueous media, all that is usually required is the addition of a nonsolvent for the stabilizing chains that is miscible with the dispersion medium. This method is also applicable to the flocculation of aqueous sterically stabilized dispersions.

5.2.2. The reversibility of flocculation

One important feature of the flocculation of sterically stabilized dispersions that sets it apart from the coagulation of electrostatically stabilized polystyrene latices, for example, is the reversibility of aggregation. It is found experimentally that if flocculation is induced by decreasing the solvency of the dispersion medium for the stabilizing moieties, spontaneous redispersion occurs on making the dispersion medium a better solvent for the stabilizing chains (Napper, 1968b; 1970a; Lambe *et al.*, 1978; Everett and Stageman, 1977; 1978a; Dawkins and Taylor, 1980; Clarke and Vincent, 1981a,b; Buscall, 1981).

Several authors, however, have reported a minor caveat in relation to the reversibility (Napper, 1968b; Everett and Stageman, 1977; 1978a; Clarke and Vincent, 1981a; Buscall, 1981). This is that if the particles are allowed to remain flocculated for relatively long periods (say, an hour or more), redispersion is no longer complete. The reasons for this aging effect are unknown. One plausible explanation for the irreversibility involves the slow diffusion of the stabilizer away from the stress zones that are created by the close approach of the particles.

5.2.3. Experimental methods

5.2.3.1. The direct turbidity method

The onset of instability in dilute sterically stabilized dispersions whose particle sizes are not too large can often be observed visually. Instability manifests itself by a dramatic increase in the turbidity of the dispersion. This is exemplified in Fig. 5.1 where the turbidity of a poly(vinyl acetate) latex stabilized by poly(oxyethylene) in 0·48 M $MgSO_4$ is plotted as a function of temperature (Napper, 1970a). At temperatures below 291 K, the turbidity is small, indicative of the long-term stability of the latex. Indeed, these dispersions are apparently thermodynamically stable. Above 291 K, however, the turbidity increases dramatically. This signifies the onset of flocculation. Note that Everett and Stageman (1978b) have shown by photon correlation spectroscopy that the onset of a visual change in turbidity coincides with the formation of low order flocs.

Figure 5.2 presents a similar plot for a poly(methyl methacrylate) latex sterically stabilized in *n*-heptane by poly(12-hydroxystearic acid). In this instance, however, the reduction in the solvency of the dispersion medium for the stabilizing moieties was achieved by adding a miscible nonsolvent (specifically ethanol) to the dispersion medium (Napper, 1968a). Flocculation was again accompanied by an abrupt increase in turbidity when a certain volume fraction of ethanol was added to the *n*-heptane. In this instance, it was possible to observe the slow flocculation of the latex particles (i.e. flocculation apparently in the presence of a small repulsive potential energy barrier at a rate slower than that predicted by Smoluchowski, 1917). It is, however, usually difficult to detect such slow flocculation because of the sharpness of the transition from stability to flocculation for sterically stabilized dispersions.

The reason for the increase in turbidity on aggregation can be grasped, at

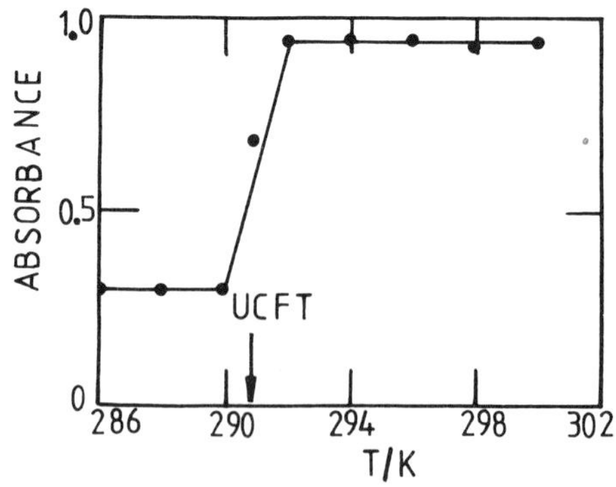

Fig. 5.1. Plot of the turbidity of a poly(oxyethylene) stabilized latex as a function of temperature in 0·48 M $MgSO_4$ (after Napper, 1970a).

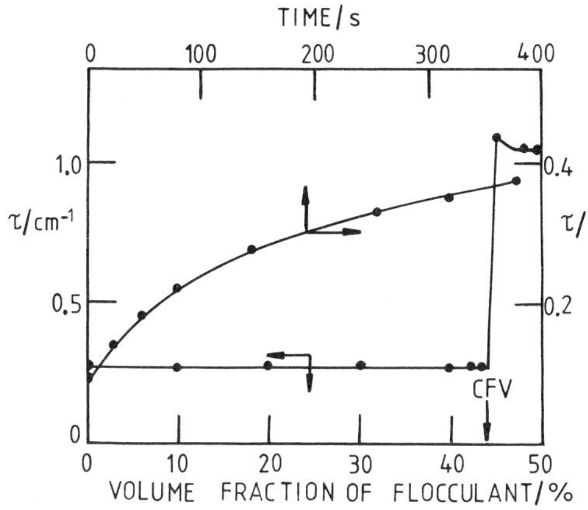

Fig. 5.2. Plots of the turbidity as a function of volume fraction of ethanol and of time for a latex stabilized by poly(12-hydroxystearic acid) in *n*-heptane (after Napper, 1968b).

least qualitatively, if the latex particles are considered to be Rayleigh scatterers (i.e. simple dipole scatterers). In that case, the turbidity, τ, is given by

$$\tau = ANV^2 \tag{5.1}$$

where A = an optical constant, N = number concentration of latex particles and V = volume of a latex particle (assumed to be monodisperse). Flocculation results in the replacement of two independent scattering centres, each of volume V, by a composite particle of volume $2V$ (also assumed to be a Rayleigh scatterer). The contribution to the turbidity of the two discrete particles is proportional to $2V^2$ whereas that of the doublet is proportional to $(2V)^2 = 4V^2$. This corresponds to a doubling of the turbidity of the dispersion if all the particles were to be converted to doublets. As flocs containing more than two particles are formed on flocculation, very large increases in turbidity may be observed. For particles larger than Rayleigh scatterers, the exponent of the volume term is progressively reduced below 2 as the particle size increases and so the sensitivity of the method necessarily declines.

5.2.3.2. The turbidity-wavelength method

Long *et al.* (1973) have devised a sensitive method for detecting low levels of flocculation in some sterically stabilized dispersions. This involves measuring the turbidity τ over a suitable wavelength range (400–600 nm) and so determining the slope of the plot of $\log\tau$ versus $\log\lambda$. This slope, n, is -4 for a Rayleigh scatterer (Kerker, 1969) in the absence of multiple scattering effects. As the particle size increases, so the absolute value of the slope, $|n|$, declines.

Consequently for particles that lie outside the Rayleigh scattering region, the value of n is strongly dependent upon the average particle size in the dispersion, at least in the size range intermediate between the Rayleigh region and the first Mie maximum. In the absence of the multiple scattering, the value of $|n|$ is also independent of the volume fraction of the latex (Cowell *et al.*, 1978).

Figure 5.3a displays some characteristic plots of $\log\tau$ *vs* $\log\lambda$ at different temperatures for poly(oxyethylene) stabilized poly(vinyl acetate) latices in 0·39 M $MgSO_4$ at different temperatures (Dodd, 1980). The derived values of $|n|$ as a function of temperature are shown in Fig. 5.3b. The sharp break in this plot denotes the onset of flocculation.

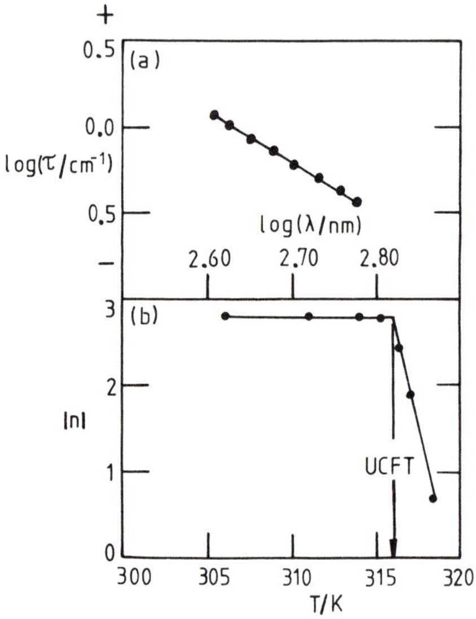

Fig. 5.3. (a) Plot of the turbidity as a function of wavelength for a poly(oxyethylene) stabilized latex; (b) plot of the turbidity-wavelength exponent as a function of temperature for a poly(oxyethylene) stabilized latex in 0·39 M $MgSO_4$ (after Dodd, 1980).

5.2.3.3. Rheological methods

The turbidity methods are unsatisfactory for large particles and/or high particle number concentrations when multiple scattering effects intrude. As shown by Hunter *et al.* (1975), rheological measurements can then be used to detect flocculation. Stable dispersions exhibit an Ostwald-type flow curve whereas flocculated systems behave in a pseudoplastic fashion. Flocculation is thus accompanied by a large increase in the Bingham yield value, τ_B, of the dispersion (see Fig. 5.4).

Cowell and Vincent (1982b) have used low shear viscometry to detect the occurrence of weak coagulation of polystyrene latices sterically stabilized by low molecular weight (1500) poly(oxyethylene). A differential capillary viscometer was used to generate exceedingly small shear rates ($<1\ s^{-1}$), the difference in height of the liquid in the two arms of the viscometer being monitored as a function of time. The (near-zero shear) viscosity was found to display a sharp break with the onset of flocculation (see Fig. 5.5).

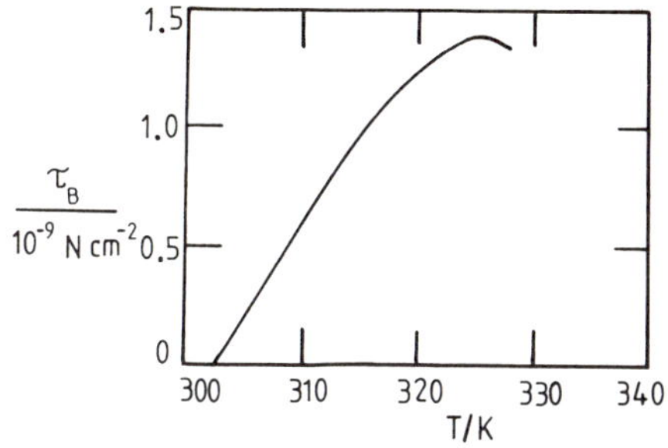

Fig. 5.4. The temperature dependence of the Bingham yield value for a poly(oxyethylene) stabilized latex in 0·45 M K_2SO_4 (after Hunter *et al.*, 1975).

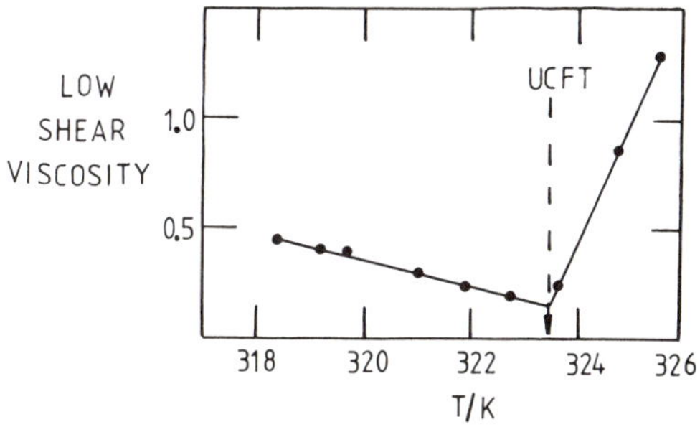

Fig. 5.5. The low shear viscosity of a poly(oxyethylene) stabilized latex in 0·26 M $MgSO_4$ (after Cowell and Vincent, 1982b).

5.3. The critical flocculation point (CFPT)

The point at which flocculation is first detected on decreasing the solvency of the dispersion medium for the stabilizing moieties is termed the *critical flocculation point* (Napper, 1968b) (see Figs 5.1–5.3). The abbreviation CFPT will be used hereafter to denote this point, which in its various manifestations will be central to much of the discussion in this and subsequent chapters.

As shown in Fig. 5.1, sterically stabilized dispersions often display an abrupt transition from long-term (indeed thermodynamic) stability to the onset catastrophic flocculation. Changing the temperature by only a few degrees kelvin is sufficient to transform a very stable dispersion into a flocculated coagulum. This dramatic temperature response contrasts sharply with the more sluggish response to temperature changes exhibited by electrostatically stabilized dispersions. Their stability is normally decreased by heating, as noted by Faraday (1857).

5.3.1. The designation of the various critical flocculation points

Sterically stabilized dispersions can be flocculated by a variety of different stimuli. Some dispersions flocculate on heating; others on cooling; and still others both on heating and cooling. Certain dispersions can also be flocculated by changing the pressure. Combinations of changes in temperature and pressure can also induce flocculation. Furthermore, almost all dispersions can be flocculated by the addition to the continuous phase of a nonsolvent for the stabilizing moieties. The critical flocculation points in these instances are differentiated according to how flocculation is induced. If a temperature variation causes instability, the critical flocculation point is referred to as a *critical flocculation temperature* (CFT) (Napper, 1968b). As some dispersions flocculate both on heating and cooling, it is imperative to distinguish between flocculation that is induced by heating and that induced by cooling. If an increase in temperature causes instability, incipient flocculation is first observed at the *upper critical flocculation temperature* (UCFT); if a decrease in temperature causes instability, incipient flocculation occurs at the *lower critical flocculation temperature* (LCFT) (Croucher and Hair, 1978). Instability that is induced by changing the pressure first appears at the *critical flocculation pressure* (CFP) (Evans *et al.*, 1975). When incipient flocculation is generated by the addition of a miscible nonsolvent to the dispersion medium, the volume fraction of nonsolvent in the continuous phase, usually expressed as a percentage, is termed the *critical flocculation volume* (CFV) (Napper, 1968b). This terminology is summarized in Table 5.1.

TABLE 5.1
Summary of the various critical flocculation points for thermodynamically limited stability

Flocculation method	Designation
temperature change	CFT
—heating	UCFT
—cooling	LCFT
pressure change	CFP
nonsolvent addition	CFV

5.3.2. Influence of the system parameters on the critical flocculation point

In order to understand the stability of sterically stabilized dispersions, it is necessary to identify the CFPT. This identification can be accomplished for many systems once the influence of a wide range of system parameters on the CFPT has been established experimentally. Such an experimental study also generates guidelines for optimizing the stability of sterically stabilized dispersions.

Apart from the solvency of the dispersion medium, the system parameters that in principle could influence the CFPT can be listed as follows:

(i) particle number concentration;
(ii) the nature of the anchor polymer;
(iii) the particle size;
(iv) the surface coverage of the particles;
(v) the nature of the disperse phase;
(vi) the molecular weight of the stabilizing moieties.

The influence of each of these factors on the CFPT will now be considered in turn.

5.3.3. The dependence of CFPT on particle concentration

5.3.3.1. Experimental

Conspectus
Whether or not the experimentally observed CFPT depends significantly upon the particle number concentration is a rather complex issue. At least five different groups have reported that for some sterically stabilized dispersions, the observed CFPT is independent of particle number concentration, within the limits of experimental error, over the number concentration range usually accessible. (Napper, 1968b; 1970a; Everett and Stageman, 1978a; Dawkins and Taylor, 1980; Clarke and Vincent, 1981b; Croucher and Hair, 1981). This unanimity is marred by the reports of three groups who have observed for other sterically stabilized dispersions the existence of a weak particle number dependence of the CFPT (Cairns and Neustadter, 1975; Everett and Stageman, 1977; 1978a; Cowell *et al.*, 1978; Cowell and Vincent, 1982b).

Evidence for a particle concentration dependence
The dependence of both the UCFT and LCFT upon the particle number concentration for poly(methyl methacrylate) latex particles stabilized by poly(dimethylsiloxane) in various alkanes was measured by Everett and Stageman (1977). Their results are shown in Fig. 5.6. Whereas the UCFT decreased with increasing particle concentration, the LCFT increased as the

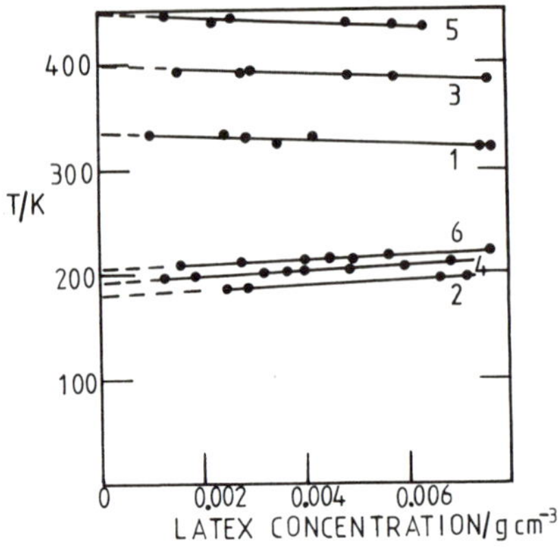

Fig. 5.6. The dependence of the CFTs (both UCFT and LCFT) upon latex particle concentration for poly(methyl methacrylate) particles stabilized by poly(dimethylsiloxane); dispersion medium: 1,2 *n*-propane; 3,4 *n*-butane; 5,6 *n*-pentane (after Everett and Stageman, 1977).

latex particle concentration increased over the latex weight fraction range studied (10^{-2}–10^{-3} w/v). Despite this seemingly contrasting behaviour, both sets of results point to the same conclusion: that dilute dispersions are marginally more difficult to flocculate than more concentrated systems. If this conclusion is correct, it implies that the configurational entropy of the latex particles can contribute significantly to their colloid stability near to the CFPT (Feigin *et al.*, 1981). This question will be taken up theoretically in Section 12.4 where the magnitude of the particle configurational entropy will be estimated.

Two further points need to be emphasised with regard to Fig. 5.6. First, the data refer to the flocculation of very large particles (ca 200 nm in diameter) stabilized by relatively low molecular weight (ca 7 000) poly(dimethylsiloxane) chains (i.e. thin steric layers). The conjunction of these two factors, specifically large particles and thin steric layers, conspire to allow the attractive London forces, operative between the core latex particles, to regulate the onset of instability. For this reason, the latex particles were observed to coagulate under significantly better than θ_L-conditions on heating, the increase in temperature apparently being necessary to reduce both the barrier layer thickness and the steric repulsion. Second, the data have been obtained with nonaqueous, nonpolar dispersion media.

Cowell *et al.* (1978) have also reported a relatively weak dependence of the UCFT of poly(oxyethylene) stabilized latices dispersed in aqueous solutions of magnesium sulphate of differing concentrations. The pattern of behaviour reported in their studies parallels that observed by Everett and Stageman (1977) for the UCFT. The UCFT decreased with increasing particle concentration over the volume fraction range 10^{-3}–10^{-5}. This implies that dilute dispersions are somewhat more stable than concentrated systems, suggesting once again that there is a significant contribution to the stability near to the CFPT arising from the configurational entropy of the latex particles. Cowell *et al.* (1978) in their studies also used large latex particles (ca 200 nm in diameter) and low molecular weight (750 and 2 000) stabilizing moieties. As a result, the van der Waals attraction between the core particles controlled stability near to the UCFT and flocculation was observed in better than θ_L-solvents. Note that in none of the dispersions discussed in this section was stability limited by thermodynamic factors.

Evidence against a particle concentration effect
The foregoing results may be contrasted with those of Croucher and Hair (1981) and Feigin *et al.* (1981). The former measured the particle concentration dependence of the UCFT of polyacrylonitrile latices sterically stabilized by polyisobutylene in 2-methylbutane (see Fig. 5.7). The UCFT was found to be insensitive to the variation of the latex volume fraction, within the limits of experimental error, over the range studied (2×10^{-3}–2×10^{-2}). Note that in these experiments, the molecular weight of the stabilizing moieties ($7{\cdot}6 \times 10^5$) was relatively high. Moreover, flocculation did not occur in better than θ-solvents. Rather, there was close agreement between the UCFT (327 K) and the θ_L-temperature (325 K).

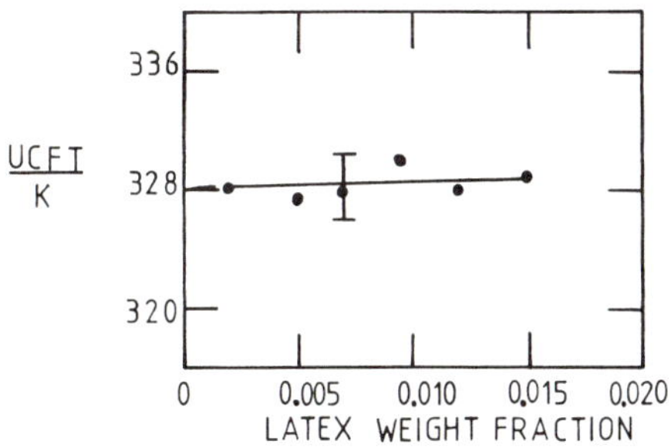

Fig. 5.7. The dependence of the UCFT upon the particle number concentration for polyacrylonitrile particles stabilized by polyisobutylene in 2-methyl butane (after Croucher and Hair, 1981).

The results obtained by Feigin *et al.* (1981) for aqueous systems complement the data of Croucher and Hair (1981) on nonaqueous systems. These are presented in Fig. 5.8. The UCFT of poly(vinyl acetate) latices stabilized by relatively high molecular weight poly(oxyethylene) (20 000 in one case and 360 000 in the other) in 0·39 M $MgSO_4$ was measured over a range of latex volume fractions (10^{-2}–10^{-4}). Once again, within the limit of experimental

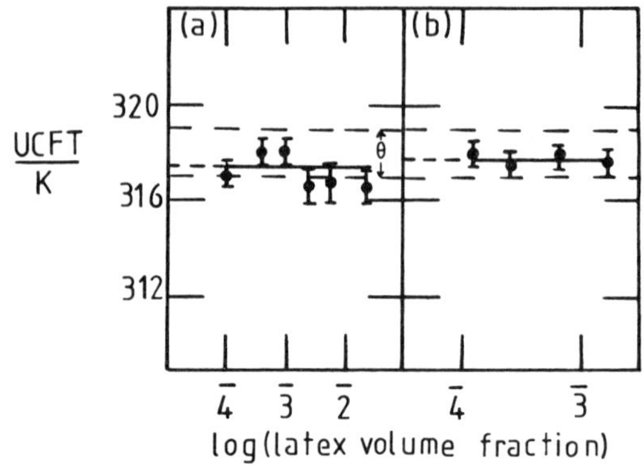

Fig. 5.8. The dependence of the UCFT upon the particle number concentration for poly(vinyl acetate) particles stabilized by poly(oxyethylene) in 0·39 M $MgSO_4$; molecular weight of poly(oxyethylene): (a) 20 000; (b) 36 000 (after Feigin *et al.*, 1981).

error, the observed UCFT was independent of the number concentration of latex particles. The UCFT values for both molecular weight stabilizing moieties were close to the corresponding θ_L-temperature.

We further note that Dawkins and Taylor (1980), whilst not presenting any quantitative results, reported qualitatively that the LCFT of poly(dimethylsiloxane) in a *n*-heptane/ethanol mixture (51 : 49 v/v) was independent of the particle volume fraction up to a value of 1×10^{-3}. The LCFT correlated strongly with the corresponding θ_U-temperature in this study.

Clarke and Vincent (1981b) also found that the LCFT of cross-linked microgel particles in *n*-butyl formate was constant over the volume fraction range 4×10^{-4} to 8×10^{-3}. Again the LCFT was very close to the θ_U-temperature, indicating that stability was controlled by thermodynamic factors.

Everett and Stageman (1978) have also reported that one of their poly(acrylonitrile) dispersions (AN 63), of radius 40 nm and stabilized by relatively high molecular weight (30 000) poly(dimethylsiloxane) moieties, exhibited no particle concentration dependence of the UCFT in propane. It is highly significant that this was the only latex studied by them whose UCFT was close to the θ_L-temperature for the stabilizing moieties under these conditions, i.e. stability was limited by thermodynamic factors rather than by van der Waals forces.

Resolution of the apparent conflict

Whilst some of the foregoing experimental results may appear at first sight to be in conflict, a definite pattern of flocculation behaviour emerges on closer scrutiny. This may be set forth as follows.

For high molecular weight stabilizing chains, the onset of flocculation occurs close to the θ-point and the CFPT is relatively insensitive to the particle number concentration. In contradistinction, for low molecular weight stabilizing moieties and large colloidal particles, flocculation occurs in dispersion media that are better solvents for the stabilizing chains than θ-solvents. This indicates that the weak van der Waals forces operative between the core particles are implicated in destabilization. The phenomenon is thus strictly one of *coagulation* rather than flocculation. The CFPT (strictly, the critical coagulation point CCPT) is found to be dependent, albeit rather weakly, on the particle number concentration. Dilute dispersions display greater stability than more concentrated systems. This suggests, as mentioned previously, that the configurational entropy of the latex particles may contribute significantly to their overall stability behaviour near to the CFPT.

It may be concluded that the apparently conflicting results reported for the particle number dependence of the CFT arises because two separate phenomena can be responsible for aggregation. Thompson and Pryde (1981) have independently reached the same conclusion, as have Cowell and Vincent (1982b). Which of these mechanisms is operative is a complex function of the particle size and its chemical nature, together with the molecular weight and conformation of the stabilizing moieties.

5.3.4. The dependence of the CFPT on the nature of the anchor polymer

One of the first questions that arises in studying the flocculation of sterically stabilized particles is whether or not the nature of the anchor polymer influences stability and thus the CFPT. The anchor polymer must, of course, be effective in attaching the stabilizing moieties to the colloidal particles. Insolubility in the dispersion medium is often sufficient to ensure effective anchoring, provided that the molecular weight of the anchor polymer is sufficiently high.

With emulsions, however, there is some evidence to suggest that the anchor polymer must not only be insoluble in the continuous phase but must also be soluble in the disperse phase to ensure optimum stability (March and Napper, 1977). March and Napper investigated the preparation of toluene emulsions in water using poly(oxyethylene) of molecular weight 79 000 as the stabilizing moieties. Anchor polymers, such as poly(vinyl acetate), polystyrene and poly(methyl methacrylate) that are soluble in toluene, generated stable emulsions that flocculated only when taken to the thermodynamic limit of stability. Polyacrylonitrile, however, which is insoluble in both water and toluene, was so ineffective as an anchor polymer that the toluene-in-water emulsions prepared with it flocculated almost immediately after preparation.

5.3.4.1. Aqueous dispersions

To ascertain the effect of anchor polymer on colloid stability, Napper (1970a) measured the UCFTs of five poly(vinyl acetate) latices stabilized by poly(oxyethylene) of molecular weight 10 000 in 0·39 M $MgSO_4$. The stabilizing moieties were covalently attached to five different anchor polymers. The results are presented in Table 5.2.

TABLE 5.2
The influence of the anchor polymer on the UCFT of some aqueous (0·39 M $MgSO_4$) dispersions stabilized by poly(oxyethylene)

Anchor polymer	UCFT/K	θ_L/K
poly(vinyl acetate)	318 ± 2	315 ± 3
poly(methyl acrylate)	314 ± 2	315
poly(methyl methacrylate)	320 ± 2	315
polystyrene	314 ± 3	315
poly(vinyl stearate)	322 ± 3	315

No major influence of the anchor polymer on the UCFT was evident, even though the anchor polymer varied from the very hydrophobic poly(vinyl stearate) to the relatively hydrophilic poly(methyl acrylate). All the anchor polymers studied, however, were nominally insoluble in water.

5.3.4.2. Nonaqueous dispersions

The results obtained with nonaqueous dispersions parallel those presented for aqueous systems. Table 5.3 displays the results obtained for the CFVs and LCFTs of poly(methyl methacrylate) latices stabilized by poly(12-hydroxystearic acid) of molecular weight 1750 in *n*-heptane (Napper, 1968b). The stabilizing moieties were attached to three different anchor polymers, each of which was insoluble in *n*-heptane. The measured CFVs and LCFTs appear to be indifferent to the chemical nature of the anchor polymer.

5.3.4.3. Conclusion

We conclude that provided the anchor polymer is attached sufficiently strongly to the colloidal particles, its chemical nature has no significant bearing on the observed CFPT. This conclusion is buttressed by the studies of Clarke and Vincent (1981a) who covalently grafted polystyrene chains on to silica particles and so obviated the need for an anchor polymer. The pattern of incipient flocculation observed for these systems conforms with that detailed in this chapter for stabilizing moieties anchored by polymer chains.

TABLE 5.3

The influence of the anchor polymer on the CFVs and LCFTs of some dispersions in *n*-heptane stabilized by poly(12-hydroxystearic acid)

		Flocculant					Flocculant				
		ethanol		*n*-propanol			ethanol		*n*-propanol		
Anchor polymer	T/K	CFV/%	θ/%	CFV/%	θ/%	T/K	CFV/%	θ/%	CFV/%	θ/%	LCFT/K
Poly(methyl methacrylate)	274	39·5 ± 0·5	39 ± 1	61 ± 1	65 ± 2	297	50·0 ± 0·5	51 ± 2	78 ± 1	84 ± 2	258
Poly(vinyl acetate)	274	38 ± 1	39	62 ± 2	65	297	52 ± 2	51	82 ± 2	84	250
Epikote 1004 epoxy resin	274	40 ± 5	39	60 ± 5	65	297	45 ± 5	51	75 ± 5	84	249

5.3.5. The dependence of the CFPT on particle size

5.3.5.1. High molecular weight stabilizing moieties

Dawkins and Taylor (1980) have measured the dependence upon particle size of the LCFT of poly(methyl methacrylate) latices stabilized by monodisperse poly(dimethylsiloxane) of molecular weight 11 200 in *n*-heptane/ethanol (51/49, v/v) mixtures. Their results are displayed in Table 5.4. It is clear that the LCFT was independent of the particle size over a five-fold variation in particle size.

Napper (1970) investigated the particle size dependence of the UCFT of poly(vinyl acetate) latices in 0·39 M $MgSO_4$ stabilized by poly(oxyethylene) of molecular weight 10 000. The results shown in Table 5.5 confirm that the UCFT was independent of the size of the latex particles, at least for an eight-fold variation in particle size. These results are consonant with those observed by Dawkins and Taylor (1980) for nonaqueous systems.

The absence of any dependence of the CFPT on the particle size suggests

TABLE 5.4
The effect of particle size on the LCFT of some nonaqueous dispersions stabilized by high molecular weight poly(dimethylsiloxane)

Particle diameter/nm	LCFT/K	θ_U/K
96	340·0	340 ± 1
130	340·0	340
250	340·1	340
480	340·2	340

TABLE 5.5
Effect of particle size on the UCFT of some aqueous (0·39 M $MgSO_4$) dispersions stabilized by high molecular weight poly(oxyethylene)

Particle diameter/nm	UCFT/K	θ_L/K
30	317 ± 2	315 ± 3
61	318	315
81	320	315
110	319	315
154	319	315
230	315	315

that the London dispersion attraction between the core latex particles was not responsible for the observed flocculation in these experiments. A radically different explanation for the attractive interactions that induced flocculation in these systems is enunciated in Section 6.3. Briefly, it is proposed that the segmental attraction between polymer chains attached to different particles in dispersion media of solvency slightly worse than θ-solvents results in flocculation.

5.3.5.2. Very low molecular weight stabilizing moieties

The results presented in this section thus far, which showed that the CFPT was independent of particle size, utilized relatively high molecular weight stabilizing moieties. It would be expected, however, that for large particles and thin steric layers, the van der Waals attraction between the core particles would control the onset of instability. This should manifest itself by a dependence of the stability on the particle size and the molecular weight of the stabilizing moieties. Moreover, coagulation should occur under better than θ-solvency conditions. As the London attraction between the particles increases with increasing particle size, it would be predicted that larger particles should be easier to coagulate than smaller particles. This is precisely what is observed experimentally. The question will be taken up in more detail in Chapter 8 but some preliminary results will be presented here to illustrate the trends involved.

The CFV results obtained with ethanol and *n*-propanol as nonsolvents for three different sized poly(methyl methacrylate) latices stabilized by low molecular weight (1750) poly(12-hydroxystearic acid) in *n*-heptane are presented in Table 5.6 (Napper, 1968b). It is apparent that with thin steric layers, the CFV was dependent upon the particle size. As the particle size increased, the CFV decreased, irrespective of the temperature or whether the flocculant was ethanol or *n*-propanol. It may be deduced that the latex particles became easier to coagulate as the particle size increased because

TABLE 5.6
The influence of particle size on the CFV of some nonaqueous dispersions stabilized by low molecular weight poly (12-hydroxystearic acid)

	Temperature							
	297 K				274 K			
	Flocculant				Flocculant			
Particle	ethanol		*n*-propanol		ethanol		*n*-propanol	
diameter/nm	CFV/%	θ/%	CFV/%	θ/%	CFV/%	θ/%	CFV/%	θ%
13	48±1	51±2	74±1	84±2	38±1	39±1	56±1	65±2
35	44	51	67	84	35	39	50	65
87	39	51	58	84	32	39	46	65

coagulation occurred in progressively better solvents than θ-solvents. In this respect, dispersions stabilized by low molecular weight moieties differ significantly from dispersions stabilized by high molecular weight species. As shown in Tables 5.4 and 5.5, the CFTs were very close to the θ-temperatures when the stabilizing moieties were of high molecular weight. No such correlation was evident with low molecular weight stabilizing moieties.

5.3.5.3. Summary

It is apparent that the CFPT is independent of the particle size when the stabilizing moieties are of high molecular weight and the particle size is not too large. This implies that the van der Waals attraction between the core particles is not responsible for flocculation in such systems. In sharp contrast, for low molecular weight stabilizing moieties, which give rise to thin steric layers, the dispersions became progressively easier to coagulate as the particle size increases. Coagulation occurs in dispersion media that are significantly better solvents than θ-solvents for these stabilizing moieties. In such systems, coagulation appears to be controlled by the London attraction between the core particles.

5.3.6. The dependence of the CFPT on surface coverage

5.3.6.1. Nonaqueous dispersions

Napper (1968b) investigated the dependence of the LCFT and the CFV of poly(methyl methacrylate) particles stabilized by poly(12-hydroxystearic acid) of molecular weight 1750 in *n*-heptane. Ethanol was used as the flocculating nonsolvent at 323 K. The results are shown in Fig. 5.9. It is clear that irrespective of whether coagulation was induced by cooling or by the addition of nonsolvent, any decrease in the surface coverage was accompanied

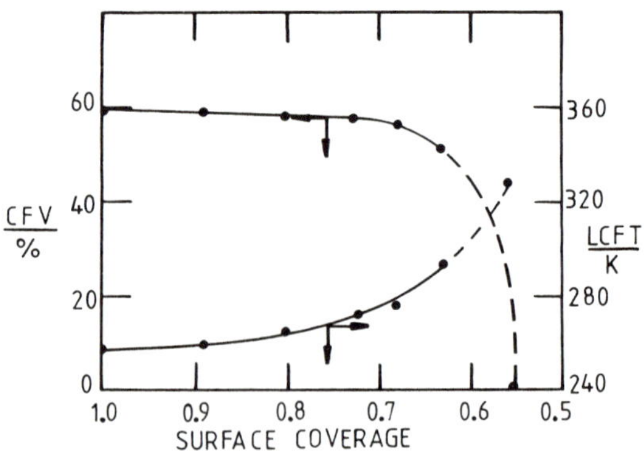

Fig. 5.9. The dependence of the CFV and LCFT upon stabilizer surface coverage for poly(12-hydroxystearic acid) stabilized latex particles in *n*-heptane (after Napper, 1968b).

by an increase in the ease of coagulation. This decrease in stability manifested itself as a reduction in the CFV (i.e. less nonsolvent was necessary to induce coagulation) or an increase in the LCFT (i.e. coagulation occurred in dispersion media of progressively better solvency for the stabilizing moieties). In contrast to the reversible flocculation observed for particles fully coated by steric stabilizer, particles only partially coated by stabilizer underwent irreversible coagulation.

5.3.6.2. Aqueous dispersions

A comparable decrease in the stability of poly(vinyl acetate) latices in 0·39 M $MgSO_4$ on reducing the surface coverage of the poly(oxyethylene) stabilizing moieties was observed by Napper (1970a) (see Fig. 5.10). The UCFT was found to decrease with decreasing surface coverage. When the surface coverage was complete, the UCFT correlated with the θ_L-temperature for

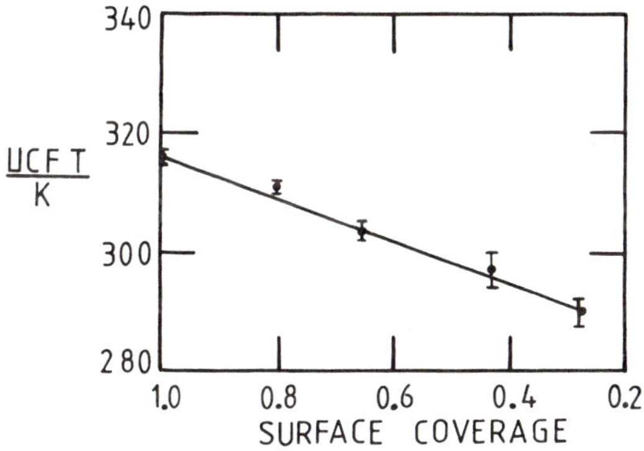

Fig. 5.10. The dependence of the UCFT of poly(oxyethylene) stabilized latex particles upon stabilizer surface coverage in 0·39 M $MgSO_4$ (after Napper, 1970a).

poly(oxyethylene) in 0·39 M $MgSO_4$. As the surface coverage decreased, so flocculation was evident in progressively better solvents than θ_L-solvents for the stabilizing moieties.

5.3.6.3. Conclusions

The similar patterns of incipient instability observed for both nonaqueous and aqueous dispersions implicate the London attraction between the core particles in the coagulation of incompletely covered dispersions. It seems likely that under the stress generated by a Brownian collision, the well-anchored stabilizing chains can move laterally on the surface of the colloidal particles ("surface migration"). This creates bare patches on the surface of the particles, allowing their close approach to proceed further. In this way, coagulation, which is driven by the London attraction between the core particles, occurs in significantly better than θ-solvents. This is an example of *displacement coagulation* (Napper, 1977).

5.3.7. The dependence of the CFPT on the nature of the disperse phase

Experiments have been performed in both aqueous and nonaqueous media in which the chemical nature of the core particles was varied (Napper, 1968b; 1970a). Some results are reproduced in Table 5.7. The experiments, however, were all conducted on latices and the core particles were all polymeric in character. The nonaqueous latices were stabilized by poly(12-hydroxystearic acid) in n-heptane. These were flocculated by the addition of ethanol. The aqueous (0·39 M $MgSO_4$) latices were stabilized by poly(oxyethylene).

It is apparent that for these latices, any variation in the nature of the polymer composing the core particles had no significant influence on the onset of instability. Notwithstanding this, it might be anticipated that with thin steric layers, the nature of the disperse phase could exert an influence on stability if its chemical nature was changed drastically (e.g. from being polymeric to being an inorganic salt or a metal). It would be expected that in the latter case, the van der Waals attraction between the core particles would control stability and that the effective Hamaker constant for the particles would be significantly altered by a drastic change in their chemical composition. This was unlikely to be the case in the experiments cited above.

5.3.8. The dependence of the CFPT upon the molecular weight of the stabilizing moieties

The presentation of the results of several experimental studies of the effect of varying the molecular weight of the stabilizing moieties in both aqueous and nonaqueous dispersion media will be deferred until after the CFPT has been more clearly identified (see Chapter 6). For the sake of completion, we mention here that the CFPT is found to be independent of the molecular weight of the stabilizing moieties provided that their molecular weight is sufficiently high (e.g. at least several thousand). For very low molecular weight stabilizing moieties, limiting stability is controlled by the London attraction between the core particles. A molecular weight dependence of the flocculation point is then observed.

5.3.9. The physical nature of flocculation

It is stressed that for many, although probably not all, sterically stabilizing

TABLE 5.7

The effects of the disperse phase on the CFPT of some aqueous and nonaqueous latices

Disperse phase	Dispersion medium	UCFT/K	θ_L/K	Disperse phase	Dispersion medium	T/K	CFV/%	θ/%
Poly(vinyl acetate)	0·39 M $MgSO_4$	318 ± 2	315 ± 3	poly(vinyl acetate)	*n*-heptane	274	39·5 ± 0·5	39 ± 1
Poly(methyl acrylate)		320 ± 2	315	poly(methyl methacrylate)		274	39 ± 1	39
Polystyrene		323	315					
				poly(vinyl acetate)	*n*-heptane	297	50·0 ± 0·5	51 ± 2
				poly(methyl methacrylate)		297	50 ± 1	51

systems, flocculation in its *physical* manifestations closely resembles the coagulation evident at high electrolyte concentrations for electrostatically stabilized dispersions. That is, the particles are observed to form aggregates (flocculum or coagulum) that sediment under the influence of gravity to the bottom of the container. The formation of two distinct phases, separated by an interface, is usually not observed. This is in sharp contrast to the coacervation (phase separation) that occurs with noncrystallizing polymers when the solvency is made significantly worse than a θ-solvent.

5.3.10. Incipient flocculation overview

Sterically stabilized dispersions may be flocculated (i.e. induced to form aggregates rather than undergo phase separation) by a variety of methods: by changing the temperature; by changing the pressure; and by adding a miscible nonsolvent for the stabilizing moieties to the dispersion medium. The common feature of these different methods is that the solvency of the dispersion medium for the stabilizing moieties must be reduced to provoke the onset of instability. The transition from long-term stability to catastrophic flocculation occurs abruptly at the CFPT.

The preceding sections elaborated patterns of flocculation behaviour at the CFPT that may be summarized as follows.

First, if the colloidal particles are small in size and fully coated by well-anchored, high molecular weight stabilizing moieties, the observed CFPT is essentially independent of

(i) the volume fraction of the disperse phase;
(ii) the chemical nature of the disperse phase;
(iii) the chemical nature of the anchor polymer;
(iv) the latex particle size;
(v) the molecular weight of the stabilizing moieties.

This pattern of behaviour is observed irrespective of whether the dispersion medium is aqueous or nonaqueous. Flocculation at the CFPT is reversible, at least if the particles are not allowed to stay in the flocculated state for too long. The experimental results suggest that for these systems, the London attraction between the core particles is not responsible for flocculation.

The CFPT is observed, however, to be dependent upon the surface coverage by the stabilizing chains. It becomes progressively easier to flocculate the dispersion as the surface coverage decreases. This may be due to the surface migration of the stabilizing chains away from the interaction zone during a Brownian collision. Coagulation occurs in dispersion media that are better solvents for the stabilizing moieties than θ-solvents. Instability appears to be induced by van der Waals attraction between the core particles, although bridging might also be implicated in some systems.

A different pattern of incipient instability emerges if the stabilizing moieties are of low molecular weight and the particle size relatively large, rather than the other way about. For these systems, stability appears to be controlled by the relatively weak London dispersion attraction between the core particles. Consequently, the CFPT is found to be a function of the particle size, larger particles being easier to flocculate than smaller particles. Moreover, the observed CFPT is found in these systems to be weakly dependent upon the number concentration of the latex particles. Coagulation at both the UCFT and LCFT (strictly the UCCT and LCCT) implies that dilute dispersions are more stable than concentrated dispersions. This suggests that the configurational entropy of the latex particles may contribute significantly to stability in sterically stabilized systems whose flocculation is controlled by the van der Waals attraction between the core particles. These anomalous systems will be considered in more detail in Chapter 8.

Chapter 6

Thermodynamically Limited Steric Stabilization: The Identification of the Critical Flocculation Point

6.1. The correlation between the CFPT and θ-point

It was shown in the preceding chapter that most sterically stabilized dispersions could be flocculated by decreasing sufficiently the solvency of the dispersion medium for the stabilizing moieties. This may be achieved using a wide range of flocculant stimuli: (i) by heating or cooling, (ii) by changing the pressure, or (iii) by adding a nonsolvent for the stabilizing moieties to the dispersion medium. The question then arises as to whether there is a general principle that encompasses all the observed CFPTs.

It was noted in Section 3.2.6.1 that the repulsion between identical linear polymer chains in free solution vanishes at the θ-point. Since steric stabilization is predicated on the occurrence of repulsion between the stabilizing moieties, it might be expected that the θ-point would represent the thermodynamic limit to stability. The attainment of θ-conditions would thus signal the onset of incipient flocculation. It has been found experimentally that for many, although certainly not all, sterically stabilized dispersions, a strong correlation is indeed observed between the CFPT and the attainment of θ-conditions for the *stabilizing moieties* in *free* solution. This correlation is evident in both aqueous and nonaqueous dispersion media, irrespective of whether flocculation is induced by changing the temperature and/or pressure, or by adding nonsolvent to the dispersion medium.

It is stressed that the correlation relates to the properties of the stabilizing moieties, not to those of the whole stabilizer (i.e. the stabilizing moieties + anchor polymer). Of course, it is highly doubtful whether the concept of a θ-solvent is applicable to a block or graft copolymer, except perhaps if the separate homopolymers which have identical θ-temperatures in the same solvent (Dawkins, 1973). This would definitely not be the case with amphipathic steric stabilizers.

6.1.1. The critical flocculation temperatures

The CFTs of a wide range of sterically stabilized dispersions have now been measured and compared with the corresponding θ-temperatures. Some of the data for aqueous systems are displayed in Table 6.1. Those for certain nonaqueous dispersions are presented in Table 6.2. For both aqueous and nonaqueous dispersion media, both the UCFTs and LCFTs have been measured. Almost all the data recorded in Tables 6.1 and 6.2 were obtained using sterically stabilized latices. The one exception is the study of Clarke and Vincent (1981a) who used silica particles on to the surfaces of which polystyrene chains had been covalently grafted. Identical results were also obtained with microgel particles (Clarke and Vincent, 1981b).

The data displayed in Tables 6.1 and 6.2 illustrate the strong correlation existing between the CFTs and the corresponding θ-temperatures for both aqueous and nonaqueous dispersion media. These results were obtained by five different groups of workers and cover some 13 different stabilizing moiety/dispersion medium systems. It is important to stress that it is not possible in general to equate the CFT with the θ-temperature.

Note that, as reported in Table 6.2, latices stabilized either by poly(α-methyl styrene) in *n*-butyl chloride or by polystyrene in cyclopentane flocculated both on heating and on cooling. Moreover, the UCFT in nonaqueous systems tends to occur as the critical temperature of the dispersion medium is approached. In aqueous systems, the UCFT can occur well below the critical temperature.

6.1.2. The critical flocculation pressures

Considerably less work has been reported on CFPs than on CFTs. There are two main reasons for this: first, the stability of sterically stabilized dispersions is less sensitive to changes in pressure than to changes in temperature; second, it is experimentally easier to vary the temperature than to vary the pressure. Of course, in many experiments where the temperature is varied, the pressure also changes significantly. In such cases, it is advisable to allow for the change in pressure by, for example, extrapolating to zero pressure (Evans and Napper, 1975).

The CFPs as a function of the temperature for a poly(vinyl acetate) latex stabilized by poly(oxyethylene) in 0·39 M $MgSO_4$ are plotted in Fig. 6.1 (Evans *et al.*, 1975). Also plotted in this diagram are the θ-pressures of

TABLE 6.1
Comparison of CFTs with θ-temperatures for some aqueous dispersions

Stabilizing moieties	Molecular weight	Dispersion medium	CFT/K	U/L	θ/K	Reference
poly(oxyethylene)	1 400	0·39 M $MgSO_4$	317 ± 2	UCFT	330 ± 10	Napper (1970a)
	4 700		320		327 ± 5	
	10 000		318		319 ± 3	
	13 500		318		315 ± 3	
	23 000		314		315 ± 3	
	49 000		316		314 ± 3	
	96 000		316		315 ± 3	
	315 000		315		314 ± 3	
	800 000		316		316 ± 3	
	1 000 000		317		315 ± 3	
poly(acrylic acid)	9 800	0·2 M HCl	287 ± 2	LCFT	287 ± 5	Evans *et al.* (1972)
	19 300		289		289 ± 5	
	51 900		283		287 ± 5	
	89 700		281		287 ± 5	
poly(acrylic acid)	16 000	0·2 M HCl	286 ± 2	LCFT	287 ± 5	Buscall (1981)
poly(vinyl alcohol)	26 000	2·0 M NaCl	320 ± 3	UCFT	300 ± 3	Evans (1971)
	57 000		301		300 ± 3	
	270 000		312		300 ± 3	
poly(acrylamide)	18 000	2·1 M $(NH_4)_2SO_4$	292 ± 3	LCFT	—	Evans *et al.* (1972)
	60 000		295 ± 5		—	
	180 000		280 ± 7		—	

TABLE 6.2
Comparison of CFTs with θ-temperatures for some nonaqueous dispersions

Stabilizing moieties	Molecular weight	Dispersion medium	CFT/K	U/L	θ/K	U/L	Reference
poly(isobutylene)	23 000	2-methylbutane	325 ± 1	UCFT	325 ± 2	θ_L	Evans and Napper (1975)
	150 000		325		325 ± 2		
poly(isobutylene)	760 000	2-methylbutane	327	UCFT	318	θ_L	Croucher and Hair (1981)
	760 000	2-methylpentane	381	UCFT	376	θ_L	
	760 000	2-methylhexane	423	UCFT	426	θ_L	
	760 000	3-ethylpentane	463	UCFT	458	θ_L	
	760 000	cyclopentane	455	UCFT	461	θ_L	
poly(α-methyl styrene)	9 400	*n*-butyl chloride	254 ± 1	LCFT	263	θ_U	Croucher and Hair (1978)
	9 400		403 ± 1	UCFT	412	θ_L	
polystyrene	110 000	cyclopentane	280	LCFT	293	θ_U	Croucher and Hair (1980a)
	110 000		410	UCFT	427	θ_L	
polystyrene	6 000	*n*-butyl formate	264 ± 1	LCFT	264	θ_U	Clarke and Vincent (1981a; 1981b)
	6 700		264 ± 1		264		
	22 600		264 ± 1		264		
poly(dimethylsiloxane)	3 200	*n*-heptane/ethanol (51:49 v/v)	340	LCFT	340 ± 2	θ_U	Dawkins and Taylor (1980)
	11 200		340		340 ± 2		
	23 000		341		340 ± 2		
	48 000		338		340 ± 2		

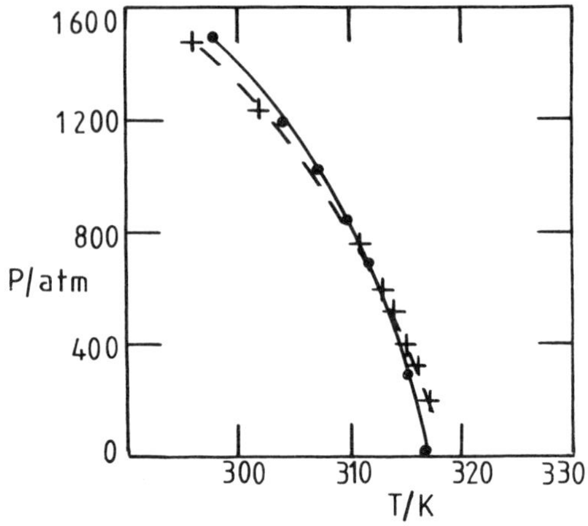

Fig. 6.1. The temperature dependence of the (+) CFP of a poly(oxyethylene) stabilized latex in 0·39 M $MgSO_4$. Also shown (●) are the corresponding θ-pressures (after Evans *et al.*, 1975).

aqueous solutions of poly(oxyethylene) under the same conditions. It is apparent that there is again a strong correlation between the observed CFPs and the corresponding θ-pressures.

6.1.3. The critical flocculation volumes

Most sterically stabilized dispersions can be flocculated by the addition to the dispersion medium of nonsolvent for the stabilizing moieties. Whilst this is a convenient practical method for inducing flocculation, the use of binary (or even ternary) mixtures as dispersion media requires caution in respect of the interpretation of flocculation data. The primary difficulty relates to the preferential interaction ("adsorption") of one of the components with the stabilizing moieties. This may have two important effects: first, the polymer may behave differently from what would normally be expected, since it is a "complexed" rather than a simple polymer; second, the preferential interaction of one component with the stabilizing moieties may deplete the continuous phase of that component and so change the composition of the dispersion medium significantly. For nonaqueous dispersion media, these effects are likely to be most acute when a polar nonsolvent is added to a nonpolar dispersion medium. For aqueous systems, a complex web of specific interactions between the stabilizing moieties and the water, together with any electrolyte present, would be the normally expected pattern of behaviour.

Despite the potential problems highlighted above, the results listed in Table 6.3 show that for the flocculation of dispersions in apolar media by the *n*-alkanols, there exists a strong correlation between the CFV and the

TABLE 6.3
Comparison of CFVs with θ-compositions for some nonaqueous dispersions

Stabilizing moieties	Molecular weight	Dispersion medium	Flocculant	Temperature K	CFV %	θ-comp %	Reference
poly(lauryl methacrylate)	25 000	*n*-heptane	ethanol	249	31 ± 1	33 ± 1	Napper (1968b)
			ethanol	274	38 ± 1	38 ± 1	
			ethanol	297	44 ± 1	45 ± 1	
			ethanol	313	51 ± 1	51 ± 2	
			n-propanol	274	56 ± 2	52 ± 2	
			n-propanol	297	70 ± 2	67 ± 2	
poly(12-hydroxystearic acid)	1 750		ethanol	274	$39{\cdot}5 \pm 0{\cdot}5$	39 ± 1	Napper (1968b)
			ethanol	297	$50{\cdot}5 \pm 0{\cdot}5$	51 ± 2	
			ethanol	313	58 ± 1	59 ± 2	
			n-propanol	274	61 ± 1	65 ± 2	
			n-propanol	297	78 ± 1	84 ± 2	
			n-butanol	274	74 ± 1	81 ± 3	
poly(isobutylene)	2 350	*n*-heptane	ethanol	294	20	20	Cairns and Neustadter (1975)
poly(dimethylsiloxane)	11 200	*n*-heptane	ethanol	340	49	49	Dawkins and Taylor (1980)
	29 800	*n*-heptane	ethanol	340	49	49	
polystyrene	—	benzene	methanol	298	25 ± 1	25	Clarke and Vincent (1981b)

corresponding θ-solvent composition for the stabilizing moieties. As with the CFTs, the CFV results were obtained by several (in this case four) different groups of workers. The results span a broad range of nonaqueous polymeric stabilizing moieties. They were all obtained with latex systems, with the exception of the data of Cairns and Neustadter (1975), who studied dispersions of carbon particles (Kosmos BB).

6.1.4. Additional experimental evidence

Additional experimental data favouring the existence of a strong correlation between the CFPT and the θ-point have already been presented in the preceding chapter. Figures 5.7 and 5.8 show that the correlation is unaffected by the number concentration of latex particles if the stabilizing moieties are of relatively high molecular weight. Tables 5.2 and 5.3 demonstrate that the correlation is not influenced by the anchor polymer. Furthermore, Tables 5.4 and 5.5 reveal that the correlation is independent of the particle size, provided that the stabilizing moieties are of a sufficiently high molecular weight. Table 5.7 shows that the nature of the disperse phase does not influence the correlation and, as discussed in Section 6.2, the correlation is observed irrespective of the molecular weight of the stabilizing moieties, provided that they are high polymers.

The foregoing conclusions are reinforced by the results of Dawkins and Taylor (1980) who measured the LCFTs of poly(methyl methacrylate) latices stabilized by polystyrene-anchored poly(dimethylsiloxane) in a *n*-heptane/ethanol (51/49 v/v) mixture. In these experiments the molecular weights of the anchor polymer and stabilizing moieties were both varied significantly, as was the latex particle size. The variation of the molecular weights of the constituent polymers composing the stabilizing chains ensured that the area per molecule occupied by the stabilizer at complete surface coverage was also highly variable. As shown in Table 6.4, despite all these changes, the LCFT was essentially constant and correlated strongly with θ_U-temperature.

6.1.5. Sterically stabilized emulsions

The thermodynamic limit to the stability of sterically stabilized emulsions was investigated by March and Napper (1977). It would be expected that in this limit, both oil-in-water and water-in-oil emulsions would obey the same general rules as are observed for sterically stabilized latices (Phillips, 1975).

Table 6.5 gives the CFTs and the corresponding θ-temperatures for several water-in-oil and oil-in-water emulsions. It is apparent that for both types of emulsions, a strong correlation exists between the observed CFTs and the θ-temperatures.

TABLE 6.4
LCFTs for poly(methyl methacrylate) dispersions stabilized by poly(dimethylsiloxane) of different molecular weights

Molecular weight Anchor polymer	Molecular weight Stabilizing moieties	Diameter/nm	Area/nm^2	LCFT/K	θ_U/K
12 700	3 200	110	6·4	339·6	340 ± 1
8 800	11 200	250	12·6	340·1	340
44 000	13 700	670	19·4	339·0	340
16 400	16 100	330	24·6	340·4	340
12 700	23 800	74	35·4	340·5	340
44 000	29 800	69	44·5	340·4	340
33 400	48 000	95	51·3	338·2	340

6.1.6. A general rule governing the thermodynamic limit to stability of sterically stabilized systems

The foregoing discussion leads to the following general rule:

> For most sterically stabilized dispersions whose particles are fully-coated by well-anchored high polymeric stabilizing moieties, the practical thermodynamic limit to stability is the θ-point for the stabilizing moieties in free solution.

The phrase "thermodynamic limit" in this context relates to the thermodynamic factors that control the segment-segment, segment-solvent and solvent-solvent interactions for the stabilizing moieties and the dispersion medium at the θ-point.

There are at least three clear exceptions to this general rule. Two further exceptions arise when the prefatory conditions are violated. All five exceptions will be considered in detail in Chapters 8 and 9.

6.2. The dependence of the critical flocculation point on the molecular weight of the stabilizing moieties

There is a simple experimental test that can be performed of the validity of the general rule enunciated above identifying the practical thermodynamic limit of the stability of many sterically stabilized dispersions as the θ-point. As noted in Section 3.2.7, the θ-point is a *segmental* property of a macromolecule that depends upon the nature of the segment/solvent interactions. Accordingly, the θ-point is independent of the molecular weight of the polymer, provided its molecular weight is sufficiently high (Flory, 1953). It follows that the CFPT should be insensitive to the molecular weight of the stabilizing

TABLE 6.5

The thermodynamic limit to the stability of sterically stabilized emulsions[a]

Anchor polymer	Stabilizing moieties	Disperse phase	Dispersion medium	Emulsion type	U/L	CFT/K	θ/K	U/L
PVAc	POE	toluene	0·39 M $MgSO_4$	oil-in-water	U	318 ± 2	318 ± 2	θ_L
PVAc	PVA	toluene	2·0 M NaCl	oil-in-water	U	297 ± 3	299 ± 2	θ_L
PVAc	PAM	toluene	2·1 M $(NH_4)_2SO_4$	oil-in-water	L	297 ± 2	298 ± 3	θ_U
PVAc	PAA	toluene	0·2 M HCl	oil-in-water	L	298 ± 2	287 ± 5	θ_U
PAA	PLM	water	*n*-pentanol	water-in-oil	L	301 ± 3	303 ± 5	θ_U
POE	PS	water	cyclohexane	water-in-oil	L	308 ± 3	307 ± 1	θ_U

[a] Key to polymers: PVAc = poly(vinyl acetate); POE = poly(oxyethylene); PVA = poly(vinyl alcohol); PAM = poly(acrylamide); PAA = poly(acrylic acid); PLM = poly(lauryl methacrylate); PS = polystyrene.

moieties. The results of some experiments that test this prediction will now be presented.

6.2.1. Dispersions

Tables 6.1 and 6.2 contain ample experimental confirmation of the validity of the prediction that the CFPT of polymer latices is independent of the molecular weight of the stabilizing moieties. The most elegant studies that have been performed to-date are those of Dawkins and Taylor (1980). These authors prepared relatively monodisperse poly(dimethylsiloxane) by anionic polymerization. The polydispersity ratio $\langle M_w \rangle / \langle M_n \rangle$ was less than 1·25 for all the stabilizing moieties. The LCFTs of poly(methyl methacrylate) latices stabilized by these monodisperse chains in a *n*-heptane/ethanol (51/49 v/v) mixture were found to be independent of the molecular weight of the poly(dimethylsiloxane) over the range 3 200 to 48 000 (see Table 6.2).

Napper (1970a) has also performed an extensive investigation of the molecular weight dependence of the UCFT. The systems studied were poly(oxyethylene) stabilized poly(vinyl acetate) latices in 0·39 M $MgSO_4$. The molecular weight was varied over ca three orders of magnitude (from $1{\cdot}4 \times 10^3$ to $1{\cdot}0 \times 10^6$). Yet the UCFT was constant within the limits of experimental error (see Table 6.1). The poly(oxyethylene) moieties used in these studies were not monodisperse, however, but rather were highly polydisperse. Inasmuch as the CFPT is independent of the molecular weight of the stabilizing moieties, the polydispersity of the chains in no way vitiated these experimental findings.

Dispersions stabilized by poly(acrylic acid) and poly(acrylamide) in aqueous solutions flocculate on cooling in the presence of suitable electrolytes. The LCFTs were found to be essentially independent of the molecular weight when the molecular weights of both polymers were varied over an order of magnitude (see Table 6.1).

6.2.2. Emulsions

March and Napper (1977) measured the effect of varying the molecular weight of the poly(oxyethylene) stabilizing moieties on the thermodynamic limit to the stability of toluene-in-water emulsions in the presence of 0·39 M $MgSO_4$. The anchor polymer was polystyrene in all cases, except for two commercial stabilizers, Pluronic F68 and Pluronic F108, for which poly(propylene oxide) was the anchor polymer. The results shown in Table 6.6 verify the independence of the UCFT on the molecular weight of the stabilizing moieties for these emulsions.

TABLE 6.6
The effects of varying the molecular weight of the stabilizing moieties of toluene-in-water emulsions

Stabilizing moieties	Molecular weight	UCFT/K	θ_L/K
Polyethylene glycol 600	600	285 ± 3	—
Polyethylene glycol 1000	1 000	320 ± 2	—
Carbowax 1540	1 400	318 ± 2	315 ± 3
Polyethylene glycol 6 000	6 000	318 ± 2	315
Carbowax 6 000	10 000	318 ± 2	315
Polyox resin WSRN-10	79 000	318 ± 2	315
Polyox resin WSRN-80	148 000	318 ± 2	315
Pluronic F68	3 500	322 ± 2	315
Pluronic F108	6 500	322 ± 2	315

6.2.3. Conclusion

We conclude that for sufficiently high molecular weights, the CFPTs of sterically stabilized dispersions are independent of the molecular weight of the stabilizing moieties. This is consistent with the identification of the θ-point as the thermodynamic limit to stability of many sterically stabilized dispersions.

6.3. The origin of the attraction at the critical flocculation point

6.3.1. Stabilization by high polymers

6.3.1.1. The unimportance of van der Waals forces

Electrostatically stabilized dispersions are readily coagulated by the addition of electrolyte. This reduces the spatial extension of the double layers to the point where instability is induced as a consequence of the London attraction between the colloidal particles. The question then arises as to whether the van der Waals forces are also responsible for the flocculation that is observed in sterically stabilized systems. A crude calculation of the magnitude of the attraction between the core particles in typical cases casts considerable light upon this question.

Consider the approach of two sterically stabilized particles, each of radius a, coated by polymer layers of thickness δ (Fig. 6.2). The van der Waals attraction, V_A, is given by

$$V_A = -A^* a / 12 H_0 \tag{6.1}$$

where A^* = effective Hamaker constant and H_0 = minimum distance between

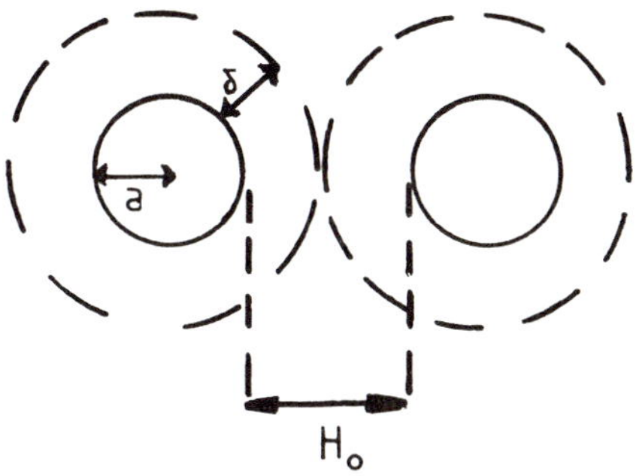

Fig. 6.2. Schematic representation of the close approach of two sterically stabilized particles; a = particle radius, δ = steric layer thickness and H_0 = minimum distance of closest approach of the particle surfaces.

the surfaces of the core particles. It will be adequate for the purposes at hand to assume that the steric layers behave as impenetrable coatings and to calculate the London attraction when the surface layers just touch (i.e. $H_0 = 2\delta$). If the well depth at this separation is several kT, coagulation of such particles would occur, despite the presence of an assumed infinitely strong steric barrier. A characteristic value of A^* for latex particles is 5×10^{-21} J (Evans and Napper, 1973c) which leads to the approximate relationship $V_A/kT \sim -0{\cdot}05(a/\delta)$. Values of the well depth calculated in this fashion are shown in Table 6.7. A typical molecular weight corresponding to the various values of the steric layer thickness is also presented in Table 6.7. This was calculated assuming that the isolated polymer chains were terminally attached (i.e. a tail). Hence $\delta \sim \alpha f l(nC_\infty)^{\frac{1}{2}} \sim \alpha f l(2MC_\infty/M_0)^{\frac{1}{2}}$, where C_∞ = limiting characteristic ratio of the polymer, l = bond length (assumed to be constant), n = number of bonds, α = intramolecular expansion factor, f = expansion

TABLE 6.7
London attraction between sterically stabilized dispersions

Molecular weight	Layer thickness nm	$\lvert V_A \rvert/kT$ Particle radius/nm 10	50	100	200	500
600	2·5	0·2	1	*2*	*4*	*10*
2 500	5	0·1	0·5	1	*2*	*5*
10 000	10	0·05	0.25	0.5	1	*2·5*
63 000	25	0·02	0·10	0·2	0·4	1
250 000	50	0·01	0·05	0·1	0·2	0·5
1 000 000	100	0·005	0·025	0·05	0·1	0·25
6 300 000	250	0·002	0·01	0·02	0·04	0·1

factor for the chain normal to the interface, M=molecular weight of the polymer and M_0=molecular weight of the repeat unit (monomer). For a typical polymer, the following values were chosen: $\alpha=1{\cdot}3, f=1{\cdot}3, l=0{\cdot}15$ nm, $C_\infty=7$ and $M_0=100$, whence $\delta \sim 0{\cdot}1 \times M^{\frac{1}{2}}$.

It is obvious from Table 6.7 that only with low molecular weight stabilizing moieties and large particles does the absolute magnitude of the attraction reach the value of several kT required for coagulation. These values appear in italics in the upper right hand corner of Table 6.7. Particles which possess larger Hamaker constants (e.g. metal sols), larger particle sizes or thinner steric layers will be more strongly attracted than indicated in this Table.

It is apparent that for most high polymers (i.e. polymers with a molecular weight exceeding 10 000) the van der Waals attraction between the core particles is insufficient to induce instability under most conditions. The origin of the attractive interaction that causes flocculation in such systems at the CFPT must be sought elsewhere.

6.3.1.2. The segmental attraction hypothesis

It was pointed out in Section 3.2.7 that in dispersion media that are poorer solvents than θ-solvents for the polymer, the macromolecular segments experience a mutual attraction. This attraction is responsible for the negative deviations from ideal behaviour displayed by polymer chains in worse than θ-solvents (see Fig. 3.3). If the solvency for the polymer is made suitably worse than that of a θ-solvent, phase separation ensues (Flory, 1953).

Flory was perhaps the first to recognize that the flocculation that occurred at the CFPT for many (although certainly not all) sterically stabilized dispersions is a consequence of the segmental attraction that occurs in dispersion media that are somewhat worse solvents than θ-solvents for the stabilizing moieties. This renders the Brownian collisions between particles sticky. As a result of the large number of segments involved in the interaction zone, the energy of attraction necessary per segment to induce flocculation (i.e. a potential energy of attraction of, say, $-10\,kT$) is very small. Consequently, flocculation occurs under solvency conditions that are only marginally worse than θ-solvents.

We stress that θ-points are rarely known with great precision. For example, the best determined θ_U-temperature is that for polystyrene in cyclohexane for which a value of 35 ± 1 °C is appropriate. The precision of most determinations of θ-temperatures is rarely better than $\pm 5-10$ °C. Consequently, it is not surprising that the CFPT, itself subject to considerable imprecision, cannot be distinguished experimentally from the θ-point in many instances.

6.3.1.3. Evidence favouring the segmental attraction hypothesis

Until recently, the evidence in favour of the hypothesis that the segments composing the stabilizing moieties become attractive in slightly worse than θ-solvents, was mainly indirect. Certainly the data presented in Tables 6.2 and 6.3 are too imprecise to establish whether the CFPT lies to the worse solvent side of the θ-point.

The fact that the second virial coefficient of polymer chains becomes negative in worse than θ-solvents is indicative of mutual segmental attraction (Flory, 1953). Moreover, as discussed in Chapter 8, when sterically stabilized particles coagulate under the action of van der Waals forces, an entirely different pattern of flocculation behaviour ensues. Specifically, flocculation occurs in dispersion media that are better solvents than θ-solvents, i.e. when the segments are undoubtedly mutually repulsive.

Klein (1980) has recently measured directly the forces between two mica plates coated by polystyrene in cyclohexane at a temperature (24 °C) below the θ_U-temperature (35 °C). These measurements showed clearly the existence of a long-range force of attraction that is in part attributable to the mutual attraction between the polystyrene segments under worse than θ-conditions. This experiment will be discussed more fully in Section 13.4.1.2.

6.3.1.4. Conclusions

We conclude that for stabilizing moieties that are high polymers, the van der Waals attraction between the core particles is too small to induce instability at the CFPT. Rather, flocculation is a consequence of the segments that impart stability in better than θ-solvents becoming mutually attractive in dispersion media that are worse than θ-solvents. The CFPT corresponds to the dispersion medium becoming a marginally worse solvent for the stabilizing moieties than a θ-solvent. This provides the basis for the general rule that the practical limit to the thermodynamic stability of many sterically stabilized dispersions is the θ-point for the stabilizing moieties in the dispersion medium.

6.3.2. Stabilization by low molecular weight polymers

Inspection of Table 6.7 shows clearly that for relatively large colloidal particles stabilized by low molecular weight stabilizing moieties, the London attraction between the core particles may be sufficient to cause coagulation, no matter how strong the steric repulsion might be. A characteristic feature of such systems should be that coagulation occurs in dispersion media that are significantly *better* solvents for the stabilizing moieties than θ-solvents. Under these conditions, the segments of the stabilizing moieties must be mutually repulsive. The CFPT (strictly CCPT) may then become a function of the number concentration of colloidal particles since the configurational entropy of the dispersed particles can now contribute significantly to stability near to the CFPT. Dilute dispersions are then inherently more stable than concentrated systems. Furthermore, the CFPT in these systems should be found to be a function of the particle size, the molecular weight and conformation of the stabilizing moieties and the solvency of the dispersion medium. A more comprehensive discussion of steric stabilization by low molecular weight moieties will be deferred until Chapter 8.

6.4. The conformations of polymers in worse than θ-solvents: the coil-globule transition

The flocculation discussed above could conceivably arise from a dramatic dimensional collapse of the stabilizing moieties (the so-called *coil-globule* transition). This would be analogous to the coagulation of electrostatically stabilized dispersions which arises from the dimensional collapse of the double layers on adding electrolyte. It is therefore necessary to review the evidence, both theoretical and experimental, supporting the existence of the coil-globule transition to assess the likelihood of its occurrence in sterically stabilized systems.

Langmuir (1926) was probably the first to suggest that van der Waals forces should cause a hydrocarbon chain molecule in a dilute gas to roll up into spheres with a density comparable to that of the liquid phase. Stockmayer (1960) extended this concept to polymer molecules in worse than θ-solvents. He recognized that the classical Flory expression for the excluded volume parameter

$$\alpha^5 - \alpha^3 = Cz \tag{6.2}$$

has no positive roots unless z exceeds some critical value [$z > z^{\text{crit}}$, where $z^{\text{crit}} = -(2/3C)(3/5)^{5/2}$]. Using the Stockmayer (1960) value for C, which is 4/3, $z^{\text{crit}} = -0{\cdot}14$, which clearly corresponds to a worse than θ-solvent, the latter demanding $z = 0 (\alpha = 1)$. For $z < z^{\text{crit}}$, only the collapsed coil is stable.

Ptitsyn *et al.* (1968) argued that the collapse process was a first order transition (i.e. the first derivatives of the Gibbs free energy with respect to both the temperature and pressure are discontinuous at the transition). They used the terms *coil* and *globule* for the two different states and likened the transition to the first order condensation of a real vapour into liquid droplets when the forces of attraction are increased sufficiently. They were led to believe that experimental observation of the transition would be very difficult, if not impossible, because intermolecular condensation would overwhelm the intramolecular contraction, leading to phase separation. Until quite recently this pessimism seemed fully justified.

de Gennes (1972) later drew an analogy between critical phenomena in a magnetic system and the self-excluded volume problem in a polymer system. Subsequently, he argued that the θ-point of a polymer of infinite molecular weight corresponds to a tricritical point (de Gennes, 1975).

6.4.1. The blob theory

Daoud and Jannink (1976) have also exploited scaling law theory (de Gennes, 1979) to explore the dramatic change in polymer conformation at temperatures somewhat below the θ-temperature. According to scaling law theory, a polymer chain can be envisaged as a succession of subunits, picturesquely referred to as "blobs" whose size is the correlation length ξ. Inside any one blob, each of which contains a number of monomeric repeat units, correlations of the excluded volume type exist. Notwithstanding this, the model assumes that at any temperature there is a characteristic length of polymer chain (blob) which obeys Gaussian statistics. This length is infinite at the θ-point and decreases away from it as a function of $|(T-\theta)|/\theta$. One upshot of this crude but simple theory is that in the vicinity of θ, the coil dimensions remain constant over a range of temperature. As noted by Akcasu and Han (1979), blob theory implies that $d\alpha_h/dT=0$ at θ (α_h=expansion factor for the hydrodynamic radius). As the temperature is further reduced below that corresponding to the region of constant dimensions, the polymer chain is predicted to undergo a dimensional collapse in a coil-globule transition.

6.4.2. Experimental evidence for coil-globule transitions

Nierlich *et al.* (1978) have measured the radius of gyration, R_g, of polystyrene of quite low molecular weight ($2{\cdot}9\times10^4$) at a concentration of $3{\cdot}4\times10^{-3}$ g cm^{-3} in deuterocyclohexane at temperatures below the θ_U-temperature, which was taken to be 38 °C. These measurements were made by small angle neutron scattering (SANS). The results were interpreted as showing that R_g decreased only slightly over the temperature range 38 °C to 20 °C. R_g decreased more rapidly, however, on cooling to 11 °C. These results were interpreted as supporting the occurrence of both a θ-region near θ and a coil-to-globule transition, as predicted by Daoud and Jannink (1976).

Slagowski *et al.* (1976) had previously measured R_g by light scattering for very high molecular weight polystyrene ($4{\cdot}4\times10^7$) at concentrations down to 2×10^{-5} g cm^{-3} in cyclohexane in the neighbourhood of the θ_U-temperature. They found abrupt changes in R_g in a narrow temperature range above and below θ_U. A small decrease (1·4 K) in temperature below $T=\theta_U$ led to a large (40%) reduction in the magnitude of R_g. Bauer and Ullman (1980) have pointed out, however, that the value of R_g measured experimentally by Slagowski *et al.* (1976) at the lowest temperature was 121 nm. This is greatly in excess of the radius of gyration (20 nm) that the equivalent collapses sphere would display (i.e. the radius of gyration of the equivalent sphere of the same weight of monomer, as Langmuir suggested).

6.4.3. Evidence against coil-globule transitions

A body of evidence has also appeared which suggests that some of the predictions of scaling law theory and the blob model in worse than θ-solvents should be treated with considerable caution. Both Bauer and Ullman (1980) and Pritchard and Caroline (1980; 1981) have studied the dimensions of polystyrene macromolecules in worse than θ_U-solvents. Pritchard and Caroline (1980), for example, measured the diffusion coefficients of monodisperse polystyrene of molecular weight 37 000 by photon correlation spectroscopy over the temperature range 6 to 60 °C. Four different concentrations in the dilute regime were studied. The derived hydrodynamic radius obtained from the diffusion coefficient at infinite dilution increased smoothly with temperature, as shown in Fig. 6.3. There was no evidence of a collapse region at low temperatures nor of a region of constant coil size around the θ_U-temperature. Bauer and Ullman (1980) reported similar experimental results in that their coils below the θ_U-temperature were very far from being collapsed globules.

The smooth change in the polymer coil dimension passing through the θ-temperature is exemplified by the data of Pritchard and Caroline (1981) shown in Fig. 6.4. Here the hydrodynamic expansion factor α_h is plotted as a function of the temperature. Contrary to the predictions of the blob theory, $d\alpha_h/dT$ is clearly non-zero at the θ_U-temperature, irrespective of the molecular weight. Its positive value is consonant with the predictions of the cluster series

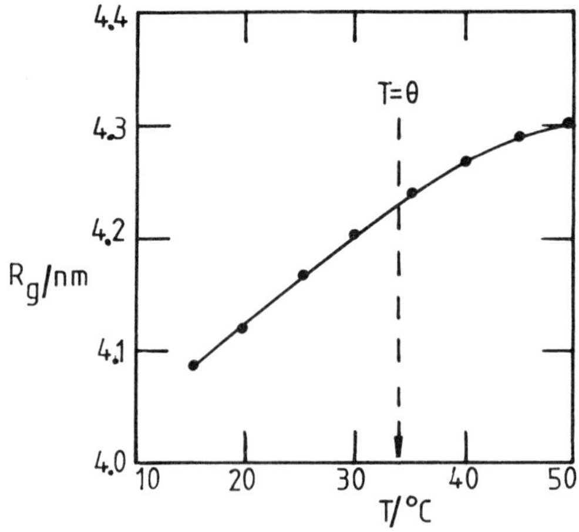

Fig. 6.3. The temperature dependence of the radius of gyration (R_g) of polystyrene in cyclohexane near to the θ-point (after Pritchard and Caroline, 1980).

expansion theory of Stockmayer and Albrecht (1958). The smooth change in chain dimensions agrees both with the mean field theory predictions of Sanchez (1979) and the results of computer simulation studies (McCrackin *et al.*, 1973; Clark and Lal, 1977).

The data shown in Fig. 6.4 also illustrate another important point. Domb (1974) has predicted theoretically that the θ-point is dependent upon the

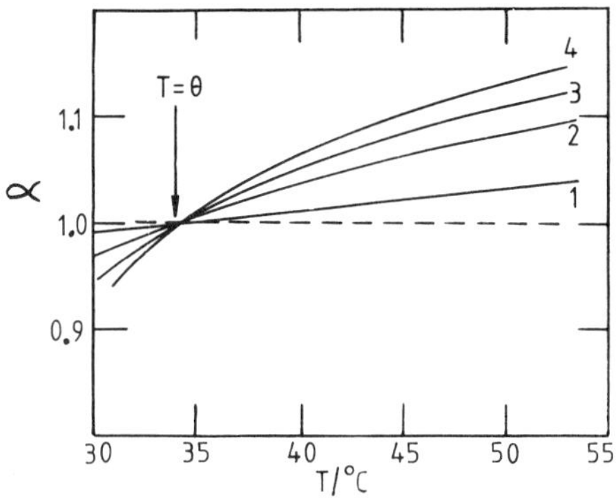

Fig. 6.4. The temperature dependence of the intramolecular expansion factor (α) near to the θ-point for polystyrene in cyclohexane; molecular weight: 1, $35{\cdot}5 \times 10^3$; 2, $6{\cdot}3 \times 10^5$; 3, $1{\cdot}25 \times 10^6$; 4, $5{\cdot}4 \times 10^6$ (after Pritchard and Caroline, 1981).

molecular weight of the polymer. The results of Pritchard and Caroline (1981) for the values of α_h for polystyrene at different temperatures in cyclohexane all converge to $\alpha_h = 1{\cdot}00$ where $T = \theta_U$ at the same temperature (35 ± 1 °C), despite a variation in molecular weight of two orders of magnitude. It may be concluded that any variation in the θ-temperature with molecular weight is insignificant.

Nose and Chu (1979a; 1979b) have also investigated the chain dimensions of monodisperse polystyrene in *trans*-decalin around the θ_U-temperature ($\theta_U = 20{\cdot}5$ °C). Again the chain dimensions were found to increase smoothly with temperature.

6.4.4. Reconciliation of the conflicting results

Swislow *et al.* (1980) have recently reported what they claim to be the first observations of the entire temperature induced coil-globule transition in a single polystyrene polymer chain in cyclohexane. These observations allow a partial reconciliation of the apparently conflicting experimental results presented above, in that they point to the reasons why other workers either did not locate a coil-globule transition or could make only incomplete observations of the transition.

The success of Swislow *et al.* (1980) stems from their use of very high molecular weight polymer ($2{\cdot}7\times10^7$) at exceedingly low concentrations (of order 1×10^{-7} g cm^{-3}). They exploited the high sensitivity of photon correlation spectroscopy to measure the hydrodynamic radius of the polymer as a function of temperature. Their results are presented in Fig. 6.5. These show clearly the collapse of the hydrodynamic radius from 125 nm to 50 nm at 32 °C (this corresponds to $\alpha\simeq0{\cdot}4$). The transition is relatively sharp, its width being about 5 °C.

Different polymer concentrations were studied to provide the data for Fig. 6.6. At the higher polymer concentrations, only the initial stages of the coil-globule transition could be observed because as the temperature was decreased, phase separation occurred. Figure 6.6 displays a portion of the phase diagram for the chosen molecular weight polystyrene in cyclohexane.

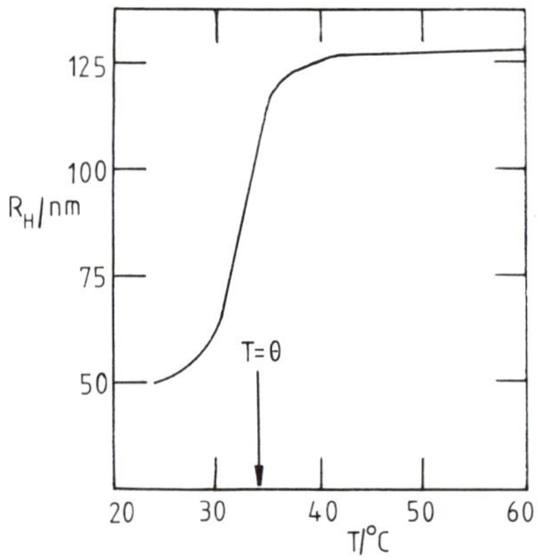

Fig. 6.5. The temperature dependence of the hydrodynamic radius of high molecular weight ($2{\cdot}7\times10^7$) polystyrene in a very dilute cyclohexane solution showing the coil-to-globule transition (after Swislow *et al.*, 1980).

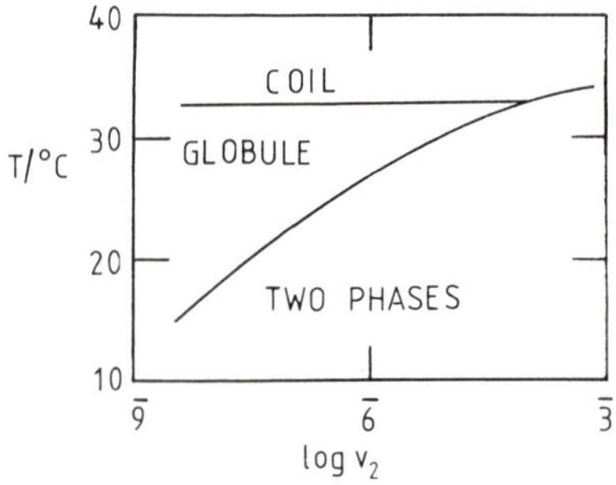

Fig. 6.6. The phase diagram for high molecular weight polystyrene in cyclohexane (after Swislow *et al.*, 1980).

The temperature of the observed coil-globule transition was, of course, independent of the polymer concentrations whereas the phase separation temperature decreased slowly with decreasing polymer concentration. The transition region intersected the phase separation boundary line in the concentration range 10^{-5}–10^{-3} g cm^{-3} for this high molecular weight sample. Other studies of the coil-globule transition used relatively high polymer concentrations so that the polymer chains underwent intermolecular segmental aggregation (i.e. phase separation) rather than intramolecular segmental aggregation (i.e. coil-globule transition). Very low concentrations indeed are necessary if the manifold of globule states that lie below the coil-globule transition line but above the coexistence curve are to be reached. It was this difficulty that afflicted several of the previous studies reviewed in the penultimate section.

Subsequently, Sun *et al.* (1980) have shown that the observed transition in polymer dimensions through the coil-globule states is in good agreement with the predictions of mean field theory.

6.4.5. Importance for steric stabilization

The above-mentioned considerations of the coil-globule transition must be borne in mind in assessing the nature of the interactions between the stabilizing moieties in dispersion media that are worse solvents than θ-solvents. The latter are responsible for the flocculation of sterically stabilized dispersions. If the stabilizing moieties were to undergo a dimensional collapse in worse than θ-solvents, London dispersion forces could conceivably be responsible for the observed flocculation at the CFPT. The experimental evidence presented previously weighs heavily against the occurrence of such a dimensional collapse. The aggregative propensities of the stabilizing moieties appear to be well satisfied at the CFPT by intermolecular associations rather than by intramolecular interactions and collapse. The possibility of a collapse cannot, however, be completely ruled out, especially in solvents very much worse than θ-solvents (i.e. far away from the CFPT).

6.5. Other proposals for the identification of the CFPT

6.5.1. The phase separation theories

Cairns and Neustadter (1975; 1978) have proposed that the CFPT occurs under conditions that lead to

> phase separation of the stabilizing moieties in free solution at the same concentration as that of the adsorbed polymer.

Unfortunately, it is not clear what is meant by the phrase "at the same concentration as that of the adsorbed polymer" because the concentration of a polymer at an interface is singularly ill-defined. It might be possible, for example, to estimate the mean density of polymer segments in a steric layer of some experimentally determined thickness. This mean value, however, would have little relevance to flocculation because the segment density in a steric barrier is usually highly variable (see Section 12.2.2.1) and it is the peripheral regions that are important in flocculation.

6.5.1.1. Experimental test of the phase separation theory

Notwithstanding these conceptual difficulties, the hypothesis put forward by Cairns and Neustadter (1975) is still susceptible to experimental test. The solvency condition required for phase separation in free solution of finite molecular weight macromolecules is always significantly worse than a θ-solvent (Flory, 1953). This follows from the θ-condition which corresponds to the limiting critical solution temperature (whether upper or lower) of a polymer of infinite molecular weight (i.e. the least soluble polymer due to the absence of an ideal contribution to mixing). Moreover, the solvency condition required for phase separation is a sensitive function of the polymer molecular weight (Flory, 1953). If the Cairns and Neustadter hypothesis were correct, the CFPT should occur in dispersion media that are significantly worse solvents than θ-solvents for the stabilizing moieties. Furthermore, the CFPT should be a sensitive function of the molecular weight of the stabilizing moieties.

The data presented in Tables 6.1 and 6.2 show unequivocally that neither of the predictions of the Cairns and Neustadter hypothesis are realized for most sterically stabilized systems (see, however, Chapter 9). The results of Dawkins and Taylor (1980) (highlighted in Table 6.8) are especially illuminating in this context. These authors stabilized poly(methyl methacrylate) particles with monodisperse poly(dimethylsiloxane) chains in a *n*-heptane/ethanol (51/49, v/v) mixture. The molecular weight of the stabilizing moieties was varied from

TABLE 6.8
The influence of the molecular weight of the poly(dimethylsiloxane) stabilizing moieties on the LCFT

Molecular weight	LCFT	θ_U/T
3 200	340	340 ± 2
11 200	340	340 ± 2
23 000	341	340 ± 2
48 000	338	340 ± 2

3 200 to 48 000. Yet the LCFTs were essentially constant at 340 ± 2 K. This is contrary to the predictions of the Cairns and Neustadter hypothesis. Moreover, incipient flocculation was not observed in these systems in significantly worse than θ_U-solvents. Instead, the LCFT correlated closely with the θ_U-temperature.

It is scarcely surprising that the hypothesis of Cairns and Neustadter leads to predictions that are at variance with experiment. Presumably the hypothesis is based upon the idea that at the CFPT, the stabilizing moieties undergo phase separation. Phase separation from solutions of noncrystallizing polymers merely involves the formation of a concentrated polymer solution as one phase (due to intermolecular segmental association in the poor solvent) and a dilute polymer solution as the other. Polymer chains which are attached to the surface of a colloidal particle and which are not compressed seem unlikely to undergo phase separation in this sense. This is especially so as phase separation demands the formation of liquid/liquid interfaces. The phenomenon that attached polymer chains usually undergo involves the formation of flocs of particles, which, just as in the solution case, is brought about by intermolecular segmental association in slightly poorer than θ-solvents. It seems unlikely that liquid/liquid interfaces are formed, although this possibility cannot as yet be definitely eliminated.

Osmond *et al.* (1975) have also asserted their preference for a phase separation approach to incipient instability in a critique of theories of steric stabilization. They argued that phase separation occurred at a temperature close to the θ-temperature. This was claimed to result from the attachment of the stabilizing moieties to the surface of the colloidal particles. Attachment removes the configurational entropy associated with the translational motion of the stabilizing moieties. Osmond *et al.* did not, however, spell out whether the phase separation envisaged by them involved the formation of liquid/liquid interfaces, although it is difficult to conceive how phase separation can proceed without their formation. Many of the objections to the hypothesis of Cairns and Neustadter raised above are equally applicable in this instance.

Everett and Stageman (1978a) have also claimed that at the CFPT, surface attached polymer chains "phase separate" from the dispersion medium and collapse on to the surface of the colloidal particle, this process occurring in dispersion media that are better solvents than θ-solvents. To-date, there is no experimental evidence to support their speculations for those systems that display a strong correlation between the CFPT and the θ-point.

6.5.2. The high molecular weight limit postulate

6.5.2.1. van der Waals attraction

Vincent (1978) has proposed that the correlation between the CFPT and the θ-

point only holds in the limit of high molecular weight stabilizing moieties. Presumably, this requirement of high molecular weight polymer, if correct, would be a consequence of the intrusion of the van der Waals attraction between the latex particles.

The results displayed in Tables 6.1 and 6.2 show indubitably that the correlation between the CFT and θ-temperature holds for latices over a wide range of molecular weights and, in some instances, for molecular weights as small as a few thousand. This is exemplified in particular by the data of Dawkins and Taylor (1980) (highlighted in the previous section in Table 6.8) for the stabilization by poly(dimethylsiloxane) moieties, and by the poly(oxyethylene) data of Napper (1970a) shown in Table 6.1.

It is not possible to say from the data presently available what is the lower limit to the molecular weight of the stabilizing moieties that still gives a correlation between the CFPT and the θ-point. The limit seems likely to vary from system to system. It would be expected to be a complex function of the chemical nature of the disperse phase and the dispersion medium, coupled with the particle size and the spatial extension of the stabilizing moieties. This conclusion is borne out by the fact that Everett and Stageman (1977; 1978a) reported coagulation in better solvents than θ-solvents with poly(dimethylsiloxane) stabilizing moieties of molecular weight 7 000. Yet Dawkins and Taylor (1980) observed the LCFT-θ_U correlation with the same moieties of molecular weight 3 200. Likewise Napper (1970a) observed the correlation between UCFT and θ_L in aqueous systems stabilized by poly(oxyethylene) of molecular weight 1 400, whereas Cowell *et al.* (1978) observed coagulation under the influence of van der Waals attraction with the same type of stabilizing moieties of molecular weight 2 000. These apparently contradictory results reported by different workers merely reflect the intrinsic complexity of the phenomenon involved.

6.5.3. Higher virial coefficients

Apart from van der Waals forces, there is a second reason for suspecting that the CFPT-θ-point correlation could break down for low molecular weight stabilizers. These give rise to high mean segment densities in the steric barrier. Under θ-conditions, the second virial coefficient, relating to the interaction of pairs of segments, vanishes. But the higher virial coefficients, relating to higher order segmental interactions, do not necessarily vanish. It is theoretically possible, therefore, for there to be residual repulsion arising from these higher order interactions in a θ-solvent.

The experimental evidence to-date suggests that the effects on stability arising from these higher order terms are negligible. This would not necessarily be the case, however, with star-branched macromolecules as stabilizers. Even in free solution, such polymers apparently display "anomalous" swelling in θ-solvents arising from higher order segmental interactions (Lifshitz *et al.*, 1978; Candau *et al.*, 1972; Cassassa, 1970; Khokhlov, 1978; 1981). de Gennes (1979) has claimed on the basis of scaling law theory that there are unusual logarithmic factors arising from subtle correlations at the θ-point associated with the three-body (and higher) segmental interactions. As they are apparently too small to be detected experimentally, it seems unlikely that they influence steric stabilization.

6.5.4. The dipolar substituent theory

Wolter (1978) has claimed that the data presented in Tables 6.1 and 6.2 are capable of a more refined interpretation. He noted that the correlation between the UCFT and the θ_L-temperature was excellent, irrespective of the molecular weight, for poly(oxyethylene) in aqueous 0·39 M $MgSO_4$ and poly(isobutylene) in 2-methylbutane. Both of these chains lack dipolar substituent groups attached to the backbone chain. An identical comment applies to the UCFT data presented in Table 6.8 for poly(dimethylsiloxane), which were unavailable to Wolter. Again, this is for a polymer lacking dipolar substituents.

In contrast, macromolecules, such as poly(acrylic acid), polyacrylamide and poly(vinyl alcohol), that possess a dipolar substituent appear to exhibit a somewhat weaker correlation between the CFT and the θ-temperature. Furthermore, whether or not a UCFT or LCFT was being observed, there appears to be a slight trend in the CFT with the molecular weight of the stabilizing moieties. This was always to the effect that an increase in molecular weight caused the CFT to be displaced further into the worse than θ-solvent domain.

Wolter regarded these systems as displaying a special type of enhanced steric stabilization (see Chapter 9): special in the sense that the degree of enhancement increased with increasing molecular weight of the stabilizing moieties whereas conventional enhanced steric stabilizations display the opposite trend. It was reasoned that a net force may perturb the conformation of a polymer molecule with dipolar substituents. As a result of a net attraction between the dipoles, the mean segment density in the stabilizing sheaths increases relative to that of the chains in free solution. This is a consequence of the loss of randomness of the orientation of the dipoles on attachment to the surface. The extent of this perturbation was postulated to be greater for higher molecular weights since these exhibit increasing conformational freedom of the chains. An appropriate analogy might perhaps be intramolecular expansion which also increases with increasing molecular weight.

A note of caution should be sounded against interpreting the data of Tables 6.1 and 6.2 in this fashion. The precision of the data is such that the trends claimed by Wolter are barely discernible and may not be experimentally significant. Note that, if as proposed in Section 6.3.1.2 flocculation at the CFPT arises from the mutual attraction of the attached segments, greater penetration into the worse than θ-solvent regime would indeed be required for higher molecular weight polymers, even in the absence of dipolar substituents, to counteract their reduced segment density. This should be less apparent in aqueous systems than nonaqueous systems. We therefore conclude that confirmation of the dipolar substituent hypothesis awaits the results of experiments specifically designed to test it.

6.6. Steric stabilization by polyelectrolytes

The stabilization of colloidal dispersions by polyelectrolytes is of considerable practical concern. We have already mentioned the early studies of Faraday (1857) and Zsigmondy (1901) wherein it was shown that proteins could protect gold sols against coagulation by electrolytes. Apart from such studies of the protective action of biopolymers, there have been few systematic investigations of the stabilizing effects of polyelectrolytes, especially those of synthetic origin.

It would be expected that in general polyelectrolytes would impart stability by a complex combination of electrostatic and steric mechanisms (i.e. electrosteric stabilization). Under certain limiting conditions, it might nevertheless be anticipated that one or other of the stabilizing mechanisms would be dominant. Thus, at low ionic strength and high degrees of ionization, electrostatic stabilization should prevail; contrariwise, at high ionic strength and/or low degrees of ionization, the principal mechanism of stabilization should be steric in origin. This latter expectation has been tested experimentally by Buscall (1981), whose results will now be considered.

6.6.1. Low degrees of neutralization

The results obtained by Buscall (1981) for the stabilization of polystyrene latices by well-anchored polyacrylate chains, of molecular weight 16 000, support the prediction that the steric mechanism is dominant at high ionic strengths. Up to a degree of neutralization of ca 0·5, the polyacrylate stabilized latices when dispersed in an aqueous solution of sodium chloride at a concentration in excess of 1 M, could be flocculated by cooling. A comparison is presented in Fig. 6.7 of the LCFTs and the corresponding θ_U-temperature for these systems as a function of the degree of neutralization. The dispersion medium was 1·6 M NaCl.

It is readily apparent from Fig. 6.7 that as the degree of neutralization increased to 0·4 (i.e. as the charge on the polyelectrolyte rose) so the LCFT declined (at least above a degree of neutralization of 0·1). There was a strong correlation between the LCFT and the θ_U-temperature over this range of neutralization. The correlation was weakened, however, when the degree of neutralization was small.

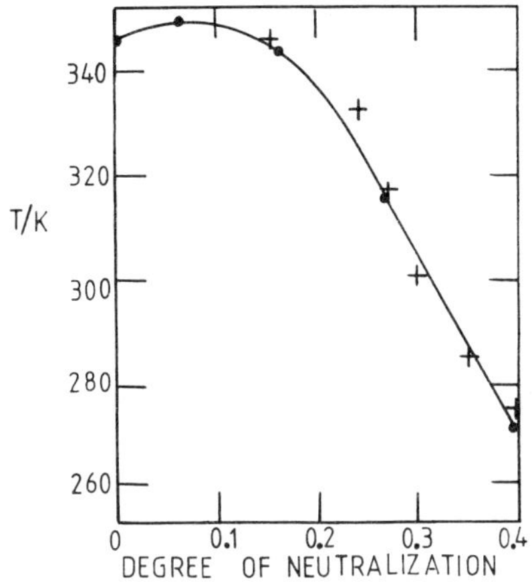

Fig. 6.7. The LCFT (+) of poly(acrylic acid) stabilized latices as a function of the degree of neutralization in 1·6 M NaCl. Also shown are the θ_U-temperatures (●) (after Buscall, 1981).

6.6.2. High degrees of neutralization

Latices stabilized by sodium polyacrylate with a degree of neutralization of 1 exhibit a complex pattern of flocculation behaviour. They were stable in saturated (ca 6 M) sodium chloride but flocculated on the addition of sodium iodide or sodium thiocyanate. Figure 6.8 shows that the LCFTs and the θ_U-temperatures correlate strongly for various concentrations of sodium thiocyanate at a degree of neutralization of unity.

The effects of sodium iodide on latex stability at 293 K was quite complicated. The dispersion was unstable when sodium iodide was present at

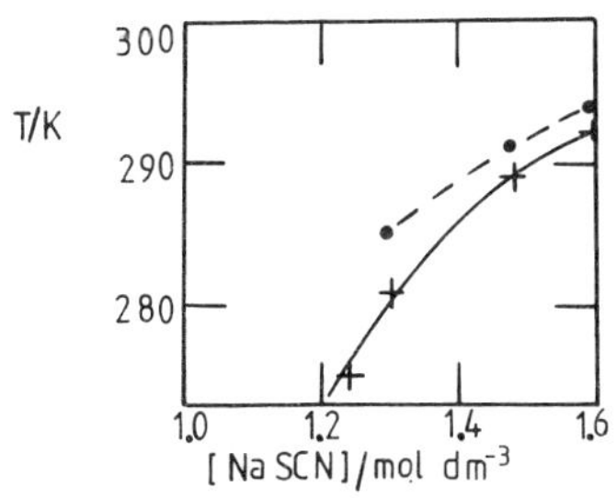

Fig. 6.8. The LCFT (+) of poly(acrylic acid) stabilized latices as a function of the concentration of sodium thiocyanate. Also shown are the θ_U-temperatures (●) (after Buscall, 1981).

concentrations in the range ca 1·5 to 2·5 M (actually a slight cooling was also necessary); above 2·5 M NaI, the latex was again stable but it flocculated when the sodium iodide concentration reached 6 M. Clearly, the segment-segment interactions must change from being repulsive to being attractive at ca 1·5 M NaI and then revert back to being repulsive at ca 2·5 M NaI. A second transformation from segmental repulsions to segmental attraction apparently occurs around 6 M NaI.

6.6.3. General discussion

The results obtained by Buscall (1981) demonstrate the importance of steric stabilization by polyelectrolytes at high ionic strengths. They are broadly in line with the conclusions of Flory and Osterheld (1954) relating to the flexibility of polyelectrolytes under such conditions. These authors demonstrated that high molecular weight poly(acrylic acid) when one-third neutralized by sodium hydroxide had a θ-temperature of 32 °C in 1·245 M sodium chloride. The dimensions of the polyelectrolyte chains under these high ionic strength conditions were identical with those measured for unionized poly(acrylic acid) in dioxane at 30 °C, also a θ-point. Such agreement would only be possible if the chain flexibility were independent of the medium. This flexibility is an essential prerequisite for a polymer chain to impart steric stabilization.

The foregoing conclusion that a polyelectrolyte chain in a medium of high ionic strength behaves in a fashion analogous to a nonionic polymer is perhaps not too difficult to understand. At high ionic strengths, the Debye length is exceedingly small (see Table 1.2). Accordingly, the interactions (usually repulsive in character) between the charged groups on the polymer chain are sharply reduced. The chain, with its closely associated counterions, therefore resembles a nonionic polymer and displays an apparent θ-point. Of course, in a ternary system, this point must be identified with the *plait point* in the limit of infinite molecular weight.

van der Schee and Lyklema (1982) have concluded on the basis of double layer experiments performed with poly-L-proline adsorbed on silver iodide that this uncharged polymer can impart steric stabilization on adsorption. This occurs despite its adsorption in a rather extended conformation, perhaps in a helical form. Apparently this chain is sufficiently flexible to allow the formation of protruding tails and loops, which can in principle impart steric stabilization. It was claimed that the adsorption of highly charged polyelectrolytes, such as the polylysines, does not lead to steric stabilization at low pH. Electrostatic repulsion between the segments was postulated to prevent the formation of loops and tails on adsorption.

6.7. Aqueous dispersions

6.7.1. The effects of electrolytes

The effects that electrolytes exert on aqueous sterically stabilized dispersions are extremely complex and still poorly understood. The potency of an electrolyte in inducing flocculation is directly related to its ability to convert water into a poorer solvent for the stabilizing moieties. Napper (1970b) has shown that for poly(oxyethylene) stabilizing moieties the order of decreasing flocculation effectiveness for anions was $SO_4^{2-} > Cl^- > NO_3^- > CNS^- > Br^- > I^-$. This parallels the classical Hofmeister series. The order of flocculation potency for the cations was found to be $Rb^+ = K^+ = Na^+ = Cs^+ > NH_4^+ = Sr^{2+} > Li^+ = Ca^{2+} = Ba^{2+} = Mg^{2+}$. This sequence is in direct conflict with the classical interpretation of the effects of these ions on hydrophilic colloids (Hedges, 1931; Alexander and Johnson, 1949). The sequence has, however, been confirmed by Boucher and Hines (1976).

The classical picture of the role of ions in aqueous polymer solutions is related to the notion that hydrophilic colloids are heavily hydrated in solution. The ions are considered to exert their influence by competing successfully with the polymer for the available molecules of water. Because of the charge on the ions, it might be expected that the ion-dipole interactions would be stronger than the dipole-dipole interactions between the water and the polymer. The classical viewpoint would therefore claim that the ions dehydrate the polymer and so induce flocculation.

The order of effectiveness for cations does not support this interpretation. The strongly hydrated lithium ion, for example, was the least potent of the monovalent cations studied. Moreover, the highly hydrated divalent cations were also relatively poor flocculants. We also note that the most effective anionic flocculant, SO_4^{2-}, is only weakly hydrated, if it is hydrated at all. The results for both anions and cations suggest that flocculation is not induced simply as a consequence of the dehydration of the stabilizing moieties.

In an attempt to clarify the mechanism whereby electrolytes influence the solvency of water for a polymer, Garvey and Robb (1979) measured the heats of dilution of poly(vinyl pyrrolidone) in aqueous electrolyte solutions. The heat of dilution was found to be negative (i.e. mixing is favoured by the enthalpy changes) so that phase separation must inevitably be controlled by entropy effects. Thermodynamics demands that in the salting-out of a polymer by a second component, this component must be negatively adsorbed from the polymer. This can be expressed mathematically by

$$d\mu_2^{\ominus} = \sum_{i=1}^{c} N^*_{si}\, d\mu_i \qquad (6.3)$$

where $\mu_2^{\ominus}$ = standard chemical potential of the polymer in the electrolyte solution and N^*_{si} = the difference in concentration of the ith component in volume elements near to and far from the nonelectrolyte. For $d\mu_2^{\ominus}$ to decrease, N^*_{si} must be negative, indicating a depletion of the electrolyte from regions immediately adjacent to the polymer.

The cause of the depletion of the ions in the vicinity of the polymer is still uncertain. Garvey and Robb (1979) suggested that the polymer constitutes a dielectric discontinuity in the medium so that the ions generate image forces, which repel the charged species. Subsequently, Robb (1982) suggested that the water structures around the polymer and certain ions are not compatible.

The idea proposed by Robb that the water structures around certain ions and the polymer is incompatible was not new. Eagland (1975) had previously advanced this concept in a somewhat different form to explain the order of effectiveness of the different anions and cations in promoting the flocculation of poly(vinyl acetate) latices stabilized by poly(oxyethylene). He pointed out that the sequence of anion effectiveness observed by Napper (1970b) was in close agreement with those found for the coagulation of electrostatically stabilized sols whose surfaces contained oxygen atoms (e.g. Fe_2O_3 and Al_2O_3 sols). It was speculated that the surface oxygen atoms play a dominant role in orienting the water molecules, producing what was termed *hydrophilic hydration*. Such oriented water molecules might be expected to be structurally incompatible with ordinary water or with the structure produced by structure-promoting anions. Incompatibility of water structure was also invoked to explain the cation effectiveness series. In this case, however, it was necessary to include in the discussion not only the hydrophilic hydration of the oxygen atom in the poly(oxyethylene) chains but also the *hydrophobic hydration* arising from the ethylene groups. Studies on proteins and polypeptides suggest that the hydration structures associated with nonpolar and polar regions are apparently different but additive. Further, the effects of the ionic constituents of electrolytes are also claimed to be additive. This is apparently a consequence of the anions interfering with the hydration of polar regions whereas cations interact with the hydration of nonpolar regions, perhaps because of the compatibility of hydrophobic hydration structure and that surrounding structure-making ions. This compatibility if applicable to poly(oxyethylene) would readily explain why the structure-making cations are the least effective flocculants. Moreover, the foregoing concepts are in harmony with the observation that the structure-forming anions are the more effective flocculants. This would be ascribed to incompatibility between the hydration structures originating from the oxygen atoms and those from the anions.

Whilst the foregoing picture is seemingly quite plausible, it is unlikely to be correct if only because the hydration about the ethylene units in poly(oxyethylene) is almost certainly not independent of that associated with the oxygen atoms. The spatial proximity of these two groups must render their hydration a cooperative phenomenon.

Napper (1970) has concluded that monatomic anions and cations are less potent flocculants if their effective primary hydration radius is large. This was considered to be a consequence of the influence that the positively hydrated ions exert on the structure of the water in their immediate neighbourhood (the so-called region B). Thus the potency of the sulphate ions is likely to be due in part to their powerful structure breaking influence. Boucher and Hines (1976) have noted that IR measurements gave the following sequences for decreasing power in destroying the structure of water: $K^+ > Na^+ > Li^+ > Cs^+$ and $NO_3^- > Cl^- > CNS^-$. This is crudely in agreement with the observed trends.

The foregoing discussion vividly illustrates the gross uncertainties associated with the understanding of the effects of electrolytes on aqueous polymer solutions.

6.7.2. The effects of temperature

Any increase in temperature would seem likely to disrupt specific interactions, such as hydrogen bonding, between the water molecules and the segments of the polymer chains. This could lead to a decrease in the solvency of the dispersion medium for the stabilizing moieties. There is some indirect evidence for this in the case of poly(oxyethylene) (Volkov, 1973; 1974; 1976) but it is far from conclusive.

Chapter 7

Thermodynamically Limited Steric Stabilization: The Thermodynamic Factors Controlling Steric Stabilization

7.1. The temperature dependence of stability

The dependence of the stability of sterically stabilized dispersions on temperature can provide some insight into the thermodynamic factors that control stability, at least near to the CFT. In fact the temperature response of stability is a generic property of sterically stabilized dispersions. Some dispersions are readily flocculated by heating, others by cooling. Some dispersions flocculate both on heating and on cooling (Napper, 1977). Indeed, it seems likely that in principle all sterically stabilized dispersions should be capable of being flocculated both by heating and by cooling, if only a sufficiently wide range of temperatures and pressures can be scanned.

7.2. A general thermodynamic treatment

Consider the close approach of two sterically stabilized particles. Let ΔG_F be the change in their free energy on close approach. Then, at constant temperature and pressure

$$\Delta G_F = \Delta H_F - T\Delta S_F \tag{7.1}$$

where T = temperature and ΔH_F, ΔS_F = the enthalpy and entropy of close approach, respectively. For the purposes at hand, the free energy of close approach is best envisaged as the *free energy of flocculation*; hence the use of the subscript F. For stability, ΔG_F must be suitably non-negative over the appropriate distances of close approach. A negative value for ΔG_F of only a few kT may, in principle, provide a potential energy well-depth sufficiently deep to induce flocculation. In practice, to impart stability it is advantageous to arrange for the free energy of flocculation to be suitably positive. This can be accomplished in three different ways. These are summarized in Table 7.1.

First, both ΔH_F and ΔS_F may be positive so that the enthalpy of close approach favours stabilization whereas the corresponding entropy term promotes flocculation. If, however, the contribution of the enthalpy term to ΔG_F exceeds that of the entropy term in absolute magnitude, ΔG_F will be positive and close approach will be disfavoured. This is termed *enthalpic stabilization* since it is the net enthalpy of close approach that ensures stability.

The second method of imparting stability is in sharp contrast to the first: ΔH_F and ΔS_F are now both negative. The enthalpy of close approach favours flocculation but the entropy term opposes it. Provided that the entropic contribution to ΔG_F exceeds, in absolute magnitude, that of the enthalpy term, stability ensues. This is termed *entropic stabilization.*

The third mechanism of imparting steric stabilization differs from the preceding two in that the enthalpy and entropy contributions to the free energy of flocculation act in concert instead of in opposition. Both now contribute to stability since ΔH_F is positive and ΔS_F is negative. The magnitudes of their respective contributions to ΔG_F are no longer important. This is termed *combined enthalpic-entropic stabilization.*

Note that certain limiting cases implicit in Table 7.1 can still result in a positive free energy of flocculation. These are $\Delta S_F = 0$ for enthalpic stabilization and $\Delta H_F = 0$ for entropic stabilization. Note, too, that if $\Delta H_F = 0$, the system is said to be *athermal.*

TABLE 7.1
Thermodynamic types of steric stabilization

ΔH_F	ΔS_F	$\lvert\Delta H_F\rvert/T\lvert\Delta S_F\rvert$	ΔG_F	Type	Flocculation
+	+	>1	+	enthalpic	on heating
−	−	<1	+	entropic	on cooling
+	−	$\gtrless 1$	+	combined enthalpic-entropic	not accessible

7.3. Identification of the stabilization type by temperature changes

7.3.1. Enthalpic stabilization

It is apparent from Table 7.1 that the requirement for enthalpic stabilization, *viz* $\Delta H_F > T\Delta S_F$, can in principle be violated by increasing the temperature. The entropic term, which promotes flocculation, becomes more competitive on heating and flocculation may ensue. Thus enthalpic stabilization is characterized by flocculation on heating. Conversely, a dispersion that is observed to flocculate on heating may be inferred to be enthalpically stabilized, *at least near to the UCFT.*

7.3.2. Entropic stabilization

With entropic stabilization, the reverse situation to that just described for enthalpic stabilization must pertain. Clearly the entropy term that promotes stability can be made less competitive by decreasing the temperature. Entropically stabilized dispersions can thus, in principle, be flocculated by cooling. A sterically stabilized dispersion that flocculates on cooling can be inferred to be entropically stabilized, *at least just above the LCFT.*

7.3.3. Combined enthalpic-entropic stabilization

As both the entropy and enthalpy terms conspire to impart stability, it is not possible to induce flocculation merely by changing the temperature in a dispersion exhibiting combined enthalpic-entropic stabilization. As discussed below, however, it may well be possible by temperature and/or pressure changes to convert the dispersion into one displaying either enthalpic or entropic stabilization. Flocculation by temperature changes could thereupon become possible.

7.4. Identification of the type of stabilization by the dilution parameters

One difficulty associated with the identification of the type of steric stabilization via responses to temperature changes is that the classification strictly only holds close to the CFT. This is brought home forcibly by recalling the behaviour of those latices that are readily flocculated both by heating and cooling. How far it is possible to deviate from the CFT before the identification becomes invalid is a complex question depending crucially upon the nature of the interactions between the segments of the stabilizing moieties and the dispersion medium. For some systems, it could be as much as several hundred degrees kelvin whereas for others it might be as little as a few degrees kelvin.

Fortunately, there is an independent, universal method that in principle allows the type of steric stabilization to be ascertained. Determination of the entropy (ψ_1) and enthalpy (κ_1) of dilution parameters for the stabilizing moieties in free solution fixes the signs of ΔH_F and ΔG_F unequivocally, provided that their conformation is not radically altered by attachment to the interface. Of course, because these thermodynamic parameters refer to dilution and not to interpenetration (i.e. concentration), their signs are reversed from those of ΔH_F and ΔS_F, as shown in Table 7.2. The allowable zero values have again been omitted from this Table. These are $\psi_1=0$ for enthalpic stabilization and $\kappa_1=0$ for entropic stabilization (the athermal case).

The dilution parameters are readily determined by a range of experimental measurements, such as the temperature dependence of the second virial coefficient. Alternatively, if the θ-temperature is measured separately, determination of the intramolecular expansion factor α, e.g. by viscometry, allows the values of ψ_1 and κ_1 to be calculated using the Flory relationship, as modified by Stockmayer

$$\alpha^5-\alpha^3=C_M\psi_1(1-\theta/T)M^{\frac{1}{2}} \tag{7.2}$$

where $\theta=(\kappa_1/\psi_1)T$, M=molecular weight and $C_M=(27/2^{5/2}\pi^{3/2})(\bar{v}_2{}^2/N_A\bar{V}_1)(\langle r^2\rangle_0/M^{-3/2}$. Here $\bar{v}_2$=partial specific volume of the polymer, $\bar{V}$=molar volume of the solvent and $\langle r^2\rangle^{\frac{1}{2}}$=rms end-to-end distance of the stabilizing moieties. Although some uncertainty attends the absolute magnitudes of ψ_1 and κ_1 determined in this fashion (arising from uncertainty in the modified Flory relationship) the signs of ψ_1 and κ_1 are unambiguously fixed by this procedure.

TABLE 7.2
The signs of the dilution parameters for different types of steric stabilization

Stabilization type	Dilution parameters κ_1	ψ_1	κ_1/ψ_1
enthalpic	−	−	>1
entropic	+	+	<1
combined enthalpic-entropic	−	+	<0

Of course, if κ_1 is negative and ψ_1 is positive, the value of θ is negative and so it is unlikely to be known independently. In these circumstances, the measurement of α for a range of molecular weights will suffice to determine both dilution parameters.

Finally, we mention that Cairns *et al.* (1976) have used an independent thermodynamic method to verify the description of polymer latices stabilized by poly(12-hydroxystearic acid) in *n*-alkanes as being entropically stabilized at room temperature and pressure. Napper (1970a) had made this assignment on the basis of the determination of the enthalpy and entropy of dilution parameters. Cairns *et al.* measured directly the enthalpy of dilution of concentrated latices in the absence of free polymer and found that the dilution of sterically stabilized latices was endothermic. This necessarily means that ΔH_F, the enthalpy of close approach, must be negative. As the system is stable, it follows that the corresponding value of ΔS_F must also be negative. Negative values for both ΔH_F and ΔS_F confirm the occurrence of entropic stabilization.

7.5. General discussion of the different types of steric stabilization

7.5.1. Generality of the classification scheme

It should first be stressed that the foregoing classification of sterically stabilized dispersions, based on whether they flocculate on heating or cooling, is quite rigorous *near to the CFT*. It is independent of any temperature dependence of ΔS_F and ΔH_F because it is always possible to write the temperature dependence of the free energy of flocculation as

$$\frac{\partial}{\partial T}(\Delta G_F) = -\Delta S_F \tag{7.3}$$

It follows that if flocculation occurs on heating, ΔG_F must change sign from positive to negative on heating. Accordingly, $\partial(\Delta G_F)/\partial T$ must be negative near to the CFPT. This immediately fixes the sign of ΔS_F as being positive. Furthermore, since stability is observed below the UCFT, ΔG_F must be positive under those conditions. This is only possible if ΔH_F is also positive.

By way of contrast, flocculation on cooling implies that ΔS_F must be negative near to the LCFT. Since flocculation is observed at the LCFT, ΔH_F must also be negative.

Finally, we note that, contrary to the claim by Osmond *et al.* (1975), the thermodynamic basis of the scheme outlined herein is independent of any model adopted to explain steric stabilization. The assignments presented in Table 7.1 have broad thermodynamic validity near to the CFT, as has been confirmed in this section using the general equation (7.3).

7.5.2. Typical examples of the classification scheme in action

Some examples of the classification of dispersions stabilized by different types of polymers *at room temperature and pressure* are listed in Table 7.3.

Enthalpic stabilization is perhaps more common in aqueous media than is entropic stabilization. Certainly, many of the commercially important

TABLE 7.3
Classification of some sterically stabilized dispersions near to room temperature and pressure

Stabilization moieties	Dispersion type	Medium example	Stabilization type	Reference
poly(oxyethylene)	aqueous	0·39 M $MgSO_4$	enthalpic	Napper (1970a)
poly(vinyl alcohol)	aqueous	2 M NaCl	enthalpic	Evans (1971)
poly(methacrylic acid)	aqueous	0·02 M HCl	enthalpic	Evans *et al.* (1972)
poly(acrylic acid)	aqueous	0·2 M HCl	entropic	Evans *et al.* (1972)
poly(acrylamide)	aqueous	2·1 M $(NH_4)_2SO_4$	entropic	Evans *et al.* (1972)
poly(oxyethylene)	nonaqueous	methanol	entropic	Napper and Netschey (1971)
polystyrene	nonaqueous	cyclopentane	entropic	Croucher and Hair (1980a)
poly(lauryl methacrylate)	nonaqueous	*n*-heptane	entropic	Napper (1968b)
poly(12-hydroxystearic acid)	nonaqueous	*n*-heptane	entropic	Napper (1968b; 1970a)
poly(isobutylene)	nonaqueous	2-methylbutane	enthalpic	Evans and Napper (1975)
poly(vinyl alcohol)	mixed	dioxan/water	combined	Napper (1969c)
poly(oxyethylene)	mixed	methanol	combined	Napper and Netschey (1971)

stabilizing moieties for aqueous systems, such as poly(oxyethylene) and poly(vinyl alcohol), function as enthalpic stabilizers in water at room temperature (Napper, 1970a). Moreover, at room temperature, entropic stabilization is markedly more common in nonaqueous media than is enthalpic stabilization. This latter ubiquity led to the serious misconception, still occasionally propagated even today, that all steric stabilization occurs by an entropic mechanism. Indeed, steric stabilizers in the past were often referred to as "entropic stabilizers". If all dispersions were entropically stabilized, as this nomenclature proclaims, then they would necessarily become more stable on heating. This is quite contrary to what is observed experimentally: many, and in principle all, nonaqueous dispersions flocculate on heating. This serious misconception of the thermodynamic processes that underpin the mechanisms whereby polymeric species impart colloid stability springs from too rigid an adherence to the conceptual framework of the early Flory-Huggins theory. This theory has now been superseded by the free volume theory which properly predicts the occurrence of enthalpic stabilization. We note that enthalpic stabilization in nonaqueous dispersion media tends to occur at higher temperatures than room temperature, indeed at temperatures approaching the critical point of the solvent. Aqueous dispersions, however, flocculate on heating at temperatures well below the critical temperature of water.

We emphasise that both enthalpic and entropic stabilization have been observed in both aqueous and nonaqueous dispersion media. Furthermore all dispersions in principle should be able to be flocculated both by heating and by cooling, if only a wide enough range of temperatures and pressures can be scanned. It is not, however, always possible to observe both critical flocculation points. Sometimes the dispersion medium freezes before flocculation at the LCFT can be reached. Alternatively, it may decompose on heating before the UCFT is reached.

Poly(oxyethylene) behaves as an enthalpic stabilizer in water and an entropic stabilizer in methanol. It will be shown in the next section that for poly(oxyethylene) stabilizing moieties, the addition of methanol to water, or vice versa, must result in passage through a region of combined enthalpic-entropic stabilization. A similar remark is applicable to poly(vinyl alcohol) in water/dioxan mixtures. Likewise, it is not possible to traverse the temperature range from the LCFT to the UCFT without passing through a region of combined enthalpic-entropic stabilization. This is illustrated in Fig. 12.12 for poly(isobutylene) in 2-methylbutane. At temperatures below ca 290 K, the stabilization mechanism is combined enthalpic-entropic; it transforms into enthalpic stabilization above ca 290 K.

7.5.3. Graphical representation of the domains of steric stabilization

Evans (1976) has shown how the various types of steric stabilization can be represented graphically by a conceptually illuminating enthalpy-entropy diagram. This is reproduced in Fig. 7.1. The axes ΔH_F and $T\Delta S_F$ represent the respective enthalpy and entropy contributions that must be subtracted in order to obtain the free energy of flocculation. The line shown with a positive slope of unity corresponds to $\Delta G_F=0$. It delineates the boundary between positive and negative values of ΔG_F. To the left of that boundary, ΔG_F is positive whereas to the right ΔG_F is negative. Thus the line with unit slope, which depicts the demarcation between long-term stability and the onset of catastrophic flocculation, is the locus of all θ-points. All the dispersions that lie in the space to the left of the line are stable, whereas those that lie to the right are unstable (van der Waals attraction is ignored here).

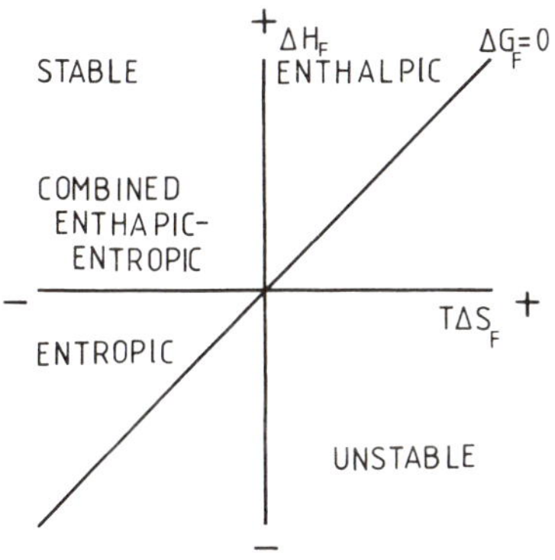

Fig. 7.1. The enthalpy-entropy diagram for sterically stabilized dispersions.

The space to the left of the θ-point line can be subdivided into three domains as shown. The upper octant is defined by both ΔH_F and ΔS_F being positive and $\Delta H_F > T\Delta S_F$. This confluence of conditions dictates the occurrence of enthalpic stabilization in that octant. The lower stabilization octant is specified by negative values for both ΔH_F and ΔS_F and also by $|\Delta H_F| < T|\Delta S_F|$. These conditions herald the onset of entropic stabilization. The intervening quadrant is specified by positive values for ΔH_F, coupled with negative values for ΔS_F. It corresponds to combined enthalpic-entropic stabilization.

It is clear from Fig. 7.1 that it is simply not possible to pass from the enthalpic stabilization octant to the entropic stabilization octant, or vice-versa, without passing through the combined enthalpic-entropic stabilization quadrant. The transition from one octant to another can be effected by a change in temperature (e.g. by heating from near the LCFT to near the UCFT) or by the addition of a different solvent to the dispersion medium. Heating an enthalpically stabilized dispersion causes the thermodynamic coordinates to translate to the right. Cooling effects the same translation with an entropically stabilized system. Addition of a nonsolvent to the dispersion medium also causes translation to the right, irrespective of which of the three domains the dispersion occupies.

7.6. The thermodynamic origins of steric stabilization in the context of free volume theory

The preceding classification of colloidal dispersions was based upon general thermodynamic considerations. This is both a strength and a weakness: a strength, because it ensures a general validity that is model independent; a weakness, because it gives no insight into why sterically stabilized dispersions display the particular generic type of stability that they do. To gain this insight, it is inevitable that the methods of statistical thermodynamics be exploited. Specific models must be postulated and the predictions of these models evaluated by standard statistical mechanical procedures to allow a comparison with experiment.

In what follows, we will adopt the free volume theory for polymer solution thermodynamics (see Section 3.3). This is a second-generation theory that retains the core features of the Flory-Huggins theory whilst grafting onto them the important concept of the difference in free volumes between the polymeric stabilizing moieties and the dispersion medium. The theory is likely to be at least qualitatively, if not semi-quantitatively, correct for nonaqueous dispersion media (Casassa, 1976).

The flocculation behaviour of many sterically stabilized dispersions can be understood from consideration of the free energy change in the *interpenetrational domain*. In what follows, this notion will be assumed to be applicable.

7.6.1. Reprise of the free volume theory

According to the free volume theory, the total free energy of close approach (or flocculation), ΔG_F, in the interpenetrational domain is the sum of three free energy contributions

$$\Delta G_F = \Delta G_F^{\mathrm{COMB}} + \Delta G_V^{\mathrm{CONT}} + \Delta G_F^{\mathrm{VOL}} \tag{7.4}$$

where $\Delta G_F^{\mathrm{COMB}}$ = combinatorial free energy of close approach, ΔG^{CONT} = free energy of close approach associated with contact dissimilarities and $\Delta G_F^{\mathrm{VOL}}$ = free energy of close approach arising from the free volume dissimilarity between the stabilizing moieties and the dispersion medium. The latter term comprehends changes in volume on mixing polymer with solvent, such changes being ignored by the Flory-Huggins theory.

7.6.1.1. The combinatorial contribution

The combinatorial free energy of interpenetration is readily understood for it is purely entropic in origin. It arises on interpenetration of the stabilizing sheaths from the demixing of polymer segments with molecules of the dispersion medium in accordance with the classical Flory-Huggins scheme (except that segment volume fractions replace polymer volume fractions). This contribution is readily recognized in the formula for the free energy of interpenetration as being the term involving the numerical coefficient $\frac{1}{2}$.

7.6.1.2. The contact dissimilarity contribution

The contact dissimilarity term is associated with the free energy difference between the solvent-polymer segment (1–2) contacts relative to solvent-solvent (1–1) and polymer segment-polymer segment (2–2) contacts. In the earlier versions of the free volume theories, the contact dissimilarity contribution for nonpolar systems was purely energetic in origin. For some systems (e.g. poly(isobutylene) in benzene), this provides an adequate description but for others (e.g. natural rubber in benzene), it is insufficient. It is unrealistic to assume that contact energy differences can arise without an entropic perturbation, however minor, to the system and some account needs to be taken of this contribution. Precise calculation of this term is precluded at present because of the intimate knowledge needed of the nature of these specific interactions. Necessarily, the magnitude of the contact dissimilarity contribution must be evaluated by force fitting the predictions of theory with experimental data on polymer chains in free solution. The introduction of a contact entropy term into the revised free volume theory does no great conceptual violence to the basic precepts of the earlier version of the theory.

7.6.1.3. The free volume dissimilarity contribution

The third contribution to equation (7.4) arises from volume changes that occur on demixing polymer segments from the molecules of the dispersion medium on interpenetration. This volume change is a consequence of the dissimilarity in free volumes between the relatively dense polymer and the more expanded dispersion medium (assumed, of course, not to be polymeric in character). The difference in density can be envisaged to arise as follows: as a liquid is heated, the van der Waals attractive intermolecular forces become relatively less important and the free volume increases; in contrast, a comparable increase in the free volume of a polymer chain is prevented by the concatenation of the polymer segments due to the much stronger covalent bonds.

We stress that free volume contributions are important in many systems even at room temperature.

7.6.2. Nonaqueous dispersions

7.6.2.1. General considerations

The signs of the contributions of the various enthalpy and entropy components to the free energy of flocculation in many nonaqueous dispersions sterically stabilized by nonpolar stabilizing moieties are displayed in Table 7.4. The same sign attributions may even apply to some aqueous dispersions (e.g. dispersions sterically stabilized by poly(acrylic acid) at low pH or polyacrylamide).

The combinatorial term, in accordance with the dictates of the Second Law of Thermodynamics, always opposes flocculation. As the steric barriers interpenetrate, molecules of the dispersion medium are forced out of the overlap volume, thus decreasing the combinatorial entropy. At constant temperature and pressure, this means that ΔG^{COMB} must be positive.

The contact dissimilarity term usually opposes stabilization in nonaqueous dispersion media (cf. aqueous dispersion media). The contact enthalpy of interpenetration is usually exothermic, as would be expected from the following elementary considerations.

Interpenetration replaces segment-solvent (1–2) contacts by segment-segment (2–2) and solvent-solvent (1–1) contacts according to the stoichiometric prescription

$$2(1-2)\rightarrow(1-1)+(2-2)$$

The sign of the contact enthalpy of interpenetration in many nonpolar systems can be established as follows. The potential energy of interaction for dispersion forces would be expected to be proportional to the product of the polarizabilities of the species involved (denoted by α_{11}, α_{22} and α_{12}). Thus the change in contact internal energy of interpenetration is proportional, according to the above prescriptions, to $-(\alpha_{11}{}^2+\alpha_{22}{}^2-2\alpha_{12}{}^2)$. Assuming that the cross-polarizability term α_{12} can be approximated by the geometric mean $(\alpha_{11}\alpha_{12})^{\frac{1}{2}}$, the contact internal energy change is simply proportional to $-(\alpha_{11}{}^2+\alpha_{22}{}^2-2\alpha_{11}\alpha_{22})$. This term must always be negative, being equal to $-(\alpha_{11}+\alpha_{22})^2$. It follows that the overall enthalpy of interpenetration arising

TABLE 7.4
The thermodynamic contributions to steric stabilization according to the free volume theory for nonaqueous (and some aqueous) dispersions

	Combinatorial		Contact dissimilarity		Free volume dissimilarity	
	ΔH^{COMB}	ΔS^{COMB}	ΔH^{CONT}	ΔS^{CONT}	ΔH^{VOL}	ΔS^{VOL}
	0	−ve	−ve	*+ve/−ve*[a,b]	*+ve*[a]	+ve
ΔG	+ve		−ve		−ve	
Favours	stabilization		flocculation		flocculation	
$\dfrac{\partial(\Delta G)}{\partial T}$	+ve		*+ve*[a]		−ve	

[a] terms of relatively small absolute magnitude are shown in italics; [b] +ve/−ve implies that either a positive or a negative sign is possible.

from contact dissimilarities (and ignoring any specific interactions) is likely to be negative and to promote flocculation. The corresponding entropy change associated with the contact dissimilarity may be positive or negative. Accordingly, it may either favour or disfavour stabilization.

The free volume entropy of interpenetration is positive and promotes flocculation. As interpenetration of the steric layers occurs, molecules of the dispersion medium in their "condensed" state in association with the segments of the stabilizing moieties are forced out of the interaction zone ("vaporized") into the bulk liquid. The greater free volume in the bulk liquid accounts for the increase in entropy of the molecules of the dispersion medium and the positive sign attributed to ΔS^{VOL} in Table 7.4. This promotes flocculation. The increase in energy associated with the new degrees of freedom of the molecules of the dispersion medium in the bulk provides an enthalpy requirement that is positive and opposes flocculation (crudely akin to the latent heat of vaporization). This causes ΔH^{VOL} to be positive in sign.

In summary, for nonaqueous dispersions, the combinatorial free energy of interpenetration favours stabilization. Both of the corresponding free energies associated with contact dissimilarity and free volume dissimilarity favour flocculation. These conclusions are represented schematically in Fig. 7.2. Since the combinatorial free energy is purely entropic in origin, it is scarcely surprising that nonaqueous sterically stabilized systems are usually found to be entropically stabilized at room temperature and pressure for it is this term that imparts stability. Anticipating the results of the next section, we stress that this does not necessarily imply that all nonaqueous dispersions are entropically stabilized at room temperature.

We note finally that both the contact dissimilarity and free volume

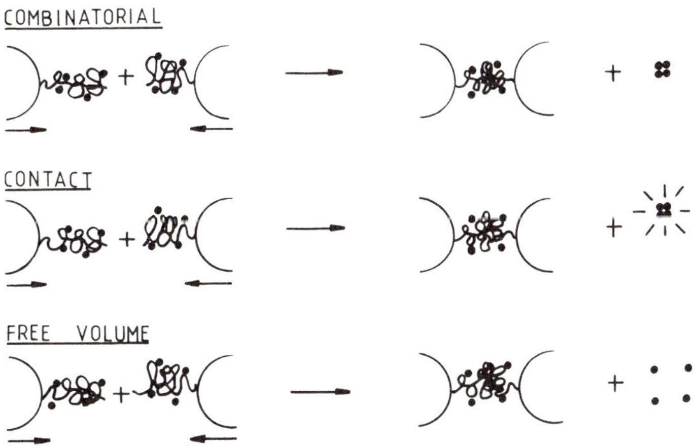

Fig. 7.2. Schematic representation of the various contributions to steric stabilization in nonaqueous dispersion media according to the free volume theory.

dissimilarity in nonaqueous media disfavour the mixing of polymer chains with dispersion medium. Both thus promote flocculation. This is in conformity with the old chemical adage *similia similibus solvuntur* ("like dissolves like"). In this context, "like" encompasses both the polarity of the polymer and solvent molecules, as well as their respective concatenated character.

7.6.2.2. The effects of temperature changes on stability

Heating

As the temperature increases, it is apparent that the combinatorial free energy $(-T\Delta S^{COMB})$ must increase in magnitude, ΔS^{COMB} being negative. However, the free volume dissimilarity contribution to the overall free energy of interpenetration increases in absolute magnitude at a faster rate. Near to the critical point of the liquid, the free volume term, which disfavours stability, becomes dominant and flocculation ensues. Thus the free volume dissimilarity is responsible for flocculation on heating.

A dispersion that flocculates on heating, as was mentioned previously, must be enthalpically stabilized just below the UCFT. For nonaqueous dispersions, the origin of the enthalpy contribution that imparts stability in these circumstances is readily pinpointed from Table 7.4. The only enthalpy contribution that is positive is that associated with the free volume dissimilarity. It is therefore this term that imparts stability just below the UCFT.

It is not possible, as was noted previously, to move from a domain of entropic stabilization to a domain of enthalpic stabilization without passing through a region of combined enthalpic-plus-entropic stabilization. Again the enthalpy term promoting stability in the combined region is that associated with the free volume dissimilarity. The entropy term that contributes to stability is the combinatorial term, just as it is in the purely entropic domain.

No mention has been made in the foregoing discussion of the temperature dependence of the contact dissimilarity contribution to the free energy of interpenetration. In light of the insensitivity of dispersion forces the temperature changes, the contact dissimilarity term would also be expected to be relatively temperature independent for nonaqueous systems. None the less, as pointed out by Croucher and Hair (1980b), the internal energy change associated with the contact dissimilarity should slowly become less negative as the temperature increases.

Cooling

Cooling an entropically stabilized dispersion decreases the contribution of the combinatorial entropy of mixing to the overall free energy of interpenetration. Whilst the free volume contribution also decreases significantly in absolute magnitude, the contact dissimilarity term is relatively unaffected by the drop in temperature, as discussed in the preceding section. It is therefore not surprising that a temperature may be reached at which the contact and free volume dissimilarity contributions, which both favour flocculation, outweigh the combinatorial free energy term. Flocculation then ensues.

7.6.3. Aqueous dispersions

7.6.3.1. General considerations

There is as yet no adequate theory to describe the thermodynamics of aqueous polymer solutions. Any discussion of the origins of the steric repulsion in aqueous dispersions must therefore be speculative. In what follows, the modifications to the free volume theory that may reasonably be required to account for the specific interactions between the water molecules and the stabilizing moieties will be enunciated. This general scheme seems pertinent to the stabilization imparted by the industrially important polymers poly(oxyethylene) and poly(vinyl alcohol). It is unlikely, however, to be relevant to the stabilization by poly(acrylic acid) at low pH or polyacrylamide. The latter stabilizers, as noted previously, appear to fall within the guidelines set out in Table 7.4.

The signs of the various contributions to the overall free energy of interpenetration are displayed in Table 7.5. Comparison with the entries in Table 7.4 for nonaqueous dispersion media show that the combinatorial and free volume terms emerge relatively unscathed when the dispersion medium is changed from being nonaqueous to being aqueous. The contact dissimilarity term, however, is radically transformed. This reflects the specific interactions (e.g. hydrogen bonding) that water molecules can undergo with suitably polar species (e.g. hydroxyl groups or the ether oxygens of the polymers cited above).

Specific interactions between the water molecules and the stabilizing moieties imply that when interpenetration occurs, water molecules are released into the bulk dispersion medium. The energy required for the new degrees of freedom (principally translational and rotational) in the bulk

TABLE 7.5
The thermodynamic contributions to steric stabilization for some aqueous dispersions

	Combinatorial		Contact dissimilarity		Free volume dissimilarity	
	ΔH^{COMB}	ΔS^{COMB}	ΔH^{CONT}	ΔS^{CONT}	ΔH^{VOL}	ΔS^{VOL}
	0	−ve	+ve	+ve	+ve	+ve
ΔG	+ve		+ve		−ve	
Favours	stabilization		stabilization		flocculation	
$\frac{\partial(\Delta G)}{\partial T}$	+ve		−ve		−ve	

dispersion medium creates a contact dissimilarity enthalpy that opposes flocculation. The increase in entropy associated with these new degrees of freedom promotes flocculation. Overall, however, the enthalpy of the contact dissimilarity term outweighs the corresponding entropy term. This results in the contact dissimilarity contribution to the overall free energy of interpenetration favouring stabilization.

Note that for aqueous systems the contact dissimilarity term favours mixing of the polymer and dispersion medium, apparently in violation of the maxim "like dissolves like". This violation is not serious, however, because it is the highly favourable nature of the interaction between the polymer and the water, as witnessed by, say, the formation of hydrogen bonds, that promotes mixing. The dissimilarity should therefore be considered to represent a strong complementarity rather than a true dissimilarity.

In summary, for some aqueous dispersions, both the combinatorial and contact dissimilarity contributions to the free energy of interpenetration favour stabilization. Only the free volume contribution apparently promotes flocculation (see Fig. 7.3).

7.6.3.2. The effects of temperature changes on stability

Heating

When the temperature is raised, the contact dissimilarity term, which favours stability, is significantly reduced. Although the combinatorial contribution to the overall free energy of interpenetration increases on heating, it would be expected, by analogy with nonaqueous dispersions, that the absolute magnitude of the free volume term (which favours flocculation) would increase. The net effect of a reduced contact dissimilarity contribution to

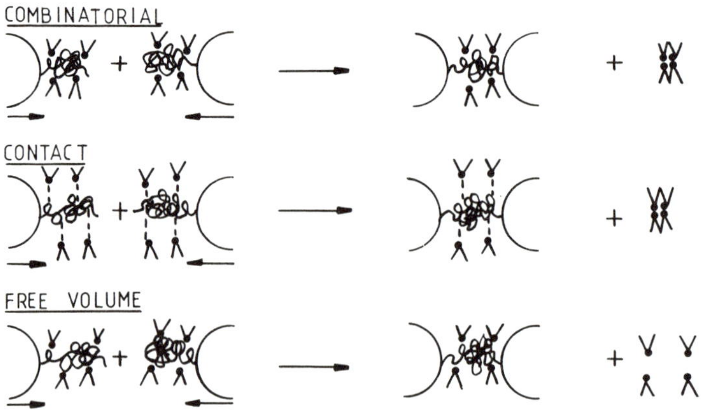

Fig. 7.3. Schematic representation of the various contributions to the steric stabilization in some aqueous systems.

stabilization and a heightened free volume contribution to flocculation is for flocculation processes to so overwhelm the stabilization effects that aggregation occurs on heating.

It is not immediately apparent from this analysis why flocculation proceeds so far removed from the critical point of the water: the θ-temperature for poly(oxyethylene) in water is ca 380 K (Burchard, 1983; Boucher and Hines, 1976; Napper, 1970a) whereas the critical point of water is 647 K at 22 MPa. One possible explanation for this huge temperature difference could reside in a greatly magnified effect of the free volume term that accompanies the H-bonding propensity of water. A second possibility would require the contact dissimilarity term to change sign and to promote flocculation quite strongly. This, however, seems rather unlikely. Further clarification of this point awaits a better understanding of the thermodynamics of aqueous polymer solutions.

Cooling

It might be expected that just below the UCFT, the enthalpies associated with the contact and free volume dissimilarities should impart enthalpic stabilization. Conversely, just above the LCFT (if accessible), the combinatorial entropy of mixing should give rise to entropic stabilization. Flocculation on cooling appears to result from the free volume contribution. This may explain why such flocculation is not always readily achieved in aqueous systems of this type.

We stress that the foregoing thermodynamic notions for aqueous dispersions are quite speculative. A better understanding awaits the formulation of theories of polymer solution thermodynamics that comprehend aqueous systems. These may well include factors that are absent from the free volume approach.

Chapter 8

Non-thermodynamically Limited Steric Stabilization: Dispersions Displaying Deviant Behaviour

8.1. Introduction

A general rule was enunciated in Section 6.1.6 that asserted that the thermodynamic limit to the stability of many sterically stabilized dispersions is the θ-point for the stabilizing moieties in free solution. It was noted at that point that there exist at least five cases which appear at first sight to be exceptions to this general rule. These occur if:

(i) the stabilizing moieties are poorly anchored or not anchored at all to the particles;
(ii) the surface coverage by the stabilizing chains is incomplete;
(iii) the steric layers are thin and/or the particle size is large;
(iv) the stabilizing moieties undergo crystallization on the surface of the colloidal particle;
(v) the stabilizing moieties are adsorbed at many points along the chain *(multipoint anchoring)*.

Closer scrutiny, however, shows that probably only cases (iv) and (v) constitute genuine exceptions to the general rule. In the other cases listed, the true thermodynamic limit is simply unattainable due to the intervention of the van der Waals attraction. In this chapter, the first four exceptions listed above will be discussed sequentially. Consideration of the fifth will be postponed until the following chapter.

Note that the use of the generic label "*non-thermodynamic limited stability*" in the title of this chapter is more a matter of convenience and concision than an accurate description of the limits to stability. Stability, in one sense, is always thermodynamically limited. The heading is meant to highlight the apparent differences between the systems discussed in the preceding three chapters and those now to be considered. The dispersions discussed previously were limited in their stability by the thermodynamics of the interactions between the segments of the stabilizing moieties and the dispersion medium near to the θ-point. For the dispersions to be considered here, this limit is either not reached or it is seemingly transgressed before the onset of incipient instability. It is in this narrow specific sense that the systems are said to be non-thermodynamically limited.

8.2. Poorly anchored stabilizing moieties

8.2.1. Copolymers

It has already been remarked that to impart optimum stability to sterically stabilized dispersions, it is necessary to ensure that the stabilizing moieties are strongly anchored to the surface of the colloidal particles. This may be achieved by means of covalent linkages (Clarke and Vincent, 1981a) or by the physical adsorption of a (usually nominally insoluble) anchor polymer (Barrett, 1975). The anchor polymer could be physically or chemically incorporated into the latex particles. Failure to provide strong anchoring usually results in displacement coagulation in dispersion media that are better solvents than θ-solvents for the stabilizing moieties. The stabilizing chains react to the stress generated during a Brownian collision with another particle by escaping from the interaction zone. This can be accomplished either by lateral movement or desorption. The colloidal particles can then approach more closely so that the London dispersion forces come into play and induce coagulation in better than θ-solvents. Such coagulation is usually irreversible. Bridging of the two particles by the polymeric chains could also play a role in certain systems (Kitahara and Hasumuma, 1972).

Evidence to support the foregoing concepts is provided by sterically stabilized emulsion systems (March and Napper, 1977). Emulsions appear to be even more demanding in their anchoring requirements than are the corresponding particulate systems (e.g. latices). This may be a consequence of the additional mobility of the stabilizing molecules at the liquid/liquid interface. March and Napper (1977) studied the stability of toluene-in-water emulsions using poly(oxyethylene) as the stabilizing moieties and polystyrene as the anchor polymer. They varied the fraction of polystyrene in the amphipathic stabilizer poly(oxyethylene-*b*-styrene), whilst holding the molecular weight of the poly(oxyethylene) constant at 79 000. The UCFT of the resulting emulsions in 0·39 M $MgSO_4$ was found to increase as the mass ratio of polystyrene to poly(oxyethylene) in the stabilizer increased (see Fig. 8.1). The UCFT reached a plateau at a mass ratio of ca 0·4, however, when the UCFT (318 ± 2 K) was found to be close to the θ_L-temperature (316 ± 3 K) of poly(oxyethylene) under these conditions.

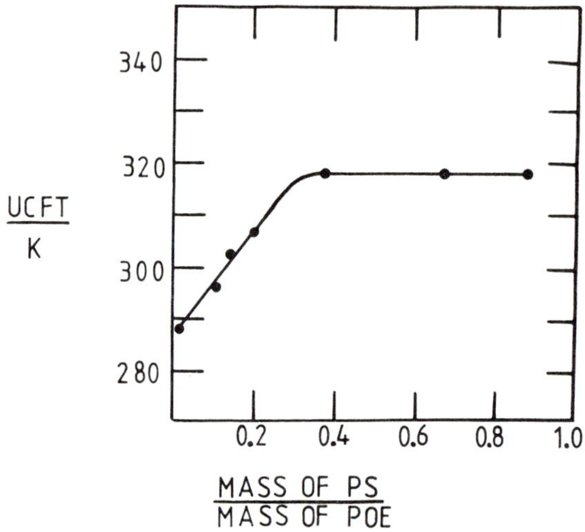

Fig. 8.1. Dependence of the UCFT of toluene-in-water (0·39 M $MgSO_4$) emulsions on the relative ratios of the polystyrene anchor polymer to the poly(oxyethylene) stabilizing moieties (after March and Napper, 1977).

8.2.2. Steric stabilization by homopolymers

The preceding discussion of anchoring provides a rationale for the relatively poor performance of unattached homopolymers as steric stabilizers. The use of homopolymers as stabilizers places conflicting demands upon the solvency of the dispersion medium. On the one hand, the dispersion medium should be a good solvent for the homopolymer to ensure that the stabilizing moieties are as extended as possible in the dispersion medium and the steric repulsion is as large as possible. On the other, the dispersion medium should be a poor solvent for the homopolymer to ensure good anchoring of the homopolymer to the colloidal particles. The latter is the more stringent requirement and apparently the more difficult to fulfil. Homopolymers usually fail as steric stabilizers because of their displacement from the interaction zone in Brownian collisions.

8.3. Incomplete surface coverage

The irreversible coagulation of sterically stabilized dispersions that occurs in better than θ-solvents when the surface coverage is incomplete, was discussed in detail in Section 5.3.6. As with poor anchoring, the displacement of the stabilizing chains from the zone of interaction in a Brownian collision allows the van der Waals forces to induce displacement coagulation. Bridging flocculation could also be implicated in some systems.

8.4. Thin steric layers

Thin steric layers normally arise from the use of low molecular weight stabilizing moieties. The consequences of this for stability are quite profound because a radically different pattern of coagulation behaviour then emerges.

It would be expected on purely intuitive grounds that the most dramatic effect arising from the use of thin steric layers should be the occurrence of incipient instability in dispersion media that are better solvents for the stabilizing moieties than θ-solvents. Coagulation might be expected to arise from the long-range van der Waals attraction between the core particles, despite the fact that the stabilizing moieties are mutually repulsive. The London attraction should generate a relatively shallow minimum in the potential energy curve for two particles at quite distant separations (see Fig. 8.2). This would be somewhat akin to the secondary minimum evident in the potential energy diagram for electrostatic stabilization, although no primary minimum need exist with sterically stabilized systems. The latter situation is especially likely to occur if the stabilizing moieties are covalently attached to the particle surface. It might be anticipated that *weak coagulation* could occur in this pseudo-secondary minimum if its depth is somewhat greater than the thermal energy (ca $3/2\,kT$) available to the particles.

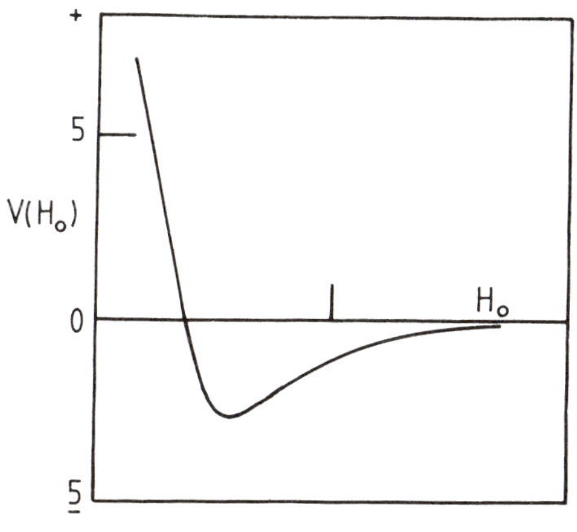

Fig. 8.2. The potential energy of interaction of two colloidal particles as a function of the distance of separation for thin steric layers.

This type of weak coagulation should depend upon the complex interaction of the multiplicity of factors that govern the London attraction between the particles. These include:

(i) the chemical nature of the disperse phase (i.e. whether metallic, inorganic or polymeric);
(ii) the particle size;
(iii) the molecular weight of the stabilizing moieties;
(iv) the conformation of the stabilizing moieties;
(v) the molecular weight of the anchor polymer;
(vi) the solvency of the dispersion medium;
(vii) the volume fraction of the latex particles.

The foregoing intuitive expectations are, by and large, realized in practice. Coagulation can be induced in better than θ-solvents for the stabilizing moieties. Incipient instability is indeed observed to be dependent upon the particle number concentration, the particle size, the molecular weight of the stabilizing moieties and the solvency of the dispersion medium. Some detailed results illustrating these effects will now be presented.

8.4.1. The effect of particle size

The influence exerted by particle size on the CFV of poly(methyl methacrylate) latices stabilized by low molecular weight (1750) poly(12-hydroxystearic acid) in *n*-heptane has already been presented in Section 5.3.5 (see especially Table 5.6). Ethanol and *n*-propanol were used as nonsolvents (Napper, 1968b). As the particle size increased, the CFV was found to decrease. It was deduced therefrom that the latex particles became progressively easier to coagulate as the particle size increased because incipient instability was manifest under progressively better solvency conditions than those of θ-solvents. This assuredly implicates the van der Waals attraction between the core particles in the coagulation process since the London attraction also increases with increasing particle size.

Very similar results have been reported by Everett and Stageman (1978a). They prepared poly(methyl methacrylate) and poly(acrylonitrile) latices in *n*-propane. The particles were stabilized by ABA block copolymers in which the stabilizing B moieties were poly(dimethylsiloxane) and the anchor A chains were polystyrene. The results for the UCFT for two different particle sizes and two different molecular weight stabilizing moieties are presented in Table 8.1. For the smallest particle size and highest molecular weight polymer chains studied (latex AN 63), good agreement between the UCFT and θ_L-temperature was apparent. However, for both molecular weights, the UCFT decreased with increasing particle size. Coagulation was again evident in better than θ_L-solvents. Larger particles were less stable than smaller particles as measured by the ease of coagulation.

TABLE 8.1
The effect of particle size on the UCFT of poly(methyl methacrylate) latices stabilized by poly(dimethylsiloxane)

Designation	Core radius (nm)	PDMS molecular weight	UCFT/K	θ_L/K
AN31	50	20 000	341	348
MM31	111	20 000	329	348
AN63	40	30 000	349	348
AN61	85	30 000	337	348

8.4.2. The effect of the molecular weight of the stabilizing moieties

The influence of molecular weight on the UCFT can be gleaned by comparing the data for latices AN31 and AN63 in Table 8.1. For the present purposes, any difference in their particle sizes can be ignored. The latex stabilized by the higher molecular weight polymer flocculated at the θ_L-point, whereas that stabilized by the lower molecular weight chains coagulated under solvency conditions significantly better than that of a θ-solvent. This suggests that whereas the former system flocculated as a result of the "stabilizing" segments becoming mutually attractive in a slightly worse than θ-solvent, the stability of the latter system near the UCFT was controlled by the London attraction between the core particles.

Additional support for this conclusion springs from the variation of the UCFT with the weight fraction of particles reported by Everett and Stageman (1978a). These are shown in Fig. 8.3. One latex (AN63) exhibited no

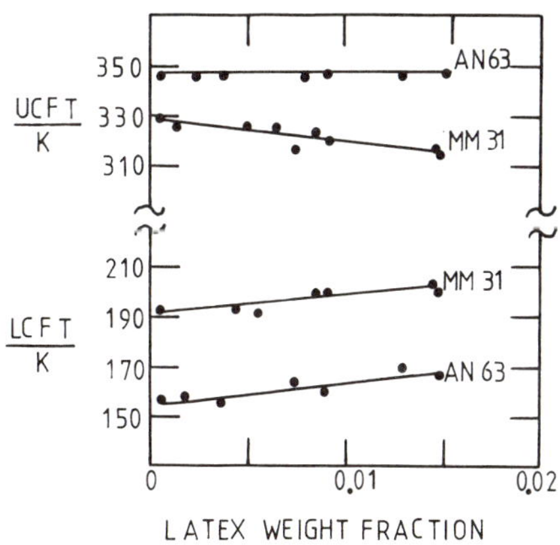

Fig. 8.3. The dependence of the UCFT and LCFT of poly(methyl methacrylate) latex particles sterically stabilized by poly(dimethylsiloxane) in *n*-propane upon the weight fraction of latex particles (after Everett and Stageman, 1978a).

significant change in the UCFT with particle number concentration. This is consistent with the observations of Croucher and Hair (1981), Dawkins and Taylor (1980), Feigin *et al.* (1981) and Clarke and Vincent (1981b). All these authors reported no variation of the CFPT with particle number concentration when the mutual attraction between the segments was apparently responsible for incipient instability. Another latex (MM31) studied by Everett and Stageman (1978a) displayed the characteristic increase in UCFT (and hence stability) with decreasing particle number concentration that accompanies coagulation that is induced by the van der Waals attraction between the core particles.

8.4.3. The chemical nature of the dispersion medium

Everett and Stageman (1978a) prepared poly(methyl methacrylate) dispersions, of radius 111 nm, stabilized by poly(dimethylsiloxane) (molecular weight 20 000) in a range of *n*-alkanes. A comparison is presented in Fig. 8.4 of the UCFT of these latices, extrapolated to zero particle number concentration, with the θ_L-temperature of poly(dimethylsiloxane) as a function of the number of carbon atoms from ethane to *n*-hexane. Also shown are the critical temperatures for these dispersion media.

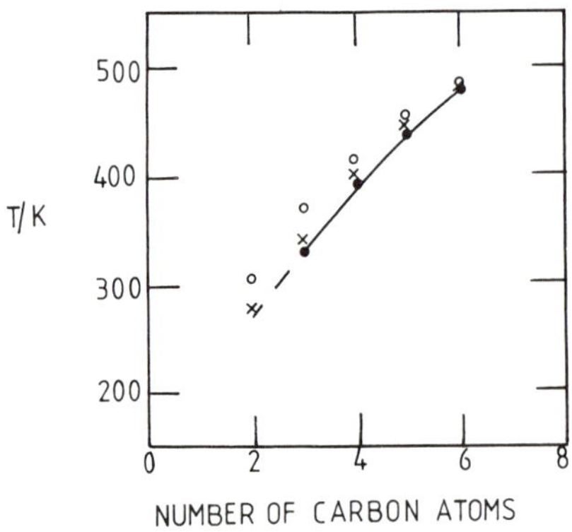

Fig. 8.4. The dependence of the CFT (●) of poly(methyl methacrylate) latex particles sterically stabilized by poly(dimethylsiloxane) in *n*-alkanes of different chain lengths. Also shown are the critical temperatures of the liquids (○) and the corresponding θ-temperatures (x) (after Everett and Stageman, 1978a).

It is evident that the thermal range of stability of the dispersions decreased with decreasing molecular weight of the liquid, as did the critical temperature of the alkanes. Moreover, the UCFT curve lies some 10–40 K below the corresponding curve for the critical temperatures.

8.4.4. The effect of particle number concentration

Cowell *et al.* (1978) have studied the dependence of the UCFT on the particle number concentration for a polystyrene latex of radius 87 nm stabilized by very low molecular weight (750) poly(oxyethylene). Their results are displayed in Fig. 8.5 for three different $MgSO_4$ concentrations.

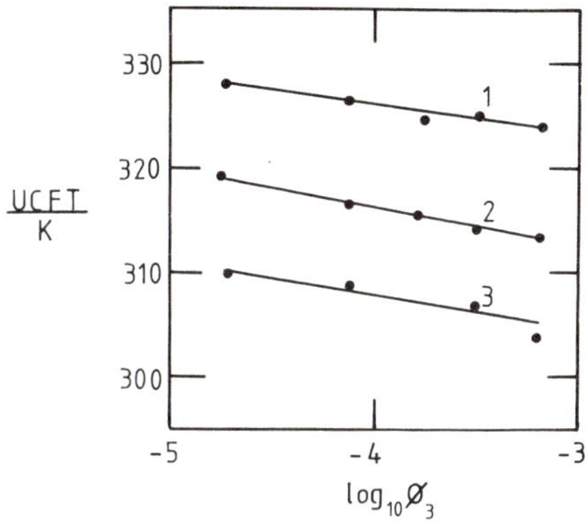

Fig. 8.5. The UCFT of polystyrene latices sterically stabilized by low molecular weight poly(oxyethylene) as a function of the volume fraction (ϕ_3) of latex particles. Concentrations of magnesium sulphate: 1, 0·065 M; 2, 0·163 M; 3, 0·26 M (after Cowell *et al.*, 1978).

In all cases, coagulation was evident under better than θ-solvency conditions and this was accompanied by a dependence of the UCFT on the volume fraction of latex particles. The slope was negative in all the systems studied, implying that the more dilute dispersions were marginally more difficult to coagulate than the more concentrated dispersions. This is in agreement with the results of Everett and Stageman (1978a) mentioned towards the end of the preceding section.

8.4.5. General discussion of weak coagulation

8.4.5.1. The physical nature of coagulation

It should first be noted that the coagulation of dispersions sterically stabilized by low molecular weight stabilizing moieties, as exemplified by the studies of Napper (1968b) and Cowell *et al.* (1978), is physically akin to the coagulation observed in electrostatically stabilized dispersions on adding electrolytes. That is to say, it is not a classical phase separation process whereby two discrete phases separated by a definite liquid/liquid interface are formed. In this sense, it departs radically from the phase separation (coacervation) evinced by noncrystallizing polymers when the solvency is made significantly worse than that of a θ-solvent. There is a strong parallelism between the physical product of aggregation of sterically stabilized dispersions, irrespective of whether the stabilizing moieties are high or low molecular weight, and that produced by the coagulation of electrostatically stabilized dispersions.

It is possible to envisage, none the less, a situation in which coacervation is the primary mode of aggregation. Thus dispersions composed of exceedingly small core particles stabilized by high molecular weight stabilizing moieties could conceivably display coacervation at the CFPT. Everett (1981) may have observed this phenomenon with gold particles stabilized by poly(dimethylsiloxane). This type of behaviour could be described, metaphorically speaking, as the tail wagging the dog. Usually the reverse pertains.

8.4.5.2. The temperature induction of weak coagulation

If the coagulation observed for particles ensheathed in thin steric layers is a consequence of the van der Waals attraction between the particles, then it is reasonable to ask why a temperature change is necessary to provoke the onset of instability. After all, the van der Waals attraction between naked core particles at a given separation is unlikely to be sensitive to temperature changes, especially in nonaqueous media.

Cowell *et al.* (1978) have proposed that at least two factors may be implicated in the onset of weak coagulation in such systems. These may be summarized as follows.

First, the major effect originates in a decrease in the range and magnitude of the repulsive steric interactions. Both diminutions occur because as the temperature is changed, the dispersion medium becomes a worse solvent for the stabilizing moieties. This may lead to a reduction in the dimensions of the stabilizing sheath and to a concomitant decrease in the range of the steric repulsion. It was argued that the change in the magnitude of the steric repulsion was more subtle. It is widely agreed that there are two contributions to the steric interaction: (a) a segment-solvent mixing term associated with the local increase in the segmental concentration in the interaction zone; (b) an elastic term arising from the loss of configurational entropy of the stabilizing moieties on compression of the steric layers by the close approach of the opposing surfaces in a Brownian collision. The mixing term is more strongly dependent upon temperature than is the elastic term, as witnessed by its change in sign at the θ-temperature. The elastic term is always repulsive. It was also claimed that with low molecular weight stabilizing moieties, the high segmental concentration and the reduced segmental motion resulted in little or no interpenetration of the sheaths in particle-particle collisions. As a result, the elastic term comes into play almost at the same point as the mixing contribution. This contrasts sharply with high molecular weight stabilizing moieties whose Brownian collisions are dominated by the mixing term.

The second factor postulated to be responsible for coagulation is an increase in the van der Waals attraction between the stabilizing sheaths arising from the increased segmental concentration that accompanies any reduction in steric layer thickness. This seems likely to be a minor effect, however.

8.4.5.3. The dependence of the CFPT upon the system parameters

The effects of particle size and molecular weight

The reasons underlying the dependence of the CFPT upon the particle size and the molecular weight of the stabilizing moieties in weak coagulation are qualitatively straightforward. The core attraction increases with increasing particle size and hence larger particles should be less stable for a given molecular weight than smaller particles. Similarly, the higher the molecular weight of the stabilizing moieties, the further apart are the core particles forced to interact in their production of the pseudo-secondary minimum since the polymer segments are repulsive in better than θ-solvents. As the London attraction decreases with increasing distance of separation, particles stabilized by higher molecular weight polymers will possess a shallower potential energy well into which they can coagulate than particles stabilized by low molecular weight polymers. Thus, particles stabilized by higher molecular weight stabilizers appear to be the more stable near the CFPT.

The effect of particle number concentration

The explanation of the dependence of the CFPT upon particle number concentration when weak coagulation takes place is not patently obvious. One clue to its origin resides in the observation that invariably dilute dispersions appear to be marginally more stable than more concentrated systems. This immediately suggests that the configurational entropy of the colloidal particles must be implicated in some fashion. Dispersed particles must lose configurational entropy when they undergo coagulation. This loss of configurational entropy can contribute to the stability of the dispersion near to the CFPT.

It has been shown theoretically (Feigin *et al.*, 1981) that the configurational entropy per latex particle can be as large as ca 5–10 kT at the usual particle concentrations of interest. If all of this configurational entropy were to be lost on coagulation, it would undoubtedly have a profound influence on any aggregation in a pseudo-secondary minimum whose depth was only a few kT. It has also been shown, as is intuitively obvious, that the magnitude of the particle configurational entropy declines with increasing particle number concentration. Hence a more dilute dispersion should appear to be more stable than a more concentrated one.

This explanation of the particle concentration dependence of the CFPT is not without its deficiencies. The most glaring defect is that the configurational entropy of the particles must become infinitely large as the particle volume fraction tends to zero. Yet none of the data published to-date show any evidence of the predicted upward curvature of the UCFT in exceedingly dilute dispersions. This may, of course, merely reflect an experimental inability to study coagulation in sufficiently dilute dispersions. Everett and Stageman (1978a) observed for nonaqueous systems that the UCFT was linear with weight fraction of latex particles. In contrast, Cowell *et al.* (1978) found that the UCFT of aqueous systems was linear with the logarithm of the volume fraction of the latex particles. The latter dependence is that predicted theoretically (Feigin *et al.*, 1981).

Resolution of the quandry whereby the configurational entropy of the particles does not apparently behave in the anticipated fashion may lie in the prediction by Long *et al.* (1973) that coagulation simply does not take place unless the particle concentration exceeds a critical value, i.e. the particle configurational entropy is so large in exceedingly dilute dispersions that it always negates the effects of the van der Waals attraction (see the next section). The nature of the flocs formed could also conceivably influence the loss of entropy on coagulation, although this would intuitively be expected to be a relatively minor effect.

Note that if the explanation advanced for the volume fraction dependence of the CFT is correct, then the common practice of extrapolating the CFT to zero particle concentration is likely to lead to error. This extrapolation procedure is presumably intended to eliminate the effects of particle configurational entropy on the CFT. Extrapolation to zero particle concentration does not seem likely to achieve that end. Rather, extrapolation in the opposite direction, i.e. to close packing where the configurational entropy is minimized, seems more likely to provide a value for the CFT that corresponds more closely to the desired quantity. This stricture must be borne in mind in reviewing the published results of Everett and Stageman (1978a), Cowell *et al.* (1978) and Cowell and Vincent (1982b).

8.4.5.4. Critical flocculation particle volume fraction

Theoretical considerations

Long *et al.* (1973) have explored, both experimentally and theoretically, the equilibrium aspects of the weak coagulation of sterically stabilized dispersions. They argued that the potential energy diagram for two colloidal particles stabilized by low molecular weight stabilizing moieties would resemble in its general shape the curve for the interaction between two molecules in the vapour phase. Just as a minimum vapour pressure (the saturation value) is necessary in order for molecules in the vapour pressure to undergo condensation, so by analogy they surmised that a minimum particle number concentration should be required for coagulation. Below this critical particle number concentration, no coagulation would be observed; but above this value, single particles would exist in equilibrium with aggregates.

The existence of a critical particle number concentration is readily comprehended in thermodynamic terms. The free energy of coagulation is given by

$$\Delta G_{\mathrm{COAG}} = \Delta H_{\mathrm{COAG}} - T\Delta S_{\mathrm{COAG}} \tag{8.1}$$

where ΔH_{COAG} and ΔS_{COAG} are the enthalpy and entropy of coagulation respectively. Coagulation is accompanied by a reduction in the potential energy of the particles and so ΔH_{COAG} is negative. In addition, the entropy change associated with coagulation is necessarily negative. The configurational entropy associated with dispersed colloidal particles, moreover, increases with decreasing particle concentration. It follows that below some critical particle concentration, the magnitude of the entropic contribution to the free energy of coagulation always exceeds the corresponding enthalpy change, i.e. $|T\Delta S_{\mathrm{COAG}}| > |\Delta H_{\mathrm{COAG}}|$. This implies that ΔG_{COAG} is positive. Coagulation will not therefore occur spontaneously below this critical particle concentration.

Long *et al.* (1973) took this analogy with vapour condensation one step further. At equilibrium, they related the dispersed and flocculated particle number concentrations (in actual fact, the particle volume fractions) through a simple Boltzmann expression:

$$\phi^d = \phi^f \exp\{-E/kT\} \tag{8.2}$$

where ϕ^y, ϕ^f = the volume fractions of particles in the dispersed state and floc state respectively, and E = energy required to remove a particle from the surface of a floc to the dispersed state. This latter energy was itself related to the depth of the potential energy well $|G_{\min}|$ by Cowell and Vincent (1982b) through

$$E \sim (z/2)|G_{\min}| \tag{8.3}$$

where z = coordination number of a particle in the centre of a floc. The factor of 2 in equation (8.3) reflects the reduced coordination number of a particle at the surface.

The critical particle concentration, according to the foregoing model, should depend upon the depth of the potential energy minimum $|G_{\min}|$ and the temperature. According to this approach, no unique value exists for the depth of the minimum that delineates the boundary between stability and instability. Rather, that simple, and widely accepted, concept must be replaced by one in which coagulation should be observable at some particle concentration, irrespective of the depth of the potential energy well. The only proviso is that the well depth must not be so small (say, less than 1 kT) that the flocs formed are too fragile to be detectable.

It is usually easier experimentally to specify the critical particle volume fraction rather than the critical particle concentration. Vincent, who has pioneered this approach, usually denotes the critical coagulation particle volume fraction by $CF\phi$. The "ϕ" in Vincent's pseudo-acronym denotes the particle volume fraction whereas the "F" denotes flocculation, rather than the term coagulation which has been preferred in this text to signify the origin of the attraction.

Experimental evidence for CFϕ

The original experiments of Long *et al.* (1973) were performed on polystyrene latex particles stabilized by an alkyl ethoxylate nonionic surfactant ($C_{16}E_6$). The very low molecular weight of the oligomeric oxyethylene chains ensured that the van der Waals attraction between the core particles was sufficiently large to provide a reasonably deep pseudo-secondary minimum in the potential energy curve. Sufficient electrolyte was added to remove any electrical double layer interactions. The optical density method outlined in Section 5.2.3.2 was used to detect coagulation. Some data obtained by Long *et al.* (1973) are shown in Fig. 8.6.

It is apparent from Fig. 8.6 that below a certain volume fraction, $CF\phi$, coagulation was not observed, the optical properties being independent of the particle volume fraction. Above $CF\phi$, however, coagulation was evident as witnessed by the dependence of the optical properties on the particle volume fraction. From the observed values of $CF\phi$, the magnitude of well depth, $|G_{\min}|$, could be calculated from the combinations of equations (8.2) and (8.3)

$$\phi^d = \phi^f \exp\{-z|G_{\min}|/2kT\} \tag{8.4}$$

subject to an assumption concerning the floc structure (e.g. random close packing), which was needed to specify z. Values of $|G_{\min}|$ determined in this manner were in reasonable agreement with those obtained by direct calculation of the van der Waals interparticle attractive energies using a double sheath model (Vincent, 1973).

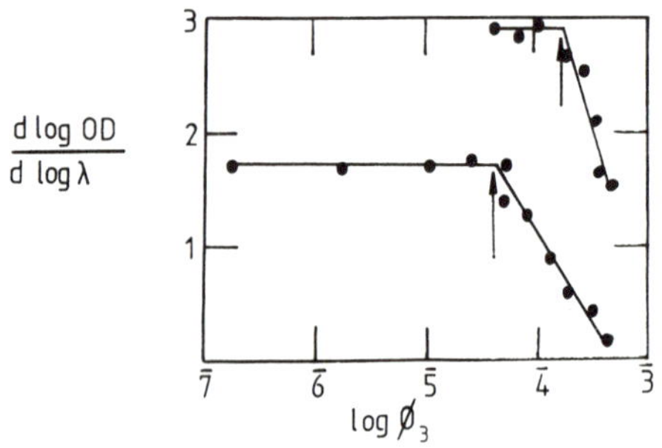

Fig. 8.6. Plots of the optical density wavelength gradient as a function of the volume fraction (ϕ_3) of particles for two different poly(oxyethylene) stabilized polystyrene latex. The arrows (↑) indicate the critical volume fraction of latex particle $CF\phi$ (after Long *et al.*, 1973).

More recently, the existence of a critical coagulation particle volume fraction has been established by Vincent *et al.* (1980) for polystyrene latices stabilized by poly(oxyethylene) terminally grafted to the particle surface and by Cowell and Vincent (1982a) for physically attached poly(oxyethylene). Bridger (1979) has found that a parallel situation pertains with silica particles sterically stabilized by terminally anchored polystyrene in ethyl benzene.

Configurational state diagrams for dispersions of weakly interacting particles

Phase rule considerations. Cowell and Vincent (1982b) have published a form for the configurational state diagrams of dispersions of weakly interacting particles. This diagram encompasses not only the dispersed particle-to-floc state transition but also the disordered-to-ordered state transition of the particles. Cowell and Vincent have actually used the term "phase" where we will prefer the terminology "configurational state". The reason for this preference is straightforward: the use of the term "phase" can easily lead to a misconception that apparently violates the Gibbs Phase Rule, given that Cowell and Vincent definitely did not observe the formation of two liquid phases.

The problem can be readily exposed as follows for latex particles to which the stabilizing moieties are terminally grafted. It is assumed that no free polymer exists in the dispersion medium. In such a system, the number of components (C) in the Gibbsian sense of being independently variable entities is 2: *viz* the particles and the dispersion medium. The number of degrees of freedom (F) is 1 as temperature is the only intensive variable that can be varied at will. Note that this assumes that the weight fraction of latex is an extensive rather than an intensive variable. It follows that the number of possible phases (P), given by $P = C - F + 2$, is 3. One of these is the vapour phase, leaving just two residual phases to be assigned.

One problem that arises in any such analysis as this is what constitutes a valid thermodynamic phase in colloidal systems. According to Findlay (1951), a phase in the Gibbsian sense must be homogeneous, physically distinct and mechanically separable, although it need not be chemically simple nor continuous. Given these criteria, it is a moot point whether a stable colloidal dispersion can be considered to be a single phase and whether a floc constitutes a separate phase. It could even be that the Phase Rule cannot usefully be applied in any simple fashion to these colloidal systems, in which case the use of the phase concept to describe coagulation is best avoided. In what follows, we will ignore these very real conceptual difficulties and attempt to interpret, albeit rather loosely, the predictions of the Phase Rule.

Experimentally it appears that at least two possible manifestations of the two residual phases predicted above can exist. These are two liquid phases separated by a definite liquid/liquid interface; or one liquid phase and one solid phase, separated by a solid/liquid interface. It is not yet known precisely what factors dictate which of these two alternatives is followed, although they presumably differ very little in Gibbs free energy. The formation of two liquid phases presumably occurs if the polymeric stabilizing moieties so dominate the association process that they undergo coacervation. Particle aggregation is a consequence of the colloidal particles and/or the stabilizing moieties providing relatively weak association interactions. What is clearly apparent from experiment is that flocculation occurs more commonly than phase separation in the systems examined thus far.

If two liquid phases are formed, one of these might be expected to be significantly more concentrated in terms of particle number than the other, by analogy with those polymer systems that undergo coacervation. The state of aggregation of the particles within those phases could not be predicted from Phase Rule considerations. If two liquid phases are not formed, then one of the two phases might be taken to be the dispersion medium, the other being the colloidal particles. This is the interpretation that seems most relevant to the studies of Cowell and Vincent. If this viewpoint is adopted, nothing can be ascertained about the state of aggregation of the particles from Phase Rule considerations. If the flocculated state were a separate phase, then there would be 3 residual phases present, not just the 2 predicted above by Phase Rule considerations. These would be the dispersion medium, the dispersed particles and the alleged flocculated phase. It seems unlikely that such a violation of the Phase Rule can be countenanced. The issue, however, is far from settled. The fact that there is a reduction in free energy, arising from the increase in the configurational entropy of the particles on dispersion, could well allow a dispersion to be regarded as a single phase. Usually, however, this free energy change is negligible and the alternative viewpoint must be preferred.

Typical state diagrams. Figure 8.7 displays the qualitative generic form of the configurational state diagram for dispersions of particles sterically stabilized by low molecular weight stabilizing moieties as proposed by Cowell and Vincent (1982b). Also shown in the Fig. are the comparable coexistence lines for dispersions stabilized by high molecular weight stabilizing moieties. The Roman numerals specify the number of existing configurational states, which are denoted by D (single dispersed particles), F (the floc state) and O (an ordered state consisting of particles in a pseudo-crystalline array). The line ab denotes the coagulation boundary line and bc represents the disorder-to-order (dispersed state→crystalline state) transition. T_c represents the limiting value for the UCFT at high volume fractions when the contribution of the particle configurational entropy is minimized. It represents a *critical temperature* for the system below which no coagulation is possible. For high molecular weight stabilizing moieties, T_c is equal to the observed UCFT, which is independent of the particle volume fraction (Fig. 8.7b).

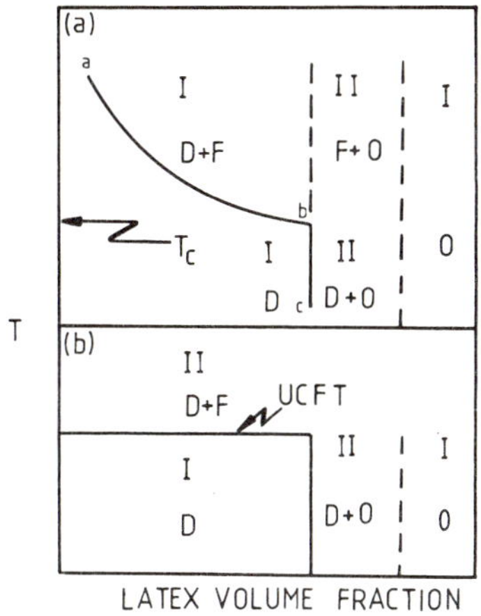

Fig. 8.7. The configurational state diagrams for dispersions of particles sterically stabilized by (a) low molecular weight and (b) high molecular weight stabilizing moieties. The Roman numerals specify the number of configurational states; D = single dispersed particles, F = the floc state and O = ordered state in a pseudo-crystalline array.

Thompson and Pryde (1981) have also published what they claim to be quantitative predictions of the form of the configurational state diagram for polystyrene particles sterically stabilized by a very low molecular weight polymer (octaoxyethylene glycol dodecyl ether $C_{12}E_8$). The extent of aggregation at equilibrium was measured by particle counting. This was related theoretically to the pairwise interaction between particles via an equilibrium constant K for the aggregation process

$$n\text{-mer} + j\text{-mer} \rightleftarrows (n+j)\text{-mer}$$

It follows that

$$K = \frac{[(n+j)\text{-mer}]}{[n\text{-mer}][j\text{-mer}]} \tag{8.5}$$

is an equilibrium constant that incorporates both enthalpic and configurational entropic contributions. The equilibrium constant can be found from the standard free energy of flocculation, *viz* $K = \exp\{-\Delta G^{\ominus}/RT\}$. It is also related to the interaction energy between the particles by a form of the second virial coefficient

$$K = 2\pi \int_R^{\infty} d^2[\exp\{-V(d)/kT\} - 1]dd \tag{8.6}$$

Here d = centre-to-centre distance and $R = 2(a+\delta)$, where a = particle radius and δ = steric layer thickness. The pair potential $V(d)$ was assumed to be composed of the van der Waals attraction and a hard sphere steric repulsion. Provided that there is only linear aggregation, the equilibrium constant can be related to the degree of aggregation by

$$N_0/N_{eq} = 1 + KN_{eq} \tag{8.7}$$

Here N_0, N_{eq} are the initial and equilibrium particle concentrations.

Equations (8.5) and (8.6) show that according to this model, the factors governing the equilibrium aggregation state of a dispersion are the particle concentration, the particle size, the barrier layer thickness, and the net interactional energy. The interactional energy that controls the state of aggregation is the absolute magnitude of the depth of the potential energy well. The calculated "phase diagram" for large (1·6 μm) particles is shown in Fig. 8.8. Note that Thompson and Pryde recognized that a true phase change was not predicted and that the use of the term "phase diagram" is contrived. The theory indicates that aggregation is more readily engendered if the particle size is large and the particle concentration is high. The former effect

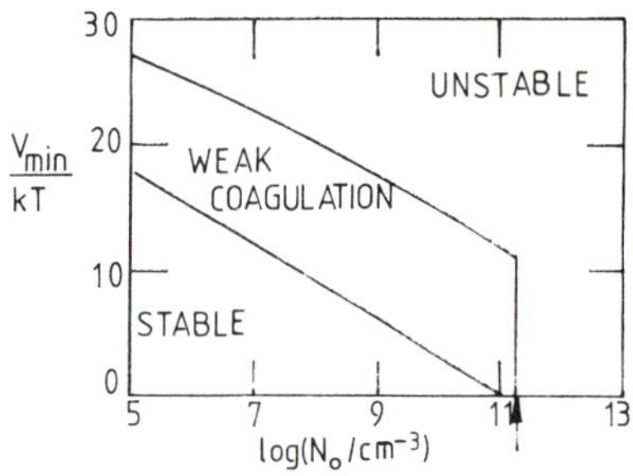

Fig. 8.8. Phase diagram for 0·6 μm diameter polystyrene latex particles sterically stabilized by low molecular weight poly(oxyethylene). The arrow (↑) indicates the onset of hexagonal close packing (after Thompson and Pryde, 1981).

arises from the enhanced van der Waals attraction between the particles as the particle size increases (an enthalpic effect) and the latter from the reduced configurational entropy at higher particle number concentrations. It follows from Fig. 8.8 that the mere existence of an energy minimum whose depth is a few kT is not in itself sufficient to cause weak aggregation. A dispersion subject to such energetics could appear either totally stable or totally unstable, or with intermediate stability.

Thompson and Pryde (1981) also claimed to have observed repulsion significantly beyond the span of the polymer chains. This was alleged to arise from ordering of water molecules beyond the physical reach of the surfactant. This finding is in accord with the conclusion of Parsegian and co-workers (Le Neveu *et al.*, 1977; Cowley *et al.*, 1978; Parsegian *et al.*, 1979), who claimed that forces of hydration between phospholipid layers can operate over separations of order several nanometres. Such claims should be treated with caution, however, because they are based on rather flimsy circumstantial evidence. As yet no direct experimental detection of such long-range solvent ordering (e.g. by neutron scattering) have been published.

The ordered configurational states of polymer latices have been investigated by Ottewill (1980), using *inter alia* neutron scattering. He has shown that a coexistence region exists between the ordered and disordered regimes for electrostatically stabilized latices. This is typified by the occurrence of single dispersed particles, which give a milky appearance to the upper latex layer, in equilibrium with an ordered iridescent layer at the bottom of the vessel.

Ottewill has further proposed that the ordered configurational states for electrostatically stabilized latices show only relatively short-range positional correlations in the volume fraction range corresponding to the coexistence region. This type of configurational state was referred to as "liquid like". It apparently owes its existence to the "softness" of the long-range electrical double layer repulsion.

Above a higher critical volume fraction of particles, the whole volume of latex displays iridescence. This was attributed to much longer-range positional correlations between the positions of the particles. Iridescence arose from the particles adopting a highly ordered "solid like" lattice, either face-centred and/or body-centred cubic.

The onset of ordered structure formation in sterically stabilized dispersions has also been reported by Ottewill (1980). Monodisperse poly(methyl methacrylate) latices, of core diameter 65 nm, were stabilized by low molecular weight poly(12-hydroxystearic acid) in dodecane. Again small angle neutron scattering provides a powerful tool to explore the structure of these ordered configurational states. It might be anticipated that the "hard" steric repulsion in these systems would narrow the coexistence region noticeably.

8.4.6. Comparison of θ-point flocculation with weak coagulation

Sterically stabilized dispersions display a very different pattern of behaviour according to whether the stabilizing moieties are high polymers or low molecular weight chains. Some of these differences are summarized in Table 8.2. It should first be pointed out that weak coagulation phenomena, if unwanted, can usually be eliminated by the simple expedient of using higher molecular weight stabilizing moieties.

If, near to the CFPT, the configurational entropy of the particles contributes to the stability of colloidal particles stabilized by low molecular weight species, then it is reasonable to ask why this configurational entropy

TABLE 8.2
Comparison of θ-point flocculation with weak coagulation of dispersions

Dispersion property	θ-point flocculation	Weak coagulation
A. Aggregate in	slightly worse than θ-solvent	somewhat better than θ-solvents
B. Origin of aggregation	strong mutual attraction between "stabilizing" segments	weak London attraction between core particles
C. Dependence of CFPT upon:		dispersions more stable if:
a. particle number concentration	none	more dilute
b. particle size	none	smaller particle size
c. molecular weight of stabilizing moieties	none	higher molecular weight
d. conformation	none	conformation is more extended
D. Critical coagulation particle volume fraction	not observed to-date	observed
E. Particle configurational entropy	unimportant	important for stability near to the CFPT
F. Reversibility	yes	yes

does not contribute correspondingly to the stability of dispersions stabilized by high molecular weight polymers. The answer, as Cowell *et al.* (1978) were the first to suggest, must surely reside in the different mechanisms whereby instability is induced.

Consider, first, the case of low molecular weight species. London dispersion forces are necessarily weak when acting at a distance, especially as they cause coagulation in somewhat better than θ-solvents. The depth of the potential energy well must therefore be shallow. It thus becomes subject to significant perturbations by the configurational entropy of the particles. On the other hand, with high molecular weight stabilizing moieties, a deep potential energy well is generated, even quite close to the θ-point, as a consequence of interpenetration. It follows that the particle configurational entropy is impotent insofar as inflicting significant changes on this well depth is concerned.

Cowell *et al.* (1978) and Cowell and Vincent (1982b) have argued that there is a significant elastic contribution to the stability of dispersions sterically stabilized by low molecular weight moieties. This is presumed to arise from the high segmental concentrations in the steric sheaths, as well as the reduced segmental motions of the somewhat extended chains. There does not appear as yet to be any direct experimental evidence to support this contention. As discussed in Section 9.3.3, it is unlikely that an elastic term arises from the high segmental concentration effect alone, although it must be admitted that any radical reduction in flexibility could contribute to the generation of an elastic contribution.

Finally, we stress that both weak coagulation and flocculation are usually readily reversible. Hence it is not possible to distinguish experimentally between these two phenomena, originating as they do in two different effects, on the simple basis of reversibility.

8.5. Crystallization of the stabilizing moieties

Everett and Stageman (1977; 1978a) have reported that poly(methyl methacrylate) latex particles stabilized by poly(dimethylsiloxane) in *n*-alkanes flocculated both on heating and on cooling. Whereas the observed UCFT was found to correlate qualitatively with the θ_L-temperature associated with the corresponding lower critical solution temperature, the LCFT occurred at a temperature far above the θ_U-temperature associated with the upper critical solution temperature. This implies that coagulation occurred under significantly better than θ-solvency conditions. Everett and Stageman (1978a) noted that the polymer phase separated from solution within a few degrees of the observed LCFT. Flocculation at the LCFT was always reversible.

Croucher and Hair (1978) postulated that the LCFT in these systems resulted from a surface-induced crystallization of the poly(dimethylsiloxane). Crystallization would be expected to allow the attractive van der Waals forces between the particles to become operative, thereby causing the particles to coagulate.

Subsequently, Croucher *et al.* (1980) and Sundararajan *et al.* (1980) used proton NMR to detect the crystallization of poly(dimethylsiloxane) from hydrocarbon solutions. The formation of a crystalline phase was confirmed by X-ray diffraction studies. Crystallization was found to occur from *n*-heptane at ca $-90\,^{\circ}$C. Moreover, the crystallization temperature was found to be insensitive to the nature of the hydrocarbon. In agreement with the observations of Everett and Stageman (1978a), poly(methyl methacrylate) and poly(vinyl acetate) latices stabilized by poly(dimethylsiloxane) in *n*-heptane coagulated at ca $-90\,^{\circ}$C as well. These results implied that crystallization was not significantly catalysed by the surface of the latex particles. It apparently took place oblivious of surface attachment.

Chapter 9

Deviant Flocculation Behaviour: Enhanced Steric Stabilization

9.1. Introduction

Four exceptions to the general rule identifying the thermodynamic limit to the stability of sterically stabilized dispersions at the θ-point were considered in the preceding chapter. They all display a common feature: coagulation occurs in dispersion media that are better solvents for the stabilizing moieties than θ-solvents. This means that coagulation takes place before the thermodynamic limit to stability is reached due to the intrusion of van der Waals forces. In many instances this premature coagulation can be circumvented by improving the quality of the steric stabilization (e.g. by better anchoring, higher surface coverage, or higher molecular weight stabilizing moieties).

A fifth exception to the general rule displays a contrasting pattern of behaviour to the exceptions discussed previously in that stability is observed in dispersion media that are significantly *worse* solvents for the stabilizing moieties than θ-solvents. This is termed *enhanced steric stabilization.* At first sight, its existence is most unexpected because the segments of the stabilizing moieties might be assumed to be strongly mutually attractive under these solvency conditions. The origins of enhanced steric stabilization are uncertain. The several hypotheses that have been advanced to explain the phenomenon will be discussed in this chapter.

9.2. Experimental evidence for enhanced steric stabilization

9.2.1. Homopolymeric stabilizing moieties

9.2.1.1. The importance of surface anchor groups

The phenomenon of enhanced steric stabilization was discovered by Dobbie *et al.* (1973). They measured the UCFTs of polystyrene and poly(vinyl acetate) latices in the presence of increasing amounts of the amphipathic stabilizers poly(styrene-*b*-oxyethylene) and poly(vinyl acetate-*b*-oxyethylene), respectively. No surface carboxylic acid groups were detectable on the surfaces of the poly(vinyl acetate) latex particles. Two polystyrene latices were studied: one possessed a high surface carboxyl content whereas that of the other was low. In addition, latex particles were prepared in which styrene was copolymerized with acrylic acid, again leading to a high incidence of surface carboxylic acid groups.

The results presented in Fig. 9.1 indicate that the poly(vinyl acetate) and the low surface carboxylic acid polystyrene latices behaved in a normal fashion. When sufficient stabilizer was present, their UCFTs correlated strongly with the corresponding θ-temperatures. At low stabilizer concentration, the

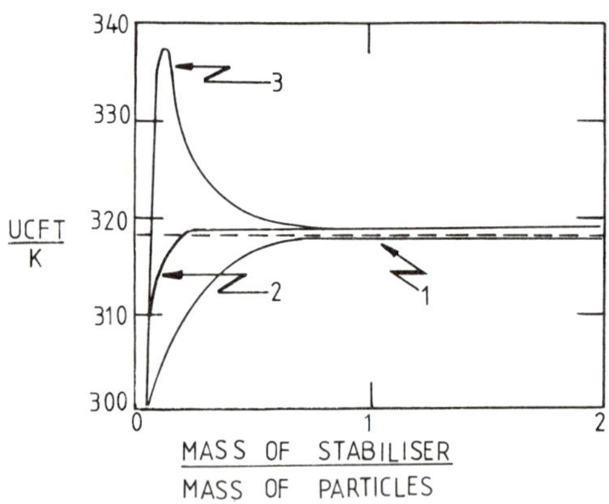

Fig. 9.1. Plots of the UCFT as a function of the relative weight of poly(oxyethylene) steric stabilizer for three different latex particles: 1, poly(vinyl acetate); 2, low carboxyl polystyrene; 3, high carboxyl polystyrene latex particles. The dashed line is θ_L (after Dobbie *et al.*, 1973).

UCFTs fell below the θ_L-temperature, indicating that coagulation was occurring in better than θ-solvents. As discussed in Section 8.2, this is a consequence of the displacement coagulation that occurs when the surface coverage by the stabilizer is incomplete. The increase in the UCFT to its plateau value presumably mirrors the adsorption isotherm of the stabilizing chains.

The behaviour of the latex particles containing significant amounts of surface carboxylic acid groups was pathologically abnormal. The UCFT in these systems increased with increasing amounts of surfactant until a maximum value was reached. Thereupon its value declined, asymptoting to that of the θ_L-temperature. The value of the UCFT at the maximum exceeded the θ_L-temperature by as much as 20 °C at this pH, well beyond the limits of experimental error (± 2 °C).

This unprecedented observation of stability in worse than θ-solvents, when the segments of the stabilizing moieties would be expected to be mutually attractive, is the enhanced steric stabilization referred to previously. Dispersions that exhibited enhanced steric stabilization appeared to flocculate irreversibly.

9.2.1.2. The influence of pH

The preceding observations demonstrate that surface carboxylic acid groups are necessary in order that enhanced steric stabilization may manifest itself in these systems. This suggests the occurrence of some form of association, presumably hydrogen bonding, between the surface carboxylic acid groups and the ether oxygens of the poly(oxyethylene) chains. To test this hypothesis, Dobbie *et al.* investigated the effect of pH on the enhancement of the UCFT above the θ_L-point using poly(oxyethylene) of molecular weight 10 000 as the stabilizing moieties. At high pH, the carboxylic acid groups, whose $pK_a \sim 4{\cdot}6$, would exist mainly in the ionized carboxylic ion (-COO^-) form and so would be unable to hydrogen bond with ether oxygens. As the pH decreases, so carboxylic acid groups (-COOH) that are capable of undergoing H-bonding appear on the particle surface (see Fig. 9.2).

The results of the pH experiments for the high carboxylic acid latex are reproduced in Fig. 9.3. Only two pH values are shown, pH 2·06 and pH 4·65.

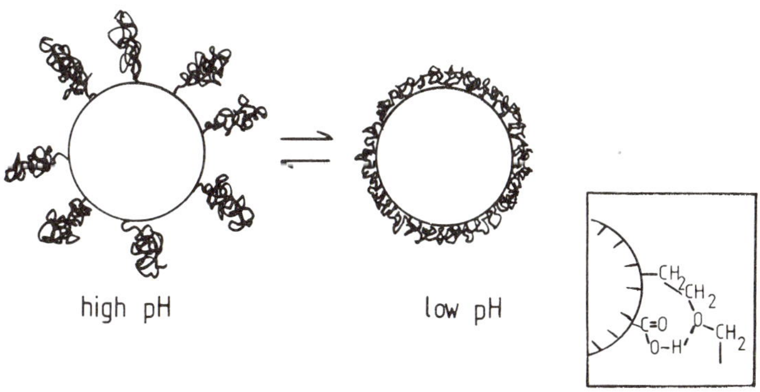

Fig. 9.2. Schematic representation of the multipoint anchoring of poly(oxyethylene) chains on high carboxyl polystyrene latex particles at low pH due to hydrogen bonding (inset).

No results are presented for the third pH studied (pH 10·6) because at the stabilizer concentrations used, the latices flocculated at temperatures well below room temperature. The maximum in the UCFT for pH 2·06 occurred at significantly higher temperatures than that for pH 4·65. Stability could be observed at the lower pH some 30 °C above the θ_L-temperature for the stabilizing moieties. Clearly, the magnitude of the enhancement increased with decreasing pH, in conformity with the concept of hydrogen bonding between the surface carboxylic acid groups and the ether oxygens in the polymer chain.

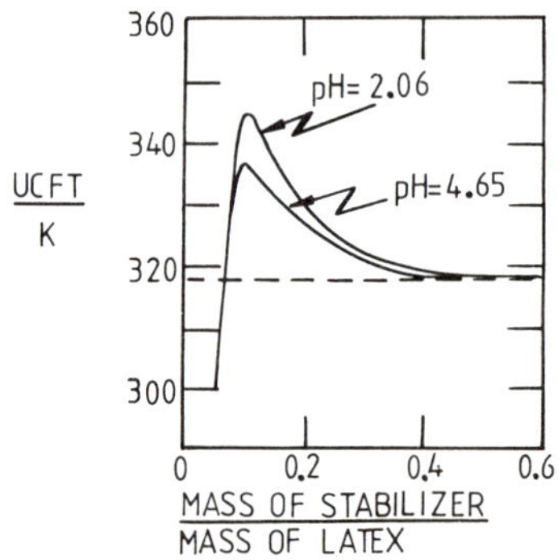

Fig. 9.3. Plots of UCFT as a function of the relative weight of poly(oxyethylene) steric stabilizer for high carboxyl polystyrene latices at two different pH values. The dashed line is θ_L (after Dobbie *et al.*, 1973).

9.2.1.3. The influence of the molecular weight of the stabilizing moieties

Dobbie *et al.* (1973) found that the molecular weight of the stabilizing moieties had a profound effect on the extent of enhancement of the UCFT above the θ_L-temperature (see Fig. 9.4).

Very high molecular weight (800 000) poly(oxyethylene) did not apparently display any measurable enhancement of the UCFT at any pH. A small enhancement was observed for a molecular weight of 96 000 and this enhancement increased as the molecular weight was further decreased. The enhancement of the UCFT above the θ_L-temperature for a molecular weight of 1540 was so large that the UCFT apparently exceeded the boiling point of the dispersion medium and so could not be observed. The data are, however, strongly suggestive of a maximum in the UCFT curve.

It is apparent that the magnitude of the enhancement in the UCFT above the θ_L-temperature was a sensitive function of the molecular weight of the polymer, at least for low molecular weight polymers. The enhancement increased with decreasing molecular weight of the stabilizing moieties.

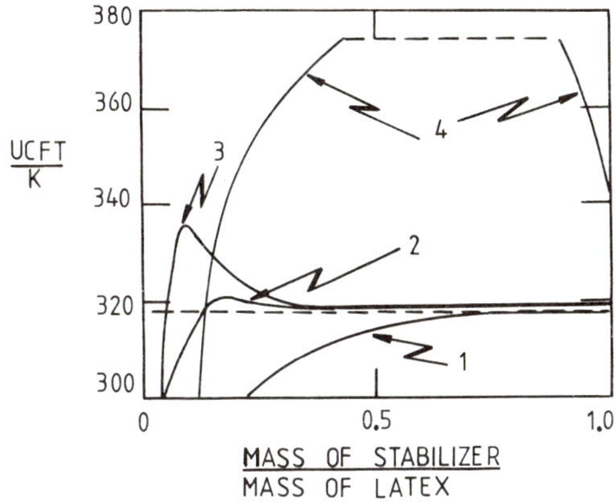

Fig. 9.4. Plots of UCFT as a function of the relative weight of poly(oxyethylene) steric stabilizer for high carboxyl polystyrene latices for different molecular weight poly(oxyethylene): 1, 800 000; 2, 96 000; 3, 10 000; 4, 1540. The dashed line is θ_L (after Dobbie *et al.*, 1973).

9.2.2. Copolymeric stabilizers

Cairns and Neustadter (1975) studied the stability of carbon dispersions stabilized by BP 45, a copolymer of methacrylic esters for which R in the ester group -COOR could be the straight chain C_4 or C_{8-12} alkyl groups, or the amino substituted ethyl group $-CH_2CH_2NH_2$. They also studied the stabilizer PV30-TEPA, a succinimide terminated poly(isobutylene). Flocculation was induced by the addition of ethanol to the dispersion medium, which was *n*-heptane. Their results, which compare the CFVs with the θ-compositions determined by the method of Cornet and van Ballegooijen (1966), are presented in Table 9.1.

The data shown in Table 9.1 confirm that it is possible to prepare dispersions in dispersion media that are worse solvents for the stabilizing chains than θ-solvents.

TABLE 9.1
Comparison of CFV with θ-composition for sterically stabilized carbon dispersions at 21°C

Stabilizing polymer	Carbon type	CFV	θ	CFV/θ
BP45	Kosmos BB	43	38	1·13
	Fiat sludge	43	38	1·13
PV30-TEPA	Fiat sludge	27	20	1·35

Lambe *et al.* (1978) studied the enhanced steric stabilization of polystyrene latices by poly(vinyl alcohol). This is included in this sub-section on copolymers because the samples studied were not fully hydrolysed. This means that the parent poly(vinyl acetate) from which they were derived was only partially (88%) hydrolysed (this is often accomplished by alcoholysis). The resultant polymer is not, however, a completely random copolymer because adjacent group effects influence the hydrolysis kinetics in such a way that some degree of "blockiness" is introduced. On average, these blocks consist of 2 ester groups to every 18 alcohol groups but blocks of average size 5–6 acetate groups are common. The chemical structure of the polymers should therefore be formally represented by a structure intermediate between poly(vinyl acetate-*b*-vinyl alcohol) and poly(vinyl acetate-*co*-vinyl alcohol) rather than poly(vinyl alcohol) as such. The random (or statistical) copolymer can be prepared by partial reacetylation of fully hydrolysed poly(vinyl alcohol).

Lambe *et al.* (1978) examined the effects of four different molecular weight fractions of poly(vinyl alcohol) on the stability of polystyrene latices. These had weight average molecular weights of 4 000, 8 000, 17 000 and 52 000 respectively. The polymer concentration was varied from 0 to 5 000 parts per million (parts/10^6) and the latex volume fraction from 10^{-3} to 4×10^{-5}.

The UCFT measured in these systems as a function of the concentration of polymer (molecular weight 8 000) added to the dispersion medium (2 M $NaNO_3$) is presented in Fig. 9.5 for three different volume fractions of latex. The UCFT in all cases increased to a maximum value and then declined. Separate measurements, using the phase separation method of Cornet and Ballegooijen (1966), suggested that in 2 M $NaNO_3$ the apparent θ-temperature for the poly(vinyl alcohol) of this degree of hydrolysis was

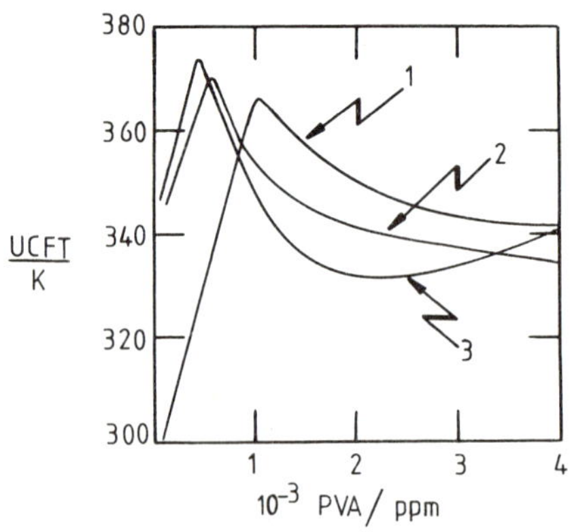

Fig. 9.5. Plots of UCFT as a function of the added concentration of poly(vinyl alcohol) (PVA) of molecular weight 8000 for different volume fractions (v_3) of polystyrene latex: 1, $v_3 = 1 \times 10^{-3}$; 2, $v_3 = 3 \times 10^{-3}$; $v_3 = 3 \times 10^{-4}$ (after Lambe *et al.*, 1978).

25 ± 5 °C. Since the maximum in the UCFT for these systems approached 100 °C, stability was observed some 70 °C above the apparent θ_L-temperature. Note, however, that the decrease in UCFT did not flatten out in these experiments at the apparent θ_L-temperature but significantly above it. A shallow minimum was also evident at the lowest volume fraction studied. The UCFT displayed a moderately strong dependence on particle volume fraction.

More recently, Tadros and Winn (Tadros, 1982) extended the results of Lambe *et al.* They confirmed that it was possible to prepare dispersions in the presence of poly(vinyl alcohol) that were stable in dispersion media (1 M KCl) which were apparently significantly worse than θ-solvents. A strong dependence of the UCFT on the molecular weight was observed at low molecular weights (see Fig. 9.6). The UCFT was found to increase to a maximum value with increasing molecular weight and then to decrease to an approximately constant value. This value of the UCFT was significantly greater (ca 20 °C) than the corresponding apparent θ_L-temperature.

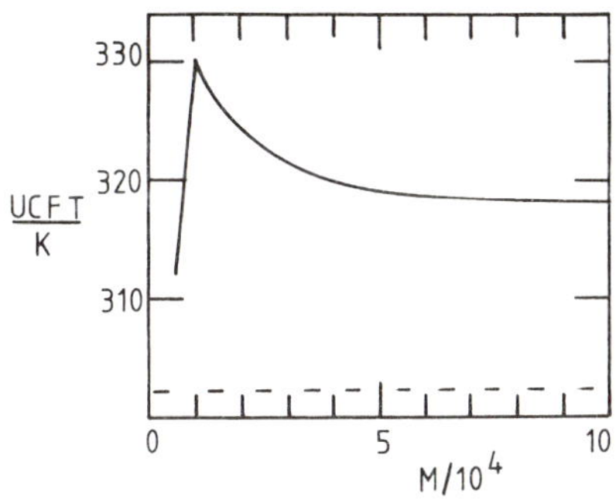

Fig. 9.6. The variation of the UCFT with molecular weight of the poly(vinyl alcohol) stabilizer for polystyrene latex particles in 1 M KCl. The dashed line indicates the claimed θ-temperature (after Tadros, 1982).

9.3. The origins of enhanced steric stabilization

As stated previously, the origins of enhanced steric stabilization are still not fully understood, although several explanations have been forthcoming. These include:

(i) the reduction of the segment density in the peripheral regions of the steric layer;

(ii) a change in the thermodynamic parameters concomitant with a large distortion of the stabilizing polymer conformation;

(iii) the intrusion of a strong elastic repulsion into the interactional free energy;
(iv) an increase in the mean segment density in the barrier layer.

These proposals will now be considered in turn.

9.3.1. The segment reduction hypothesis

Dobbie *et al.* (1973) observed that enhanced steric stabilization was evident under conditions when the interaction between groups located at the particle surface and the stabilizing moieties were favoured. If sufficient in number, such interactions would be expected to perturb the conformation of the stabilizing moieties significantly (Smitham and Napper, 1979). They argued that the existence of a large number of points of attachment along the stabilizing chain (so-called *multipoint anchoring*) would result in *loopy* adsorption (Silberberg, 1962a,b; 1967; 1968). Each polymer chain is then composed of three different types of segments (see Fig. 4.7): first, those in trains, which are short lengths of polymer that abut the surface of the particle without appreciable extensions into the dispersion medium; second, those in loops, which are sections of the chain that project substantially into the dispersion medium; finally, segments in tails. Tails are terminal sections of polymer chains that project into the dispersion medium. They differ from loops in that they are attached to the surface of the particle by a single, terminal anchoring point.

Segments in trains do not contribute directly to the steric repulsion, which arises from the segments composing the tails and the loops. The trains, however, are essential for stability to be imparted by homopolymers because they provide the anchoring so necessary to counteract the stress generated by a Brownian collision.

The poly(oxyethylene) stabilizing moieties used in the studies of Dobbie *et al.* (1973) would normally be attached to the particle surface as single loops or, less probably, single tails when surface coverage was complete. However, for surfaces containing carboxylic acid groups with which the ether oxygens of the poly(oxyethylene) can H-bond, the conformation is likely to be markedly flattened, at least at low surface coverages. This was shown schematically in Fig. 9.2. The conformational change maximizes the polymer-surface enthalpic interactions, consistent with the minimization of the loss of conformational entropy.

The occurrence of strong H-bonding between carboxylic acid groups, which act as H-bond donors, and ether oxygens, which act as H-bond acceptors, is well recognized in classical chemistry (Pimental and McClellan, 1960). It is confirmed in the case of poly(acrylic acid) (to which the particle surface is crudely analogous) and poly(oxyethylene) by their mutual precipitation from aqueous solution at suitable pH. Smitham and Napper (1979) postulated that multipoint anchoring transforms dramatically the shape of the segment density distribution function. This is illustrated in Fig. 9.7 for poly(vinyl alcohol) of molecular weight 17 000. The steric layer thickness, L, was assumed to be constant at 9·8 nm. It is readily apparent from the diagram, where the normalized segment density distribution function $\hat{\rho}$ is plotted as a function of the distance from a flat interface, that for a tail $\hat{\rho}$ has a maximum value at about half the thickness of the steric layer. Where multipoint anchoring occurs, theory predicts that an approximately exponential distribution function pertains (Silberberg, 1962a,b; 1967; 1968; Scheutjens and Fleer, 1980). This has its maximum value at (or near) the surface.

The qualitative change in the form of the segment density distribution function concomitant with strong surface interactions and loopy adsorption has recently been confirmed experimentally by small angle neutron scattering and pulsed nmr studies, which have allowed the segment density distribution functions to be determined directly (Barnett *et al.*, 1981a,b; Cosgrove *et al.*, 1982). They will be more fully discussed in Section 12.3.1.1.

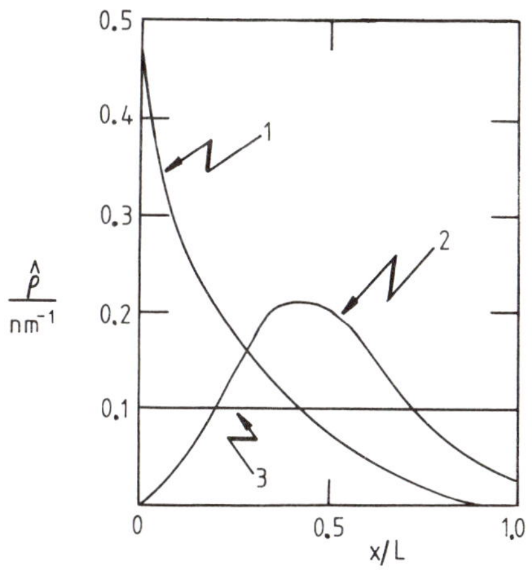

Fig. 9.7. The distance dependence of the normalized segment density distribution function for 1, an exponential function, 2, a radial Gaussian function and 3, a constant segment density function (after Smitham and Napper, 1979).

Figure 9.7 shows that loopy adsorption dramatically reduces $\hat{\rho}$ in the outer regions ($x/L > 0{\cdot}5$) of the steric layer compared with that for tails, for example. Physically, this can be readily understood: the energetically favoured surface interactions result in segments being displaced from the peripheral regions of the steric layer and relocated in closer proximity to the surface.

It was indicated previously that for many systems the CFPT correlates strongly with the corresponding θ-point. Flocculation in these systems results from the segmental attraction that is operative between the steric layers attached to opposing particles in marginally worse than θ-solvents. The magnitude of this attraction is a sensitive function of the segmental density, varying roughly as $\hat{\rho}^2$. Any reduction in $\hat{\rho}$ drastically decreased the magnitude of the steric interactional energy. This is exemplified in Fig. 9.8 which displays the relative free energies for loopy-type adsorption and for tails in the interpenetrational domain (Smitham and Napper, 1979). The interactional energy for loopy adsorption is reduced by almost an order of magnitude compared with that for tails. Adsorption in the form of single loops would presage a similar pattern of behaviour to that of tails.

What the severe reduction in interactional energy that accompanies multipoint anchoring means for flocculation can now be discerned. In order to generate a potential energy well depth sufficiently large to induce flocculation, it is necessary to compensate for the reduction in segment density by a greater penetration into the worse than θ-solvent regime. It follows that with low segment densities in the peripheral regions, stability can be apparent in significantly worse than θ-solvents.

The marked effect of molecular weight observed by Dobbie *et al.* (1973) is also explicable. Very high molecular weight chains were not significantly influenced by interactions with the surface, presumably because most segments were rather distant from the surface. No significant enhancement was therefore observed. On the other hand, with very low molecular weight chains, almost all the segments were perturbed and a very large enhancement was evident. A simple calculation, based on the surface group concentration obtained by potentiometric titration, showed that the average loop size

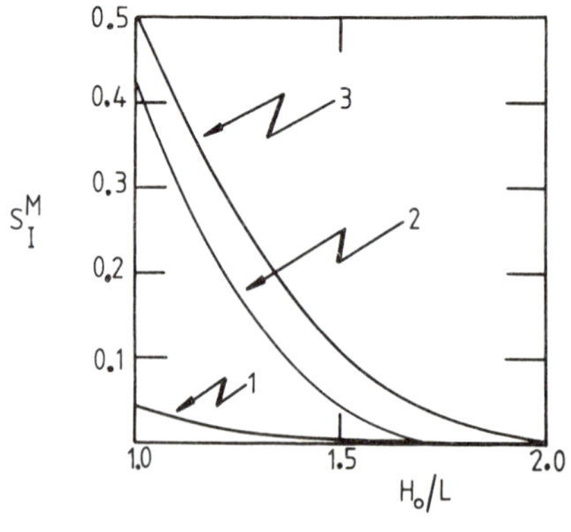

Fig. 9.8. The distance dependence of the steric interaction function in the interpenetrational domain for two spheres with different segment density distribution functions: 1, exponential function; 2, radial Gaussian function; 3, constant segment density function (after Smitham and Napper, 1979).

between successive anchor points with polymer of 10 000 molecular weight could be as small as five to ten oxyethylene groups.

9.3.2. Change in thermodynamic parameters

Smitham and Napper (1979) have suggested that enhanced steric stabilization may arise in part from a thermodynamic origin. The interaction between the surface groups and the segments of the polymer chains in multipoint anchoring could distort the conformation of the stabilizing moieties drastically. Specifically, the flexibility of the chain could be significantly reduced. Adsorption would be expected to render the conformational motion of the segments more difficult, at least in the vicinity of the trains. This might be expected to alter, to some degree, the thermodynamic parameters that govern segment-solvent interactions.

Silberberg (1976) has presented semiquantitative arguments that are in line with this motion. He has claimed that the range of unlimited mixing of highly folded chains with solvent is greatly enlarged over that for random chains. This conclusion, which is intuitively appealing, can be reached from a different approach. Folded chains should follow more closely the Bragg-Williams model (Moore, 1972) for molecular mixing rather than the Flory-Huggins model if access to the folded regions of the polymer is denied to the solvent. The θ-point for a Bragg-Williams molecule corresponds to $\chi_1 = 2$ rather than the usual Flory-Huggins value of $\chi_1 = \frac{1}{2}$ (Tompa, 1956). Loopy adsorption might be visualized in this context as representing a degree of folding that is intermediate between the Flory-Huggins random coil model and the Bragg-Williams folded chain model. It is therefore feasible that the effective θ-point could be shifted, at least for low molecular polymers, to values of χ_1 significantly greater than $\frac{1}{2}$. This would correspond to the worse than θ-solvent domain for the unfolded chain in free solution.

One difficulty with the foregoing postulate is that it requires significant exclusion of the molecules of the dispersion medium from the domains of the steric barriers. This seems to be physically unlikely. Flory (1970) has stressed that the thermodynamic properties of polymer solutions should be essentially independent of the polymer conformation.

9.3.3. The elastic repulsion hypothesis

The elastic repulsion hypothesis has arisen from the studies using partially hydrolysed poly(vinyl alcohol) as stabilizer. In this instance, no interactional groups are apparently necessary on the surface of the colloidal particles, presumably because the blocks of poly(vinyl acetate) segments, which are nominally insoluble in water, anchor the poly(vinyl alcohol) stabilizing moieties to the surface. Of course, poly(vinyl alcohol) is a decidedly atypical polymer. Its highly idiosyncratic behaviour, which presumably arises from the hydroxyl groups, sets it apart from normal polymers.

Lambe *et al.* (1978) have extended the theory of Cowell *et al.* (1978) for low molecular weight stabilizing moieties to embrace enhanced steric stabilization. They have postulated that with multipoint anchoring the adsorbed layer thickness is relatively small. Consequently, the steric layers do not undergo significant interpenetration in Brownian collisions. Instead, they are postulated to undergo an elastic compression. This elastic compression, arising as it does from the loss of configurational entropy of the polymer chains, is always repulsive, irrespective of the solvency of the dispersion medium. The presence of the elastic repulsion is postulated in worse than θ-solvents to prevent the much weakened segmental attraction from forming a potential energy well of sufficient depth to cause flocculation. This theory is really an elaboration of the *denting* mechanism, first proposed by Bagchi (1974a,b).

The validity of the elastic repulsion hypothesis is uncertain. It seems unlikely that the concentration of segments, even for quite low molecular weight chains, is ever sufficiently high to prevent interpenetration under θ- or worse than θ-conditions for normal polymers [this, of course, excludes poly(vinyl alcohol)]. Flory (1976), in his Nobel Prize address, stressed that the dense packing of polymer segments may appear at first glance to be an impossibly difficult task. Yet thorough examination of the problem shows that flexible macromolecules are readily able to meet this challenge without departures from their natural proclivities. The reason for this is simple: the overwhelmingly large number of conformations accessible to flexible macromolecules guarantees that there are numerous ways of satisfying the mutual exclusion engendered by the finite (real) size of the polymer segments. The large number of conformations of a polymer chain is reflected in the conformational partition function, which for many carbon backbone polymers is of order 2^n, where n = number of bonds in the chain (Flory, 1969). Even for small n (e.g. $n = 100$), this is an incomprehensibly large number (10^{30}).

It is further remarked that compression, involving as it does the exclusion of the segments from complete volume elements in space, is unlikely to be preferred to interpenetration wherein the segments have access to most (usually more than 70%) of those volume elements. If there is an elastic contribution to steric stabilization, it must surely arise, not directly from a high segment density, but from the loss of flexibility concomitant with multipoint anchoring at the surface.

The foregoing conclusion is borne out by the data presented in Tables 6.1 and 6.2, wherein it is shown that for certain systems at least, the CFT and the θ-temperature are strongly correlated even for stabilizing moieties of molecular weight as low as a few thousands. This immediately excludes any significant elastic interactions in these systems because such effects are always repulsive and relatively insensitive to the solvency of the dispersion medium. Elastic repulsion, if present, must weaken any correlation between the CFPT and the θ-point and lead to stability in worse than θ-solvents.

It is possible that an elastic contribution to steric stabilization could have a kinetic, rather than a thermodynamic, origin. The polymer segments in the stabilizing moieties may relax too slowly to allow interpenetration to proceed in the time span of a Brownian collision. A denting mechanism would then be operative in order to transmit the repulsive stress to the particles. This kinetic mechanism for the introduction of an elastic repulsion would predict that the enhancement should be strongly dependent upon the particle size which determines the duration of a Brownian encounter. No experimental evidence for such a particle size dependence has apparently been published.

It is possible, in one sense, to reconcile the elastic repulsion hypothesis with the reduced segment density hypothesis. The intrusion of a more significant elastic repulsion could be one consequence of a reduced segment density. To see this, we utilize the results of Section 3.2.6 wherein the excess mixing free energy was shown to be roughly proportional to the square of segmental concentration. In contrast, the elastic term is approximately linear in the same parameter. Hence the mixing term is more profoundly decreased on segmental reduction than is the elastic term (Smitham and Napper, 1979).

There is, of course, a qualitative difference between the addition of a bona fide block copolymer of the type studied by Dobbie *et al.* (1973) and that of the partially hydrolysed poly(vinyl alcohol) investigated by Lambe *et al.* (1978) and Tadros and Winn (Tadros, 1982). Specifically, at complete surface coverage, the anchor polymer in the experiments of Dobbie *et al.* apparently completely shielded the surface carboxylic acid groups from interaction with the ether oxygens of the poly(oxyethylene) chains. The UCFT was then close to the θ_L-temperature. With partially hydrolysed poly(vinyl alcohol), however, the cause of multipoint anchoring, the unhydrolysed poly(vinyl acetate) segments, is present at all surface coverages. The UCFT is thus unlikely to decrease to the θ_L-temperature at high surface coverages but, rather, to remain significantly above it, as was apparently observed.

We note in passing that it is by no means certain that enhanced steric stabilization has actually been observed with partially hydrolysed poly(vinyl alcohol) as the stabilizer. First, it is questionable whether the concept of a θ-temperature is applicable to copolymers in general, except in rather special circumstances: e.g. it is certainly applicable to alternating copolymers and perhaps to some random copolymers; but it is not applicable to block or graft copolymers, except perhaps to those composed of blocks with identical θ-temperatures in the same solvent. Second, the putative θ_L-temperature determined for partially hydrolysed poly(vinyl alcohol) by the phase separation method of Cornet and van Ballegooijen (1966) applies to the *whole* stabilizer, not just to the poly(vinyl alcohol) moieties. Yet it is these latter moieties which presumably impart stability. The presence of the nominally insoluble poly(vinyl acetate) segments would be expected to depress the observed value of the θ_L-temperature. The conclusion emerges that just because the observed UCFT is greater than the putative θ_L-temperature of the copolymer, this does not necessarily imply that the true θ_L-temperature of the homopolymeric poly(vinyl alcohol) moieties is exceeded. It follows that the existence of enhanced steric stabilization with partially hydrolysed poly(vinyl alcohol) has not yet been established unequivocally, although its occurrence does seem highly likely. Unfortunately, it is not possible to use the method of Cornet and van Ballegooijen (1966) to determine the actual θ_L-temperature of fully hydrolysed poly(vinyl alcohol) because its crystallinity and insolubility renders the method both inoperative and invalid.

Finally, we note that the maximum in the UCFT observed with poly(vinyl alcohol) is readily explicable without the elastic repulsion hypothesis. It merely requires the adsorbed, partially hydrolysed polymer to change from a relatively flat conformation at lower concentrations to a conformation more extended normal to the interface, and thus exhibiting reduced multipoint anchoring, at higher concentrations. This explanation is closely analogous to that proposed by Dobbie *et al.* (1973) for poly(oxyethylene) attached to polystyrene latices containing surface carboxylic acid groups.

9.3.4. Increased mean segment density hypothesis

Feigin (1980) has pointed out that a significant increase in the mean segment density in the barrier layer could cause stability to be apparent in the worse than θ-solvent regime. He assumed that each coating could be approximated by a layer of constant segment volume fraction v_2. The free energy of mixing in a volume V of such a layer is

$$\Delta G_{\mathrm{MIX}} = (kT\,V/V_1)[\ln(1-v_2)+\chi_1 v_2] \tag{9.1}$$

where χ_1 = interaction parameter and V_1 = volume of a dispersion medium molecule. Penetration of two identical polymer coatings causes the two volume elements to telescope into one of volume V whilst another volume V of dispersion medium is expelled into the bulk dispersion medium. This is accompanied by a free energy change

$$\Delta G_1 = (kT\ V/V_1)\ [(1-2v_2)(\ln\{1-2v_2\}+2\chi_1 v_2) - 2(1-v_2)(\ln\{1-v_2\}+\chi_1 v_2)] \tag{9.2}$$

Flocculation would be expected for suitably negative values of ΔG_1. As only a few kT would be required for this, we can find the critical values of χ_1 to a reasonable approximation by setting $\Delta G_1 = 0$. The values so obtained, denoted by χ_1^{crit}, are presented in Table 9.2 for different values of the polymer volume fraction.

TABLE 9.2
Critical values for the interaction parameter as a function of the polymer volume fraction

v_2	0·001	0·01	0·02	0·05	0·10	0·20	0·30
χ_1^{crit}	0·5005	0·5051	0·510	0·527	0·557	0·632	0·738

It is immediately apparent that χ_1^{crit} increases with increasing segmental concentration. Whereas $\chi_1^{\mathrm{crit}} \sim \frac{1}{2}$ for small values of the volume fraction, χ_1^{crit} exceeds $\frac{1}{2}$ for larger values. The latter would correspond to dispersion media of worse solvency than θ-solvents whilst the former corresponds to near θ-conditions. The occurrence of enhanced steric stabilization would thus be explained by an increase in the segment density in the steric barrier.

At first sight, it seems that an overwhelming argument against this proposal is that it should be universal: that is, it should predict that the CFPT even in the absence of multipoint anchoring should be a strong function of the segment volume fraction. As discussed in Section 6.5.1, this manifestly is not so. Closer scrutiny, however, leads to the recognition that the segment volume fraction in the peripheral regions of the steric barriers may be extremely low ($<0{\cdot}01$, say), irrespective of the molecular weight. If this is correct, the correlation between the CFPT and the θ-point would still stand. Only with multipoint attachment would the segment density in the steric barriers lead to $\chi^{\mathrm{crit}} > \frac{1}{2}$, i.e. to enhanced steric stabilization.

We note parenthetically that the absence of a molecular weight dependence of the CFPT with many dispersions is a strong argument against surface phase separation at the CFPT. Surface phase separation would be described by a closely analogous theory. It would lead inexorably to a relatively strong sensitivity of the CFPT to the molecular weight of the stabilizing moieties. It must be admitted, however, that χ_1 is often a reasonably strong monotonically increasing function of the segmental volume fraction. This would tend to dampen the sensitivity of the CFPT to the molecular weight.

9.5. Conclusions

There seems little doubt that it is possible to prepare sterically stabilized dispersions in worse than θ-solvents. The origins of this enhanced steric stabilization have yet to be firmly established. Very low segment densities in the peripheral regions of the steric layers concomitant with multipoint anchoring may be implicated. Other factors, such as elastic effects and changed thermodynamic parameters, cannot as yet be ruled out as possible contributors to enhanced steric stabilization.

Chapter 10

Stabilization by Attached Polymer: Theories of Steric Stabilization

10.1. Classification of theories

There have been several attempts to develop theories for predicting the interaction between sterically stabilized particles. Attacks on this problem to-date can be classified under two headings.

(i) *Ab initio theories.* These ambitious theories endeavour to predict the conformation of polymer molecules at the interface and, from the resulting segment density distribution functions, to predict the steric interaction. As we shall see, none of the theories published to-date is really a genuine *ab initio* theory.

(ii) *Pragmatic theories.* These recognize the difficulties associated with trying to develop a general theory describing the conformation of polymers at a crowded interface. They therefore take as their starting point the segment density distribution functions determined experimentally (e.g. by small angle neutron scattering). Alternatively, a mathematical form is assumed that is scaled to the measured barrier layer thickness. From this can be calculated the steric interaction.

Neither of these two approaches is very satisfactory. The conformation of a polymer molecule at an interface, which determines the segment density distribution function, is critically dependent upon a manifold of factors that include:

(i) the chemical constitution of the polymer;
(ii) the chemical nature and geometrical shape of the interface;
(iii) the chemical composition of the solvent;
(iv) the mode of attachment of the polymer chains to the surface;
(v) the surface density of the polymer molecules at the interface.

It is the subtle interaction of these diverse factors that determines the conformation of a polymer at an interface and thus the steric interaction in a particular system.

The quantitative prediction of steric interactions from first principles includes two classical problems in polymer science. The first is the conformational problem alluded to already. The second is the excluded volume problem. For stabilizing moieties at a crowded interface, this encompasses both intermolecular and intramolecular excluded volume. Given the unresolved difficulties that beset these two questions, and ignoring the problem of attempting to predict how a polymer molecule of a particular chemical composition interacts with a surface of a different composition in a given solvent, it is scarcely surprising that no satisfactory quantitative theory of steric stabilization has yet emerged. Nor is one in sight. But a start has been made on assembling the methodologies that will be required ultimately to generate a general theory.

One further difficulty not touched upon in the foregoing discussion is the absence of a truly quantitative theory describing polymer solution thermodynamics. Even a second generation theory, such as the equation-of-state theory, probably only represents a qualitative or, at best, a semi-quantitative theory of polymer solution thermodynamics (Casassa, 1976). In the absence of a fully quantitative theory, it seems justifiable to make do with the classical Flory-Huggins theory, provided that the cracks that have appeared in its superstructure are papered over. These include using the concentration dependent interaction parameter χ that is determined experimentally. Most of the theories of steric stabilization that have been developed to-date have unfortunately been based upon a concentration independent interaction parameter (see Table 10.1), although there are some exceptions (see, e.g. Evans and Napper, 1977).

10.2. General considerations of the steric interaction

10.2.1. The three domains of close approach

It is convenient from a conceptual viewpoint to delineate three domains of close approach for sterically stabilized colloidal particles (Evans and Napper, 1973a). These zones delimit domains in which different phenomenology is clearly recognizable. They are determined by the relative sizes of the spans of the attached polymer chains (assumed for simplicity to be monodisperse) and the separation between the particles. The *span* can be defined in this context as specifying the conformational average distance normal to the surface that the furthermost segment meanders from the surface. The span thus sets the absolute limit to the average conformational thickness of the steric layers.

The three domains of close approach can be envisaged with the aid of Fig. 10.1, which depicts two flat parallel plates coated by stabilizing moieties of span L_S and separated by a distance d.

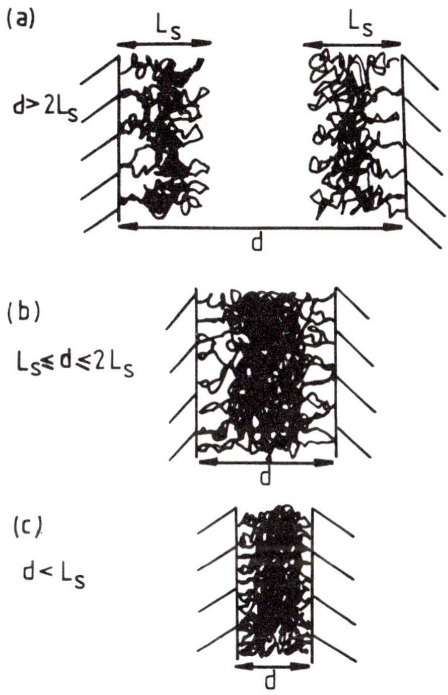

Fig. 10.1. Schematic representation of the three domains of close approach for sterically stabilized flat plates: (a) the noninterpenetrational domain ($d > 2L_S$); (b) the interpenetrational domain ($L_S \leq d \leq 2L_S$); (c) interpenetrational-plus-compressional domain ($d < L_S$).

10.2.1.1. The noninteractional domain ($d \geq 2L_S$)

Provided that the distance between the plates is greater than twice the span of the stabilizing moieties, no significant interaction can take place between the steric layers (see Fig. 10.1a). This ignores the trivial interaction that could occur for conformations that extend instantaneously beyond the average span. This regime, which is defined algebraically by $d \geq 2L_S$, might be termed the *noninteractional domain*. We ignore here any claims that the polymer chains can interact beyond their spans via perturbation of the solvent structure.

10.2.1.2. The interpenetrational domain ($L_S \leq d \leq 2L_S$)

If the distance between the plates is intermediate between one and two spans of the stabilizing moieties, the polymer chains attached to one plate can interpenetrate those attached to the opposite plate (see Fig. 10.1b). This regime, defined by $L_S \leq d \leq 2L_S$, is appropriately called the *interpenetrational domain*.

10.2.1.3. The interpenetrational-plus-compressional domain ($d < L_S$)

When the separation between the plates decreases to less than the span of the stabilizing moieties, the moieties on one surface are compressed by the opposing impenetrable surface. Interpenetration of the segments attached to one surface with the segments of the opposing compressed chains can still occur. This zone may therefore be termed the *interpenetrational-plus-compressional domain* (see Fig. 10.1c).

10.2.2. The components of the steric free energy

The preceding division of the region of close approach of two sterically stabilized particles into three domains leads quite naturally to a discussion of the steric interaction in terms of two basic components: the mixing free energy and the elastic free energy.

10.2.2.1. The mixing free energy

In the interpenetrational zone, the dominant process that generates the steric interaction is the increase in the segment density of the stabilizing moieties. This is effected at the expense of the molecules of the dispersion medium, which are excluded into the bulk continuous phase. It is apparent that in a good solvent, interpenetration, almost by definition, must be thermodynamically disfavoured. Conversely, in a worse than θ-solvent, interpenetration is thermodynamically favoured. Thus the interpenetrational free energy depends in a sensitive way upon the solvency of the dispersion medium. Interpenetration produces a demixing of polymer segments from molecules of the dispersion medium. Polymer solution theories are traditionally formulated in terms of the reverse of this process, i.e. mixing. It is therefore appropriate to call the overwhelming free energy process in the interpenetrational domain a *mixing free energy* ΔG_{MIX} (perhaps demixing would be more logical).

Hesselink *et al.* (1971) have preferred the designation *osmotic* for the major free energy change in the interpenetrational domain. This terminology accurately portrays the consequences of the steric interaction. The interaction of the steric layers generates a difference in the chemical potential between molecules of the dispersion medium in the interactional zone and those in the bulk dispersion medium. This chemical potential difference could, of course, be related to an excess osmotic pressure.

One prime difficulty confronts the use of the designation "osmotic" in this context: most, though perhaps not all, methods of imparting colloid stability generate a chemical potential difference between the molecules of the dispersion medium in the bulk continuous phase and those in the interactional zone. Langmuir (1938), for example, pointed this out in the case of electrostatic stabilization and the identical comment is appropriate to depletion stabilization. To eliminate this obvious ambiguity between various types of "osmotic" stabilization, we will follow the precedent established by Meier (1967) who was the first to apply Flory's original term "mixing" to steric stabilization.

Osmond *et al.* (1975) have claimed that there is a significant additional entropic contribution to the total steric free energy in the interpenetrational domain. This is said to spring from the loss of configurational entropy of the stabilizing moieties attached to one particle due to the physical presence of the *polymer chains* attached to the other. Of course, this term, which is in essence an elastic-type contribution, would vanish in a θ-solvent, unlike the genuine elastic contribution to be discussed in the next section. The latter arises from the compression of the stabilizing moieties by the opposing particle surfaces on very close approach.

Whilst Osmond *et al.* (1975) are strictly correct in drawing attention to the existence of an elastic free energy in the interpenetrational domain, the magnitude of this term is likely to be trivially small. The reason for this is that the segmental volume fraction for most polymer chains is usually quite small ($<0{\cdot}05$, say) and so the volume from which the incoming polymer chains are excluded is also relatively small. This conclusion is corroborated by the results of Dolan and Edwards (1975) that are reproduced in Fig. 10.2. Here the polymer segment density distribution functions both in the presence and

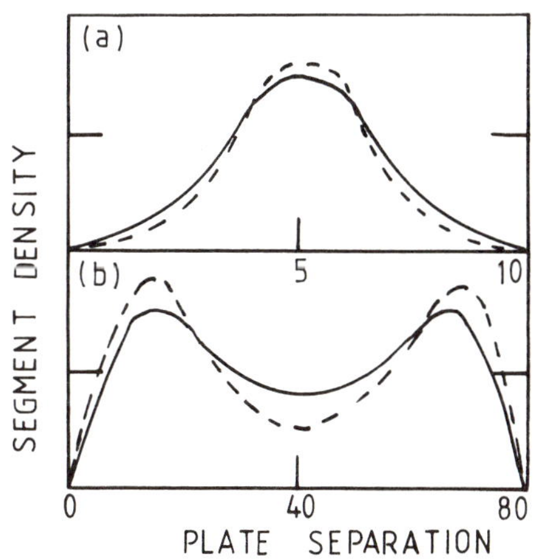

Fig. 10.2. The segment density distribution function for two flat plates stabilized by tails for an excluded volume parameter ($u=5$) according to Dolan and Edwards (1975). The dashed line is for zero excluded volume ($u=0$).

absence of excluded volume effects are displayed at two different distances of separation. These refer to very good solvency conditions with an intramolecular expansion factor $\alpha = 1{\cdot}3$. Such minor perturbations in the segment density distribution functions as arise from the presence of the opposing chains would probably not be measurable experimentally. The steric interaction in the interpenetrational domain is also relatively insensitive to such minor changes (Smitham and Napper, 1979).

We conclude that no serious error is inflicted upon the theoretical predictions in the interpenetrational domain by ignoring the minor conformational readjustments that the polymer chains undergo in the presence of the opposing chains. It follows that we may write without serious error

$$\Delta G_I = \Delta G_{\mathrm{MIX}} \qquad (10.1)$$

where the subscript I denotes the interpenetrational domain and MIX refers to the mixing term.

10.2.2.2. The elastic free energy

Once the separation between the particles becomes less than the span of the stabilizing moieties, the spatial extension of the chains normal to the particle surface is diminished by the compression imposed by the presence of the opposing surface. This loss of configurational entropy generates an interaction that is always repulsive, irrespective of the goodness of the solvency of the dispersion medium. In this sense, it differs dramatically from the mixing free energy, the sign of which is sensitive to the quality of the solvent. Of course, the mixing term is also operative in the interpenetrational-plus-compressional domain so that we may write

$$\Delta G_{I+C} = \Delta G_{\mathrm{MIX}} + \Delta G_{\mathrm{EL}} \qquad (10.2)$$

where the subscript $I+C$ denotes the interpenetrational-plus-compressional domain and the subscripts MIX and EL refer to the mixing and elastic contributions respectively.

This separation of the total free energy into an elastic contribution and a mixing contribution has been criticized by Osmond *et al.* (1975), Dolan and Edwards (1975) and Levine *et al.* (1978). Such criticism is fully justified in the sense that it is well-known that these two contributions are not independent: obviously, the elastic term must depend upon the conformation adopted and this will be determined *inter alia* by the mixing term.

In defence of the artifice under criticism, it can be argued that the uncertainty introduced by this artificial separation is significantly less than that stemming from the simulation of a real polymer chain by the primitive random flight model, which doesn't even recognize the necessity for such a separation. It should also be recalled that this separation was exploited by

Flory in his now classic theory of the intramolecular expansion factor α. Despite the expenditure of considerable effort, Flory's theory, or at least the Stockmayer (1960) modified version of it, has scarcely been improved upon in later years if comparison with experiment forms the judgemental basis (Yamakawa, 1971). In any event, the high segment densities often generated in the interpenetrational-plus-compressional domain are difficult to comprehend using the Flory-Huggins theory, especially if the interaction parameter is assumed to be concentration independent.

Both Meier (1967) and Hesselink *et al.* (1971) have referred to the elastic contribution as a *volume restriction* term. Whilst not declaiming this terminology, it needs to be pointed out that volume restriction in the interpenetrational-plus-compressional domain also influences the mixing of polymer segments with solvent and so contributes to the mixing term. In these circumstances, it seems preferable to retain in its entirety the original terminology used by Flory (1953), who favoured the terms "elastic" and "mixing".

10.2.3. Interpenetration and flocculation

The incipient flocculation behaviour of thermodynamically limited sterically stabilized dispersions, not subject to compression, can be understood primarily in terms of what happens in the interpenetrational domain. This greatly simplifies the interpretation of experiments of this type. Dispersions that display deviant flocculation behaviour may possibly require some consideration of the interpenetrational-plus-compressional domain. Certainly, dispersions that undergo compression at high solids content could well pass into that domain and require consideration of the elastic contribution to the total free energy change. The elastic contribution is also important in the macroscopic mica plate experiments, especially in θ- or worse than θ-solvents (Klein, 1982). In a θ-solvent, the primary repulsion (apart from that arising from the virial coefficient terms higher than the second) originates in the elastic contribution. Elastic effects also provide the dominant repulsion for sterically stabilized particles that are dispersed in melts of the stabilizing moieties. Under those conditions, the mixing term is negligible in all domains, leaving only the elastic contribution in the interpenetrational-plus-compressional domain to provide the repulsion so necessary for stability. Even if the dispersion medium is the melt of a polymer different from the stabilizing moieties, the mixing terms are still relatively small.

The interpenetrational-plus-compressional domain is also important in heterosterically stabilized dispersions where the different stabilizing sheaths are composed of compatible polymers. Elastic repulsion can give rise to heterosteric stabilization.

10.3. A general thermodynamic formalism for steric stabilization

A general thermodynamic formalism for steric stabilization has been presented by Everett (1978c). This is an elaboration of that set forth by Ash *et al.* (1973), which will be discussed in some detail in Section 17.5 in connection with the stabilization by polymers free in solution. As Everett (1978c) has pointed out, the procedure propounded by Ash *et al.* (1973) is inapplicable to steric stabilization because equilibrium is not maintained between adsorbed polymer and polymer in the bulk solution. An alternative approach must therefore be devised.

Consider two parallel flat plates coated by irreversibly attached stabilizing moieties. These identical plates, each of area A, are held apart at a constant distance h by an externally applied force J. This force is given a positive sign when the plates repel one another and a negative sign when they are attractive. It is related to the force per unit area f by $J = fA$. The system is capable of work of both the PdV and $Afdh$ types. Everett compares this system, denoted by I, with a reference system (II) that is identical except that it lacks the plates.

The total differential of the internal energy U of system I may be written

$$dU = TdS - PdV + \mu_1^\sigma dn_1 + \mu_2^\sigma dn_2 + 2\sigma dA - Afdh \qquad (10.3)$$

where μ = chemical potential, σ = interfacial tension between the liquid phase and the solid surface, n = amount of substance and the subscripts 1 and 2 refer respectively to the dispersion medium and the polymer. Equation (10.3) presumes that the contribution of the plates to U arises solely from surface effects.

The analogous equation for system II is

$$dU^\circ = TdS^\circ - PdV + \mu_1^\sigma dn_1^\circ + \mu_2^\sigma dn_2^\circ \qquad (10.4)$$

the superscript o denoting the reference system. The concentration of both the dispersion medium and the polymer in the reference system is taken to be uniform throughout its volume. System II, however, may contain different amounts of polymer and solvent due to their interaction with the surface.

Subtraction of equation (10.4) from equation (10.3) yields

$$d(U - U^\circ) = Td(S - S^\circ) + \mu_1^\sigma d(n_1 - n_1{}^\circ) + \mu_2^\sigma d(n_2 - n_2{}^\circ) + 2\sigma dA - Afdh \qquad (10.5)$$

Let the surface excess amounts, given by $n_i - n_i{}^\circ$ for $i = 1, 2$, in system I of the dispersion medium and polymer be $n_1{}^\sigma$ and $n_2{}^\sigma$ respectively. Then equation (10.5) can be recast into the form

$$d(U - U^\circ) = Td(S - S^\circ) + \mu_1^\sigma dn_1^\sigma + \mu_2^\sigma dn_2^\sigma + 2\sigma dA - Afdh \qquad (10.6)$$

Introducing the Gibbs-PV-free energy ($G = U + PV - TS$), we find that

$$d(G - G^\circ) = -(S - S^\circ)dT + \mu_1^\sigma dn_1^\sigma + \mu_2^\sigma dn_2^\sigma + 2\sigma dA - Afdh \qquad (10.7)$$

Cross-differentiation of equation (10.7) results in the relationships

$$\frac{\partial^2(G-G^o)}{\partial h \partial n_2^\sigma} = -A\left(\frac{\partial f}{\partial n_2^\sigma}\right)_h = \left(\frac{\partial \mu_2^\sigma}{\partial h}\right)_{n_2^\sigma} \tag{10.8}$$

Integration of these relationships yields

$$f - f^o = -(1/A)\int_{n_2^\sigma=0}^{n_2^\sigma}\left(\frac{\partial \mu_2^\sigma}{\partial h}\right)_{n_2^\sigma} dn_2^\sigma \tag{10.9}$$

which can be rewritten as

$$f - f^o = -2\int_{\Gamma_2=0}^{\Gamma_2}\left(\frac{\partial \mu_2^\sigma}{\partial h}\right)_{\Gamma_2} d\Gamma_2 \tag{10.10}$$

where $\Gamma_2(=n_2^\sigma/2A)$ is the surface concentration (a real excess) of polymer. If, alternatively, n_2^σ is expressed in terms of the surface excess amount of the polymer segments, μ_2^σ can be identified with the chemical potential of the segments. When the macromolecule assumes its equilibrium conformation, μ_2^σ must be uniform throughout the polymer.

Integration of equation (10.10) yields an expression for the difference in the potential energy of interaction in the presence and absence of polymer at a separation h

$$V_P(\Gamma_2, h) - V_P(o, h) = -\int_{\infty}^{h}(f - f^o)dh = 2\int_{\Gamma_2=0}^{\Gamma_2}[\mu_2^\sigma(h) - \mu_2^\sigma(\infty)]d\Gamma_2 \tag{10.11}$$

This formula shows that in general the steric repulsion between two flat plates is determined by the dependence of the segment chemical potential upon the distance of separation (h) and the segment surface excess amount (n_2^σ). In many theories of steric stabilization, it is tacitly assumed that the potential energy is a linear function of n_2^σ so that the integral becomes $[\mu_2^\sigma(h) - \mu_2^\sigma(\infty)]\Gamma_2$. This will not be true in general, however.

10.3.1. The constant segment density

Everett (1978) exemplified the application of equation (10.11) for two parallel plates by considering the case where the segment density in the steric layer, of span L_S, is constant. In those circumstances, the average volume fraction of the stabilizing segments ($\langle v_2^\sigma \rangle$) between the plates is given by

$$\langle v_2^\sigma \rangle = n_2^\sigma \bar{V}_2/Ah = 2\bar{V}_2\Gamma_2^s/h \tag{10.12}$$

where $\bar{V}_2$ = the segmental molar volume and the superscript s is a reminder that it is the segmental surface excess that is relevant here. The factor of 2 accounts for the presence of 2 identical plates.

The Flory-Huggins expression for segment chemical potential (Flory, 1953) for unattached polymer is

$$\mu_2^\sigma = \mu_2^{\sigma,\ominus} + (RT/x)\ln\langle v_2^\sigma\rangle - RT([x-1]/x)(1-\langle v_2^\sigma\rangle) + RT\,\chi_1(1-\langle v_2^\sigma\rangle)^2 \tag{10.13}$$

where χ_1 = interaction parameter, x = number of lattice sites occupied by each polymer chain and the superscript $\ominus$ denotes the standard state.

Inserting equation (10.12) into equation (10.13) gives to good approximation

$$\mu_2^\sigma = \mu_2^{\sigma,\ominus} + (RT/x)\ln(2\,\bar{V}_2\,\Gamma_2^s/h) - RT(1-2\,\bar{V}_2\,\Gamma_2^s/h) + RT\chi_1(1-2v_2\,\Gamma_2^s/h)^2 \tag{10.14}$$

This is the expression for the segmental chemical potential that Everett claimed was appropriate in order to evaluate equation (10.11).

Everett, following Evans and Napper (1973a), distinguished between three domains of close approach: (i) $h \geq 2L_S$, when no direct interaction between the steric layers is possible; (ii) $L_S \leq h < 2L_S$, when interpenetration of the two steric layers can occur; and (iii) $h < L_S$, when not only the interpenetration of the two steric layers can occur but also the compression of the layers by the close approach of the opposing plates takes place. Everett confined his attention to the interpenetrational domain (ii).

The change in the potential energy of interaction between the plates in moving from a separation of $2L_S$, when the interaction just commences, to one of h ($L_S \leq h < 2L_S$) is readily calculated from

$$[V(\Gamma_2^s,h) - V(o,h)] - [V(\Gamma_2^s,2L_S) - V(o,2L_S)] = 2\int_0^{\Gamma_2^s} [\mu_2(h) - \mu_2(2L_S)]\,\mathrm{d}\Gamma_2^s \tag{10.15}$$

This integral can be evaluated after substituting equation (10.14) into the integrand:

$$\begin{aligned} V_P(\Gamma,h) - V_P(o,h) = 2\int_0^{\Gamma_2^s} &\{(RT/x)\ln(2\,\bar{V}_2\Gamma_2^s/h) \\ &-(RT/x)\ln(2\,\bar{V}_2\Gamma_2^s/2L_S) - RT(1-2\,\bar{V}_2\Gamma_2^s/h) + RT(1-2\,\bar{V}_2\Gamma_2^s/2L_S) \\ &+ RT\chi_1(1-4\,\bar{V}_2\Gamma_2^s/h + 4\,\bar{V}_2^2[\Gamma_2^s/h]^2 \\ &- RT\chi_1(1-4\,\bar{V}_2\Gamma_2^s/2L_S + 4\,\bar{V}_2^2[\Gamma_2^s/2L_S)]^2)\}\mathrm{d}\Gamma_2^s \\ = -2RT\Gamma_2^P\ln(h/2L_S) &+ (\tfrac{1}{2}-\chi_1)RTx^2\,\bar{V}_2(2\Gamma_2^P)^2(1/h - 1/2L_S) \\ &+ (1/3)\chi_1 RTx^3\,\bar{V}_2^2(2\Gamma_2^P)^3(1/h^2 - 1/[2L_S]^2) \end{aligned} \tag{10.16}$$

where Γ_2^P ($=\Gamma_2^s/x$) refers to the whole polymer molecules. In solvents that are better than θ-solvents, all three terms are repulsive and contribute to stability. Note carefully that the potential is nonlinear in Γ_2^P.

10.3.1.1. The first virial coefficient term

The first term (J_1) in equation (10.16) has an osmotic character since its contribution to the force between the plates, obtained from $-(\partial J_1/\partial h)$, is $2\,RT\,\Gamma_2^P/h = RTc_2$, where c_2 = mean polymer concentration between plates. Dolan and Edwards (1974; 1975) obtained a similar term but, due to an algebraic error, they presented the term with the opposite sign, implying not a repulsive interaction but rather an attraction.

The presence of this osmotic term for fixed flat plates to which the stabilizing moieties are irreversibly attached is questionable. No comparable term appears in the theories of Meier (1967) and Hesselink, Vrij and Overbeek (1971) and for good reasons: this term is a consequence of using the segment chemical potential for unattached polymers, rather than the correct value for attached polymers. The term $(RT/x)\ln\langle v_2^\sigma\rangle$ in equation (10.13) arises from the $n_2^\sigma \ln\langle v_2^\sigma\rangle$ contribution to the parent free energy of mixing, as does the term associated with -1 in the next contribution to equation (10.13) [i.e. $+(RT/x)(1-\langle v_2^\sigma\rangle)$]. For high polymers, both terms are relatively unimportant to the totality of the free energy of mixing and the question of their validity is essentially an academic matter. Nevertheless, attachment of the polymer chains to the fixed flat plates eliminates the $n_2^\sigma \ln\langle v_2^\sigma\rangle$ term from the parent free energy of mixing term because this term expresses the configurational entropy associated with the movement of the centres of mass of the polymer chains. Flory (1953) has suggested the alternative, but entirely equivalent, viewpoint that $(RT/x)\ln\langle v_2^\sigma\rangle$ is negligible because it is as if x has tended to infinity. Whichever viewpoint is adopted, inclusion of this term for fixed plates appears improper.

The question of whether the same term is warranted for chains irreversibly attached to dispersed particles, which are capable of translational motion, is not so easily resolved. The centres of mass of the attached chains possess translational movement by virtue of the motion of the particles. However, the chains attached to one particle undergo correlated movement, quite different from their uncorrelated behaviour in free solution. Hence, even for dispersed particles, this term is unlikely to be correct but it is not immediately obvious how the configurational entropy of the sterically stabilized particles can be properly introduced. Finally, we note parenthetically that the presence of a term associated with the osmotic pressure of the dispersion medium may be appropriate at high compressions.

10.3.1.2. The higher order virial coefficient terms

The second term of equation (10.16) contains the thermodynamic factor $(\frac{1}{2}-\chi_1)$. Accordingly, it vanishes under θ-conditions when $\chi_1=\frac{1}{2}$. In a dispersion medium that is a better solvent for the stabilizing moieties than a θ-solvent, $\chi_1 < \frac{1}{2}$ and the second term contributes a repulsive interaction. When $\chi_1 > \frac{1}{2}$, i.e. for a worse than θ-solvent, this term will result in attraction. If an estimate of x in equation (10.16) is required, it may be approximated by the ratio of the molar volume of the polymer to that of the dispersion medium.

10.4. Models for polymer chains

Many of the putative *ab initio* theories of steric stabilization that have been advanced to-date have been cast using the random flight chain as the model for the polymer molecule (see Table 10.1). The random flight chain, also termed the freely jointed chain, is by far the most primitive model of a polymer molecule. Although incorporating the connectivity that distinguishes a macromolecule from a minimolecule, the random flight chain nonetheless ignores completely the precise details of this articulation inasmuch as it disregards the fixed valency angles and the barriers to free rotation about bonds. These structural details, which determine the architecture of the macromolecule, distinguish one polymer molecule from another.

Reviewing this type of approach, it can be stated that the most pedantic viewpoint would assert that the random flight chain is quite incapable of providing *quantitative* predictions about the behaviour of real polymer molecules, although it is well placed to elaborate the qualitative features of polymeric phenomena. A less restrictive viewpoint would allow the predictions of the random flight chain to be applied to real polymer molecules, subject either to appropriate scaling procedures or to an equivalent reduction procedure. Scaling in this context refers to the establishment of a 1:1 correspondence between a given number of real bonds and an equivalent number of freely jointed bonds. For this scaling procedure to be valid, it would have to be proven that the same correspondence embraces all properties of the polymer chain (e.g. if the rms end-to-end distance of the model chain is equated to that of the real chain in free solution then the steric repulsion generated by the two chains must necessarily be the same). No proof of this 1:1 correspondence between all the properties of a real polymer molecule of arbitrary length and those of an equivalent random flight chain has been forthcoming. Indeed, it is unlikely to be generally valid, although it is more likely to be applicable to the thermodynamic than the constitutive properties of a polymer.

A simpler approach to steric and depletion stabilization is to use the predictions of the random flight chain at the same reduced distance, say, $H_0/\langle r^2 \rangle^{\frac{1}{2}}$. The chain dimensions $\langle r^2 \rangle^{\frac{1}{2}}$ are used here as an arbitrary reduction parameter. The validity of this simple procedure can only be assessed after an exact theory has been elaborated. Unfortunately, such an

TABLE 10.1
Comparison of the various pseudo-*ab initio* theories of steric stabilization

Theory	Model	Solution theory	Interaction parameter	Separate elastic	Combined mixing + elastic
Mackor (1951)	inflexible rod	nil	—	elastic only	—
Clayfield and Lumb (1966–1974c)	random flight	nil	—	elastic only	—
Jäckel (1964)	gel layer	—	—	elastic only	—
Meier (1967)	random flight	Flory-Huggins	concentration independent	elastic + mixing	
Hesselink, Vrij and Overbeek (1971)	random flight	Flory-Huggins	concentration independent	elastic + mixing	
Dolan and Edwards (1974; 1975)	random flight	Flory-Huggins	concentration independent	—	combined
Gerber and Moore (1976; 1977)	random flight	Flory-Huggins	concentration independent	—	combined
Levine *et al.* (1978)	freely rotating	Flory-Huggins	concentration independent	—	combined

evaluation is not yet possible for steric stabilization in the absence of an exact theory.

The arbitrary scaling procedures set forth in the preceding paragraphs mean that none of the theories of steric stabilization advanced to-date are genuine *ab initio* theories. They make no endeavour to predict the actual dimensions of the chains, merely scaling to the experimentally observed dimensions. For this reason, we shall refer to such theories as being *pseudo-ab initio*.

The reason why the random flight model has proved so popular theoretically stems from its simplicity, which offers hope for the development of analytic solutions. The problem can usually be cast in the form of a diffusion-like or a Schrödinger-wave-equation-like differential equation, the solutions of which are reasonably well explored. A tendency has developed in recent times to apply extremely sophisticated mathematical procedures to what are really very primitive models for polymer chains (see, e.g. Levine *et al.*, 1978). Whether the ends merit the means in such instances cannot yet be assessed objectively. A strategy that might be more productive in terms of the development of a practical theory for steric stabilization is to aim for a simpler mathematical description of more complex models of polymer chains. It should also be borne in mind in developing *ab initio* theories that a simple model that may well suffice in polymer solution thermodynamics may be quite inadequate for the simulation of the conformational properties of polymers. Polymer solution thermodynamics seem to be relatively insensitive to molecular architecture *per se* whereas the conformation of a polymer chain is extremely sensitive to it.

10.5. The early theories of steric stabilization

10.5.1. The loss of configurational entropy theories

10.5.1.1. Mackor

Mackor (1951) was perhaps the first to endeavour to calculate the free energy of repulsion between sterically stabilized particles. This work was instigated after van der Waarden (1950; 1951) had shown experimentally that aromatic molecules with long-chain aliphatic substituents could have a profound effect on the stability of carbon black particles dispersed in a paraffin (see Section 2.4.2). For this reason, Mackor adopted a model in which he assumed that the aromatic nuclei were adsorbed onto the carbon black particles in a flat configuration, thus anchoring the alkyl chains to the surface. These chains were assumed to project into the dispersion medium and were modelled as rigid rods, of length L, flexibly attached to the particle surfaces by ball joints.

The repulsion was envisaged to arise solely from the loss of configurational entropy of the rods on close approach of the opposing surfaces (see Fig. 10.3). It is therefore possible to write for a separation h between the plates:

$$\Delta G_R = G_R(h) - G_R(\infty)$$
$$= -TS_R(h) + TS_R(\infty) \qquad (10.17)$$

where G = Gibbs free energy and S = entropy. The Boltzmann relationship, $S = k \ln W$, permits the loss of configurational entropy to be evaluated. The number of accessible configurations W of the rigid rod can be set proportional to the area on the surface of the hemisphere, centred on the ball joint and of radius L, that is accessible to the rods. If the particles, assumed for convenience to be flat parallel plates, are closer than the length of the rods, the area on the surface of the hemisphere that is accessible to the rods is simply $2\pi Lh$. The accessible area at infinite separation is $2\pi L^2$ and so

$$\Delta G_R = kT \ln\{W(\infty)/W(h)\}$$
$$= -kT \ln(h/L) \qquad (10.18)$$

For values of h not too far removed from L, the logarithm can be expanded to give:

$$\Delta G_R = kT(1 - h/L) \qquad (10.19)$$

where terms of the expansion of higher order than the first have been

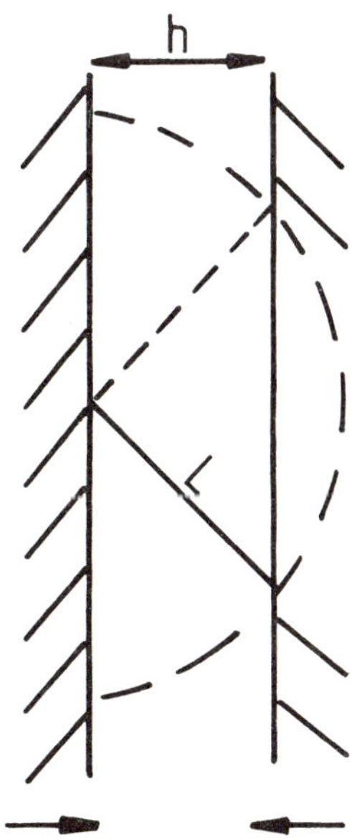

Fig. 10.3. Diagrammatic representation of the loss of configurational entropy that causes steric stabilization according to the theory of Mackor (1951).

neglected. This result is readily derived by writing $h/L = 1 - \varepsilon$, where $\varepsilon = (L-h)/L$. The value of ΔG_R given by equation (10.19) refers to just one molecule. For a single surface coated by $N_S\Theta$ stabilizing molecules per unit area, where Θ = fractional extent of coverage, the total repulsive free energy

$$\Delta G_R = N_S \Theta k T(1 - h/L) \tag{10.20}$$

This result, which Mackor derived by an alternative route, is only applicable at small Θ since all rod-rod encounters (involving terms in Θ^2 and higher) are ignored.

Of course, any theory that models even an alkyl chain, let alone a polymer molecule, as a rigid rod is foredoomed to quantitative failure. Nevertheless, this primitive model allowed Mackor to calculate crude potential energy *vs* distance curves. These demonstrated that, at least in principle, the loss of configurational entropy of the stabilizing chains could be sufficiently large to outweigh the van der Waals attraction and so lead to stability. Subsequently, Mackor and van der Waals (1952) extended the theory to higher surface coverages and dumbbell shaped molecules by using a quasi-lattice model for the liquid state. Interpenetration of the surface layers was thus incorporated into the theory.

Nowadays, apart from the assumed lack of flexibility of the stabilizer, an additional major criticism that would be levelled at the Mackor approach would be its complete neglect of the nature of the solvent, a quantity that has been shown experimentally to be of crucial importance.

Two results of the Mackor analysis, which is now merely of historic interest, still linger on today. The first is the misconception that the overall repulsion in steric stabilization is *always* the consequence of the loss of configurational entropy of the stabilizing moieties. If this were really true, no sterically stabilized dispersion could be flocculated by heating, which perforce favours entropic effects. Yet almost all sterically stabilized dispersions can be so flocculated. The second misconception is that the potential energy diagrams for sterically stabilized particles always resemble those of an electrostatically stabilized system in that they exhibit a primary maximum, which is what Mackor found. As we shall see, this is not generally correct.

10.5.1.2. Clayfield and Lumb

Clayfield and Lumb (1966a,b; 1968; 1974a,b,c) in an extensive series of papers carried out Monte Carlo simulations of polymer chains attached to the surfaces of particles and their effects on colloid stability. These numerical calculations were in the spirit of the Mackor approach. However, the flexibility of the polymer chains was now incorporated into the theory by simulating random flight chains. These chains were placed on a cubic lattice, connoting a valency angle of 90°. Both terminally attached homopolymers and random copolymers that gave a loopy type of adsorption were studied.

The results of Clayfield and Lumb relate entirely to the loss of configurational entropy of the polymer chains on close approach of the particles, due either to the presence of the impenetrable surface of the opposite particle or the polymer chains that are attached to that particle. In the early papers, the effect of the solvent on the conformation of the macromolecules was ignored but an attempt was made to include the role of solvency in some of the later publications. Notwithstanding this, essentially what Clayfield and Lumb calculated was the elastic contribution to the repulsive free energy of interaction between sterically stabilized particles. As such, their results are manifestly unable to explain the observed flocculation of sterically stabilized particles that is induced by decreasing the solvency of the dispersion medium. Even if only for this reason, the assertion by Osmond *et al.* (1975) that the Clayfield and Lumb theory was the best available at that time is clearly untenable.

Clayfield and Lumb (1974a) purported to show by simulation that polymer chains do not interpenetrate substantially and that the neglect of interpenetration of the stabilizing layers did not introduce serious errors into their theory of steric stabilization. Feigin (1980) has pointed out several serious flaws in their approach. The most important is that the volume fraction of segments in the steric sheaths was assumed by Clayfield and Lumb to be ca 0·5. This is an unrealistically high value, being perhaps an order of magnitude too large. Even a χ_1 value of close to 0·5 can correspond to an extremely good solvent at this segmental concentration. It might be added in passing that there is a substantial body of experimental evidence, reviewed in Section 12.2.3.2, which shows that, as predicted by scaling law theory, polymer interpenetration is favoured over the alternative possibility (*viz* compression) in the semi-dilute range of concentration.

10.5.1.3. The elastic modulus theory of Jäckel

Jäckel (1964) has proposed that protective layers around a colloidal particle can be considered to be similar to a cross-linked gel network and that the *mechanical* properties of these gel layers determine their protective action, rather than their physicochemical properties. If the modulus of the layers is sufficiently large, deformation in a Brownian collision (see Fig. 10.4) is such as to prevent the van der Waals attraction from coming into play. Any viscous drainage losses of molecules of the dispersion medium from the gel layer would weaken its elastic response and so reduce the repulsion, as would any viscous processes in the layer that dissipate collisional energy as heat energy.

Jäckel calculated the elastic energies of compression (${}^{S}V^{\mathrm{EL}}$) that would typically be generated in Brownian collisions between such gel layers from an equation originally developed by Hertz (1888):

$$ {}^{S}V^{\mathrm{EL}} = 0{\cdot}75E(\delta - H_o/2)^{5/2}(a+\delta) \qquad (10.21) $$

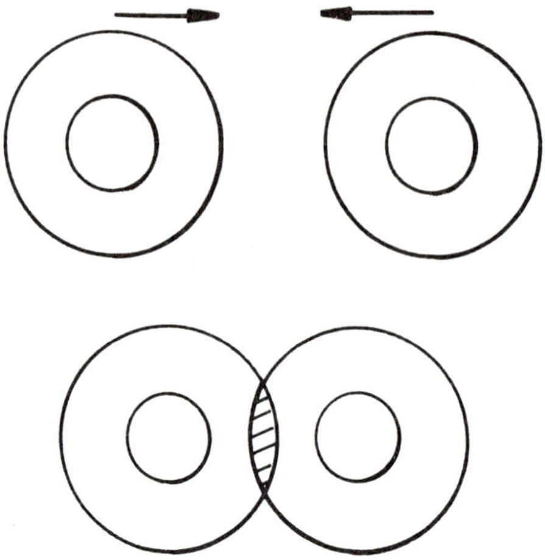

Fig. 10.4. Schematic representation of the close approach of two sterically stabilized particles (not to scale).

where E = elastic modulus, δ = steric layer thickness, H_o = minimum distance of surface separation so that $(\delta - H_o/2)$ is the barrier layer compression, and a = particle radius. These compressional energies were such that stability could be imparted. This is illustrated by Fig. 10.5, which was published by Sonntag (1974) ostensibly to model steric stabilization by poly(vinyl alcohol).

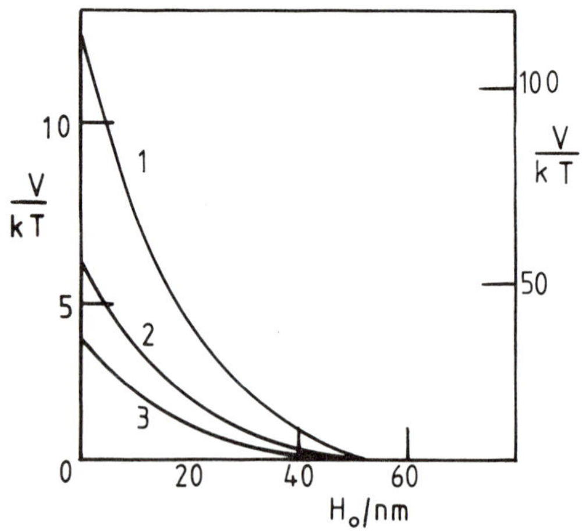

Fig. 10.5. The distance dependence of the potential energy of interaction of latex particles sterically stabilized by poly(vinyl alcohol) in water. The different particle radii were (1) 500 nm, (2) 100 nm and (3) 10 nm. The left-hand ordinate corresponds to an elastic modulus of $1{\cdot}4 \times 10^3\ \mathrm{N\,m^{-2}}$ whereas that of the right-hand side corresponds to $1{\cdot}2 \times 10^4\ \mathrm{N\,m^{-2}}$ (after Sonntag, 1974).

Actually, the elastic modulus measured by Sonntag may conceivably be a factor of 10 too small according to the measurements of Cain *et al.* (1978). Accordingly, the scaling on the right hand ordinate has been amended to present an alternative energy diagram.

Perhaps the strongest argument that can be marshalled against the gel-layer approach is that for many facets of steric stabilization, it can be demonstrated experimentally that it is the *physicochemical* properties of the steric layers that control stability, not the mechanical properties. There is also considerable doubt as to whether all polymer layers behave like cross-linked gels, although it might be anticipated that some atypical steric layers (e.g. poly(vinyl alcohol) moieties in water) would display gel-like behaviour (Cain *et al.*, 1978) (see also Israelachvili *et al.*, 1980). Jäckel's interesting approach may well be resurrected when the dynamic properties of sterically stabilized dispersions are investigated, e.g. by rheological measurements.

Chapter 11

Pseudo-'*Ab Initio*' Theories of Steric Stabilization

11.1. Introduction

The early theories of steric stabilization reviewed in the previous chapter were primitive and phenomenologically incomplete. Specifically, the quality of the solvency of the dispersion medium for the stabilizing moieties did not enter directly into these theories. This grave deficiency was remedied, however, in the second and later generation theories. In this chapter, we will consider the pseudo-*ab initio* theories of steric stabilization that incorporate the solvency of the dispersion medium.

11.2 The second generation '*Ab Initio*' theories

11.2.1. The germinal theory of Meier for isolated chains

The first successful application of the modern theories of polymer solution thermodynamics to steric stabilization was accomplished by Fisher (1958) (see Section 12.2.1) in terms of the mixing contribution to the free energy. Fisher, however, did not attempt, to predict the conformational properties of the stabilizing moieties and so his approach is not classified as an *ab initio* theory. The first pseudo-*ab initio* theory of steric stabilization, which set forth correctly the principal causes of the steric interaction, was published by Meier (1967). This clearly recognized for the first time the mixing and the elastic contributions to the steric free energy. The theory was couched in terms of a random flight chain that was terminally attached to a planar interface. The conformation of an isolated chain was predicted by solving the diffusion equation with the boundary conditions relevant to the parallel impenetrable plates.

11.2.1.1. The diffusion equation

As pointed out by Yamakawa (1971), the diffusion equation is readily derived from the Fokker-Planck equation. This equation, which is one of the cornerstones of the theory of transport processes, can be written

$$\frac{\partial W}{\partial n}\Delta n = -\ \nabla\cdot(\langle \mathbf{l}\rangle W) + (\tfrac{1}{2})\ \nabla\ \nabla:(\langle \mathbf{ll}\rangle W) \tag{11.1}$$

where W = the probability associated with the end-to-end distance of a chain of n links, each of length l, ∇ = the gradient ($\mathbf{i}\,\partial/\partial x + \mathbf{j}\,\partial/\partial y + \mathbf{k}\,\partial/\partial z$), and both of the moments $\langle \mathbf{l}\rangle$ and $\langle \mathbf{ll}\rangle$ are either proportional to Δn or vanish. The random flight chain is characterized by

$$\langle \mathbf{l}\rangle = 0 \tag{11.2}$$

which simply expresses the complete randomness of the chain. Moreover,

$$\langle \mathbf{ll}\rangle = (1/3)\langle l^2\rangle \mathbf{I} = (l^2/3)\mathbf{I} \tag{11.3}$$

where $\mathbf{I}$ = unit tensor. For sufficiently large values, n may be taken to be a continuous variable. Substitution of these relationships, together with $\Delta n = 1$, into equation (11.1) yields

$$\frac{\partial W}{\partial n} = (l^2/6)\ \nabla^2 W \tag{11.4}$$

This is the diffusion equation as used by Meier and later workers.

11.2.1.2. The free energy contributions

Meier (1967) calculated the segment density distribution functions for polymer chains attached to parallel flat plates. Even for one plate, the chains extended from the plate to infinity. This, of course, is completely unphysical. It must be said, however, that the values predicted for the segment densities beyond the contour length of the chain are minuscule. This mathematical artifact is not trivial because it meant that Meier did not make any distinction between the various domains of close approach. At all distances of separation, both an elastic and a mixing contribution were deemed to be present in the steric interaction. Of course, Meier made no allowance in his calculation for the excluded volume effect arising from the chains attached to the opposite surface, only for the compression resulting from the opposite surface itself.

The mixing free energy

The mixing free energy change on close approach can be calculated as detailed below. This derivation deviates significantly from that of Meier but it is entirely equivalent.

The Gibbs free energy of mixing of polymer segments and solvent in a volume element δV according to the Flory-Huggins theory may be written as

$$\delta(\Delta G) = \delta(\Delta G^*_{\mathrm{MIX}}) + \delta(\Delta G_{\mathrm{SP}}) \tag{11.5}$$

where $\Delta G^*_{\mathrm{MIX}}$ = free energy of mixing calculated solely from external considerations of mixing segments and solvent, and ΔG_{SP} = free energy change due to all other sources, including specific interactions between neighbours. The first term in equation (11.5) can be evaluated (see Section 3.2.1) as

$$\delta(\Delta G^*_{\mathrm{MIX}}) = kT(\delta n_1 \ln v_1 + \delta n_2 \ln v_2) \tag{11.6}$$

where δn_1, δn_2 = number of molecules of the dispersion medium (1) and polymer (2) respectively, and v = volume fraction.

The second term in equation (11.6) derives from the configurational entropy of the centres of mass of the polymer molecules. This term is always relatively small but attachment renders it negligible. It is as if the stabilizing moieties were of infinite molecular weight. Hence

$$\delta(\Delta G^*_{\mathrm{MIX}}) = kT\delta n_1 \ln v_1 \tag{11.7}$$

According to the extended Flory-Huggins theory, the second term in equation (11.5) is given (see Section 3.2.2) by

$$\delta(\Delta G_{\mathrm{SP}}) = kT\chi_1(\delta n_1)v_2 \tag{11.8}$$

Here χ_1 is the interaction parameter measured experimentally at volume fraction v_2 such that $kT\chi_1$ represents the excess free energy associated with the transfer of one solvent molecule from pure polymer to a polymer solution at volume fraction v_2.

The foregoing considerations yield

$$\begin{aligned}\delta(\Delta G_{\mathrm{MIX}}) &= kT(\ln v_1 + \chi_1 v_2)\delta n_1 \\ &= kT(\ln[1 - v_2] + \chi_1 v_2)\delta n_1\end{aligned} \tag{11.9}$$

Since $v_2 = \rho_2 V_S$, where ρ_2 = number of polymer segments per unit volume and V_S = volume of a segment, and

$$\delta n_1 = \frac{(1 - \rho_2 V_S)\delta V}{V_1} \tag{11.10}$$

it follows from equation (11.9) that

$$\begin{aligned}\delta(\Delta G) = kT(V_S{}^2/V_1)\,(\tfrac{1}{2} - \chi_1)\rho_2{}^2\delta V \\ - kT(V_S/V_1)\,(\tfrac{1}{2} - \chi_1)\rho_2\delta V \\ - \tfrac{1}{2}kT(V_S/V_1)\rho_2\delta V\end{aligned} \tag{11.11}$$

where V_1 = volume of a molecule of the dispersion medium. Note that the expansion of the logarithmic term in equation (11.9) was truncated at the quadratic term to obtain this result.

Equation (11.11) is the basic dissolution equation for amorphous polymer chains attached to the particles. The origins of the three terms are readily identified. The third term on the right hand side represents the dissolution of randomly oriented attached segments in a θ-solvent. The middle term is the free energy change engendered by changing the quality of the solvent from that of a θ-solvent to a better solvent with $\chi_1 < \frac{1}{2}$, whilst retaining the quasi-random coil θ-solvent conformation. The remaining term formally allows for the changes in the polymer conformation on changing the solvent to a better than θ-solvent.

To find ΔG^M, the mixing free energy change for the approach of two sterically stabilized particles from infinite separation to a separation d, the total mixing free energy of the initial state ($\Delta G(\infty)$) must be subtracted from that in the final state $\Delta G(d)$:

$$\Delta G^M = \Delta G(d) - \Delta G(\infty) \tag{11.12}$$

This is accomplished by integrating equation (11.11), term by term, over the entire available space. In so doing, any linear term of the type $\int_V \rho_2 \mathrm{d}V$ must be constant, independent of the separational distance, since it represents the total number of segments present. The result is that on taking the difference required by equation (11.12), all terms in equation (11.11) that are linear in ρ_2 vanish and only the squared terms remain

$$\Delta G^M = kT(V_S{}^2/V_1)\,(\tfrac{1}{2} - \chi_1)[\textstyle\int_V \rho_2{}^2(d)\mathrm{d}V - \int_V \rho_2{}^2(\infty)\mathrm{d}V] \tag{11.13}$$

This is the basic equation that is used to derive the mixing contribution to the steric interaction. To simplify the formalism of the volume integrals, we will use the shorthand notation whereby $\rho_2(d) = \rho_d$.

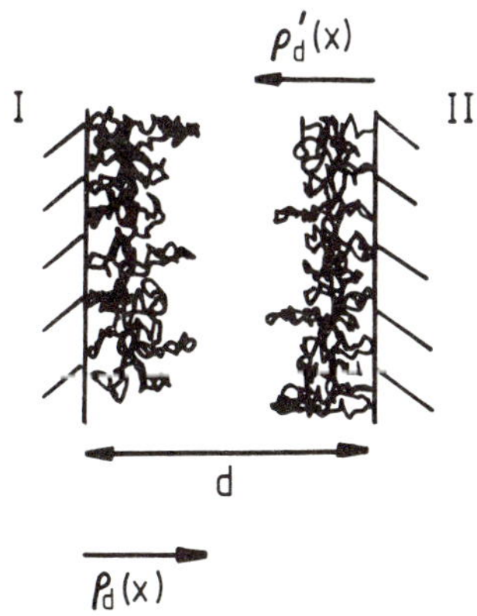

Fig. 11.1. Schematic representation of two sterically stabilized flat plates separated by a distance d.

Flat plates. The simplest geometry to handle in this context is that of two parallel flat plates that are coated by identical steric layers. Because of its ubiquity in practical dispersion systems, we will assume that the surface coverage is high. This means that the segment density parallel to the flat plates is constant to good approximation. The volume integrations in equation (11.13) are then transformed into integrals normal to the surface. Suppose $\rho_d(x)$ is the segment density at a distance x from the surface for the polymer attached to plate I and $\rho_d'(x)$ is the corresponding quantity for the polymer attached to plate II. We note parenthetically that $\rho_d'(x)=\rho(d-x)$. This allows equation (11.13) to be recast to read

$$\Delta^{FP}G^M=kT(V_S^2/V_1)\,(\tfrac{1}{2}-\chi_1)\left[\int_0^d(\rho_d+\rho_d')^2\mathrm{d}x-\int_0^\infty(\rho_\infty+\rho_\infty')^2\mathrm{d}x\right] \qquad (11.14)$$

where the superscript FP denotes a flat plate potential.

It is convenient to rewrite equation (11.14) in terms of the normalized segment density $\hat{\rho}_d$, which is proportional to ρ_d but normalized such that $\int_0^d\hat{\rho}_d\mathrm{d}x=1$. To find the value of $\Delta^{FP}G^M$ per unit area of plate, we recall that $\int_0^d\rho_d\mathrm{d}x=\upsilon i$, where $\upsilon=$the number of polymer chains per unit area of surface and $i=$the number of solvent sized segments in each polymer molecule (assumed for convenience to be monodisperse). It follows that $\rho_d=\upsilon i\hat{\rho}_d$ and $\rho_d'=\upsilon i\hat{\rho}_d'$. Equation (11.14), which gives the free energy per unit area, can then be written as

$$\Delta^{FP}G^M=2kT(V_S^2/V_1)\,(\tfrac{1}{2}-\chi_1)\upsilon^2 i^2 P^M \qquad (11.15)$$

where the segment density integral for two flat plates (P^M) is given by

$$P^M=(\tfrac{1}{2})\left[\int_0^d(\hat{\rho}_d+\hat{\rho}_d')^2\mathrm{d}x-\int_0^\infty(\hat{\rho}_\infty+\hat{\rho}_\infty')^2\mathrm{d}x\right] \qquad (11.16)$$

The evaluation of the mixing free energy at a given plate separation thus boils down to the evaluation of P^M. Note that the steric interactional free energy according to equation (11.15) can be written as the product of a thermodynamic term ($\mathcal{T}$), incorporating the factor $(\tfrac{1}{2}-\chi_1)$, and a segment density term P^M

$$\Delta^{FP}G^M=\mathcal{T}P^M \qquad (11.17)$$

Other particle geometries sometimes lead to

$$\Delta G^M=\Gamma\mathcal{T}S \qquad (11.18)$$

where $\Gamma=$a geometrical factor and $S=$the analogue of P for the new geometry. In general, however, the particle geometry will influence the segment density distribution function and so Γ and S will not be readily separable in this simple fashion, except perhaps in limiting cases.

A. The interpenetrational domain. If $L_S\leq d\leq 2L_S$, the polymer chains interpenetrate but are not compressed by the opposing surface. The expression for the segment density integral becomes

$$P^M=\tfrac{1}{2}\left[\int_0^d(\hat{\rho}_d^2+2\hat{\rho}_d\hat{\rho}_d'+\hat{\rho}_d'^2)\mathrm{d}x-\int_0^\infty(\hat{\rho}_\infty^2+2\hat{\rho}_\infty\hat{\rho}_\infty'+\hat{\rho}_\infty'^2)\mathrm{d}x\right] \qquad (11.19)$$

The term $\hat{\rho}_\infty\hat{\rho}_\infty'$ must be zero since at infinite separation the steric layers on the opposing surfaces are quite unable to interact. This simplifies P^M to

$$P^M = \tfrac{1}{2}\left[\int_0^d (\hat{\rho}_d{}^2 + 2\hat{\rho}_d\hat{\rho}'_d + \hat{\rho}'^2_d)\mathrm{d}x - \int_0^\infty (\hat{\rho}_\infty{}^2 + \hat{\rho}'_\infty{}^2)\mathrm{d}x\right] \tag{11.20}$$

which is generally valid. For the interpenetrational domain, the segment density at a separation d is identical with that at infinite separation (ignoring the small intermolecular excluded volume perturbation). This simplifies the expression for P^M to

$$P_I^M = \int_0^d \hat{\rho}_d\hat{\rho}'_d \,\mathrm{d}x \tag{11.21}$$

where the subscript I denotes interpenetration. The logical sequence in the development of this relationship should be carefully noted. It is not possible to set $\hat{\rho}_d = \hat{\rho}_\infty$ and $\hat{\rho}'_d = \hat{\rho}'_\infty$ in the original equation (11.16) since P_I^M would then be zero. This is manifestly incorrect. Paradoxically, it is possible to set $\hat{\rho}_d = \hat{\rho}_\infty$ and $\hat{\rho}'_d = \hat{\rho}'_\infty$ in the final expression for P_I^M.

B. The interpenetrational-plus-compressional domain. For $d < L_S$, the steric layers are compressed and $\hat{\rho}_d \neq \hat{\rho}_\infty$. The simplifying relationships in this domain are $\int_0^d \hat{\rho}_d{}^2 \mathrm{d}x = \int_0^d \hat{\rho}'^2_d \mathrm{d}x$ and $\int_0^\infty \hat{\rho}^2_\infty \mathrm{d}x = \int_0^\infty \hat{\rho}'_\infty{}^2 \mathrm{d}x$. These equalities are demanded by symmetry since the steric interaction must be independent of which of the plates we designate I and II. Accordingly, we find in this domain, denoted by the subscript $I+C$

$$P_{I+C}^M = \int_0^d \hat{\rho}_d{}^2 \mathrm{d}x - \int_0^\infty \hat{\rho}_\infty{}^2 \mathrm{d}x + \int_0^d \hat{\rho}_d\hat{\rho}'_d \mathrm{d}x \tag{11.22}$$

Note that P_{I+C}^M must always be positive.

The three terms in equation (11.22) are readily identifiable. The first two terms can be roughly considered to be associated with the *intramolecular* mixing of segments with solvent at the two different separational distances. The terms are, of course, different because of the compression of the steric layer on one plate by the close approach of the surface of the opposing plate. Of course, if $d > L_S$, these two terms are identical and cancel one another. The third term is the *intermolecular* interpenetrational term at the separation d, the corresponding term for infinite separation being identically equal to zero.

The elastic free energy

In the interpenetrational-plus-compressional domain, the mixing free energy calculated as described above must be supplemented by the elastic free energy. Meier (1967) expressed this by the Boltzmann entropy relationship

$$\Delta G^{EL} = -T\Delta S^{EL} = -2\upsilon kT \ln P_n(d) \tag{11.23}$$

where $P_n(d)$ = the probability that all chain elements are within a distance d of the surface to which they are attached when the other surface is at infinity. Note that the factor of 2 allows for the two surfaces. Alternatively, we may write

$$\Delta G^{EL} = -2\upsilon kT \ln \{\Omega(d)/\Omega(\infty)\} \tag{11.24}$$

where $\Omega(d)$ = the configurational thermodynamic probability at a particle separation d.

Meier solved the diffusion equation with the appropriate boundary conditions. Curiously, as first pointed out by Di Marzio (1965), this corresponds to the placement of *adsorbing* barriers at $x=0$ and $x=d$, even though the physical surface at $x=0$ corresponds to an impenetrable reflective surface. Meier obtained rather unwieldy expressions for the segment density distribution functions, which will not be reproduced here. Results were, however, obtained for both low and high surface coverages.

Hesselink (1969) has argued that Meier's derivation for tails incorporates a procedural error that invalidates his quantitative results for the mixing free energy. Meier formulated a probability distribution function $W_k(x,d)$, which describes the probability that the terminal (kth) bond of the tail lies at a distance in the range x to $x+dx$ from the surface when the plate separation is d. The segment density distribution function was then evaluated by summing the end-to-end probability functions for all subchains of k bonds over the total number of bonds n

$$\rho_d = \sum_{k=1}^{n} W_k(x,d) \tag{11.25}$$

The W_k were calculated by Meier assuming that all segments were contained within the plates.

Hesselink has argued that whilst this procedure ensures that bonds 1 to k do not transgress the interfaces, it places no restrictions on the bonds $(k+1)$ to n. The summation in equation (11.25) thus includes some conformations that are forbidden by the presence of the impenetrable barriers. Since the smaller is k, the more likely are the subsequent segments to cross the plates, the counting of these forbidden conformations should lead to a segment distribution that is biased towards conformations close to the interface, i.e. the segment distribution is compressed too close to the interface. The result of this artifact is that the Meier theory predicts values for the interaction between plates that are usually ca 20–30% too low.

11.2.2. Hesselink, Vrij and Overbeek's extension of the Meier theory

Hesselink *et al.* (1971) have corrected the error that crept into Meier's theory for steric stabilization by isolated identical tails. They have also extended the theory to encompass stabilization by identical loops, and by homopolymers where every segment of the chain has *a priori* an equal chance of being adsorbed at the interface. Copolymers with anchor groups randomly distributed along the chains have also been treated. The following discussion will be restricted to loops and tails. This limitation is adopted because homopolymers are not very good steric stabilizers, unless they can interact strongly with well-spaced adsorption sites on the particle surface. In the absence of such sites, homopolymers place conflicting demands on the solvency of the dispersion medium: the dispersion medium must be a poor solvent to generate good anchoring but a good solvent to impart robust stability.

Hesselink *et al.* based their theory on random flight statistics appropriate for a six-choice cubic lattice (Hesselink, 1969; 1971). In the absence of a second interface, the segment density of the isolated chain was calculated much in the spirit of the Meier approach but using a procedure that correctly eliminates the conformations that transgress the barrier (Hoeve, 1965; Hoeve *et al.*, 1965). This method was subsequently extended to allow for the presence of the second impenetrable interface (Hesselink, 1971).

11.2.2.1. Total free energy

The total free energy of steric interaction ($\Delta^{FP}G^{TOT}$) obtained by Hesselink *et al.* for two parallel flat plates coated by steric layers is

$$\Delta^{FP}G^{TOT} = 2(2\pi/9)^{3/2}\upsilon^2 kT(\alpha^2-1)\langle r^2\rangle M(d) + 2\upsilon kTV(d) \qquad (11.26)$$

Here the first term is the mixing free energy and the second term is the elastic free energy. Note carefully that a crude scaling assumption is implicit in any specific practical application of equation (11.26). The random flight dimensions $(nl^2)^{\frac{1}{2}}$ were equated to the experimentally determined rms end-to-end distance of the chain in free solution $\langle r^2\rangle^{\frac{1}{2}}$. The expansion factor α was taken into account by using $\langle r^2\rangle^{\frac{1}{2}} = \alpha\langle r^2\rangle_0^{\frac{1}{2}}$, the subscript zero referring to the unperturbed dimensions. It should be recognized that there is as yet no theoretical justification for this arbitrary scaling procedure. Its necessity transforms an *ab initio* theory into a pragmatic theory whose validity remains to be established, either by experiment and/or by theory. Note, too, that Flory's unmodified expression for the expansion factor α was used in this derivation:

$$\alpha^5 - \alpha^3 = \frac{(9/2\pi)^{3/2} V_S{}^2(\frac{1}{2}-\chi_1)}{V_1(nl^2)_0{}^{3/2}} \qquad (11.27)$$

This is known to be too large by a factor of ca 2, at least near to the θ-point (Stockmayer, 1960).

The segment density distribution functions

The mixing free energy term depends critically upon the mode of attachment since this determines the segment density distribution functions. For identical isolated tails, Hesselink *et al.* derived the following expression

$$\hat{\rho}_d(x)=3/(nl^2R_1)\left\{\sum_{m=-\infty}^{+\infty}\left[m\int_0^d(\exp\{-q(md-d+x-b/2)^2\}-\right.\right.$$

$$2\exp\{-q(md-b/2)^2\}+\exp\{-q(md+d-x-b/2)^2\})db$$

$$\left.+\int_x^d(\exp\{-q(md+b/2)^2\}-\exp\{-q(md+x-b/2)^2\})db\right] \quad (11.28)$$

In this equation,

$$R_1=\sum_{m=-\infty}^{\infty}[\exp\{-6m^2d^2/nl^2\}-\exp\{-3(2m+1)^2d^2/2nl^2\}] \quad (11.29)$$

and $q=6/(nl^2)=1/\langle s^2\rangle$, where $\langle s^2\rangle$ = rms radius of gyration of the tails in free solution. The analogous result for loops is

$$\hat{\rho}_d(x)=12/(nl^2R_2)\sum_{m=-\infty}^{\infty}[(m+1)\,(md+x)\exp\{-q(md+x)^2\}-$$

$$m^2d\exp\{-qm^2d^2\}] \quad (11.30)$$

where

$$R_2=\sum_{m=-\infty}^{\infty}(1-12m^2d^2/nl^2)\exp\{-6m^2d^2/nl^2\} \quad (11.31)$$

Homopolymers yielded

$$\hat{\rho}_d(x)=bR_3[\exp\{-xb/2\}-\exp\{-bd+bx/2\}]^2/(1-\exp\{-bd\})^2 \quad (11.32)$$

with

$$R_3=[1-\exp\{-bd\}]^2/[1-2bd\exp\{-bd\}-\exp\{-2bd\}] \quad (11.33)$$

where $b=2(6)^{\frac{1}{2}}\,t/nl$ and $t\sim 0{\cdot}7$. Note that for $d\rightarrow\infty$, $R_3\rightarrow 1$ and $\hat{\rho}_d(x)$ decreases exponentially with x.

In the interpenetrational domain (specified by $d\geq(nl^2)^{\frac{1}{2}}$), the normalized segment densities can be approximated by the following analytical expressions (Hesselink, 1977)

$$\textit{tails}:\ \hat{\rho}_d(x)=(24/5\langle r^2\rangle)x\exp\{-12x^2/5\langle r^2\rangle\} \quad (11.34)$$

and

$$\textit{loops}:\ \hat{\rho}_d(x)=(12/\langle r^2\rangle)x\exp\{-6x^2/\langle r^2\rangle\} \quad (11.35)$$

The distance dependent mixing functions

Flat plates. The foregoing segment density distribution functions allow the distance dependent mixing function $M(d)$ in the expression for the steric interactional free energy [equation (11.26) to be evaluated numerically using equations (11.28) and (11.30)]. The values obtained by Hesselink *et al.* (1970) as a function of the reduced distance $d/(nl^2)^{\frac{1}{2}}$ are listed in Table 11.1.

Again it is possible to obtain analytic expressions if $d > (nl^2)^{\frac{1}{2}}$

$$\textit{equal tails}: M(d) \sim (6\pi/5)^{\frac{1}{2}}(4\langle r^2\rangle/15d^2 - 1)\exp\{-6d^2/5\langle r^2\rangle\} \quad (11.36)$$

and

$$\textit{equal loops}: M(d) \sim (3\pi)^{\frac{1}{2}}(6d^2/\langle r^2\rangle - 1)\exp\{-3d^2/\langle r^2\rangle\} \quad (11.37)$$

Spheres. One advantage of these analytical formulae is that they allow simple analytic results to be established for spheres by means of the Derjaguin (1934) integration procedure. This permits flat plate potentials to be converted into sphere potentials. The mixing free energy for spheres obtained in this fashion is

$$\Delta^S G^M = [(2\pi^{5/2}/27]\upsilon^2 kTa\langle r^2\rangle^{3/2}(\alpha^2 - 1)S^M \quad (11.38)$$

where a = particle radius and

$$S^M = \int_{d/\langle r^2\rangle^{\frac{1}{2}}}^{\infty} M(d)dd$$

Some typical values of S^M obtained in this fashion are presented in Table 11.2. These were calculated using the following expressions for the distance dependent S^M function

$$\textit{equal tails}: S_T^M = (1{\cdot}2\pi d^2/\langle r^2\rangle)^{\frac{1}{2}}\exp\{-1{\cdot}2d^2/\langle r^2\rangle\} \quad (11.39)$$

and

$$\textit{equal loops}: S_L^M = (3\pi d^2/\langle r^2\rangle)^{\frac{1}{2}}\exp\{-3d^2/\langle r^2\rangle\} \quad (11.40)$$

TABLE 11.1
Values for the distance dependent functions $M(d)$ and $V(d)$ for flat plates

$d/(nl^2)^{\frac{1}{2}}$	$M(d)$ Tails	$M(d)$ Loops	$V(d)$ Tails	$V(d)$ Loops
0·6	3·72	2·97	3·00	1·48
0·8	2·40	1·72	1·28	0·34
1·0	1·59	0·84	0·58	0·06
1·2	1·04	0·31	0·26	
1·4	0·67	0·09	0·11	
1·6	0·41	0·02	0·04	
1·8	0·23		0·02	
2·0	0·13			
2·5	0·02			

TABLE 11.2
The distance dependent functions for tails and loops attached to spheres

$d/(nl^2)^{\frac{1}{2}}$	S_T^M	S_L^M
1·0		0·153
1·1		0·0895
1·2	0·416	0·0490
1·3		0·0251
1·4	0·249	0·0120
1·5		0·0054
1·6	0·141	0·0023
1·7		
1·8	0·075	
2·0	0·038	
2·2	0·018	
2·4	0·007	

The elastic free energy
Hesselink *et al.* have also evaluated the distance dependence function $V(d)$ [see equation (11.26)] for the elastic free energy contribution to steric stabilization. The result obtained for tails is the same as that published by Meier (1967)

$$V(d) = -\ln \sum_{m=-\infty}^{\infty} [\exp\{-6m^2d^2/nl^2\} - \exp\{-3(2m-1)^2d^2/2nl^2\}] \tag{11.41}$$

The result for identical loops is

$$V(d) = -\ln \sum_{m=-\infty}^{\infty} [(1-12m^2d^2/nl^2)\exp\{-6m^2d^2/nl^2\}] \tag{11.42}$$

Values obtained for $V(d)$ by numerical evaluation of these formulae have already been presented in Table 11.1.

The foregoing expressions reduce to simple analytic forms for $d \gtrsim (nl^2)^{\frac{1}{2}}$ (i.e. in the interpenetrational domain). Only the leading terms, corresponding to $m=0, \pm 1$, need be considered as the summations converge rapidly. Expansion of the logarithms then yields for tails the simple exponential form

$$V(d) \sim 2\exp\{-3d^2/nl^2\} \tag{11.43}$$

whereas for loops

$$V(d) \sim 2(12d^2/nl^2 - 1)\exp\{-6d^2/nl^2\} \tag{11.44}$$

Again a Derjaguin integration can be used to evaluate these expressions for the elastic repulsion for spheres. As demanded by physical considerations, their values in the interpenetrational domain are usually negligible compared with the mixing free energy and so that exercise is scarcely warranted.

11.3. The potential energy curves for sterically stabilized dispersions

11.3.1. Good solvency conditions

Typical potential energy curves calculated by Hesselink *et al.* for the steric interaction between parallel flat plates in a good solvent for the polymer ($\alpha = 1 \cdot 2$) are displayed in Fig. 11.2. These include the van der Waals attraction between the opposing plates. What is interesting about these curves is that they are radically different from those calculated for electrostatically stabilized dispersions, the typical shape of which is shown in the insert to Fig. 11.2. Whereas electrostatically stabilized dispersions usually display a primary maximum and a primary and, sometimes, a secondary minimum, sterically stabilized dispersions merely display a *pseudo-secondary minimum.* The minimum is termed "pseudo-secondary" because there is usually no primary minimum. The depth of this minimum, which in principle must always be present for steric layers of finite extent, is negligibly small compared with thermal energy for stable dispersions. The steric repulsion at the larger distances of separation (i.e. in the interpenetrational domain) arises from the mixing interaction. The elastic term only becomes important on closer approach (i.e. in the interpenetrational-plus-compressional domain) when its magnitude increases sharply.

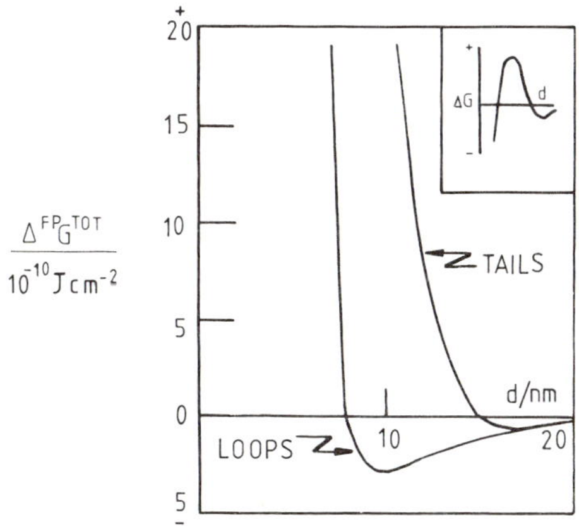

Fig. 11.2. The potential energy diagrams for two flat plates sterically stabilized by tails and loops. Inset shows the potential energy diagram for electrostatically stabilized particles (after Hesselink *et al.*, 1971).

Hesselink *et al.* concluded that the main parameters that determine the total free energy of steric interaction are:

(i) the number (and hence mass) density of the stabilizing species;
(ii) the molecular weight of the stabilizing moieties;
(iii) the mode of attachment of the stabilizing moieties to the particle surface;
(iv) the intrinsic quality of the solvency that the dispersion medium displays towards the stabilizing moieties;
(v) the particle geometry;
(vi) the net London dispersion attraction between the particles.

11.3.2. The effect of the solvency of the dispersion medium on stability

Figure 11.3 presents the original theoretical predictions of Hesselink *et al.* for the effects on the boundary between stability and instability of altering the solvency of the dispersion medium at different molecular weights. The change in the solvency is manifest by the varying value of the intramolecular expansion factor α (see curve 1). The larger the value of α, the better is the solvency of the dispersion medium. Note that these calculations were performed for an extremely low number density of stabilizing chains. This led Hesselink *et al.* to the conclusion that the θ-solvent (corresponding to $\alpha = 1{\cdot}0$) has no special significance with regard to incipient instability. Stability was predicted in dispersion media that were both better and worse than θ-solvents. Moreover, incipient instability was predicted to be a strong function of the molecular weight of the stabilizing moieties.

The physical bases of these theoretical predictions are quite straightforward. At very low surface coverages, the steric interactional energy is

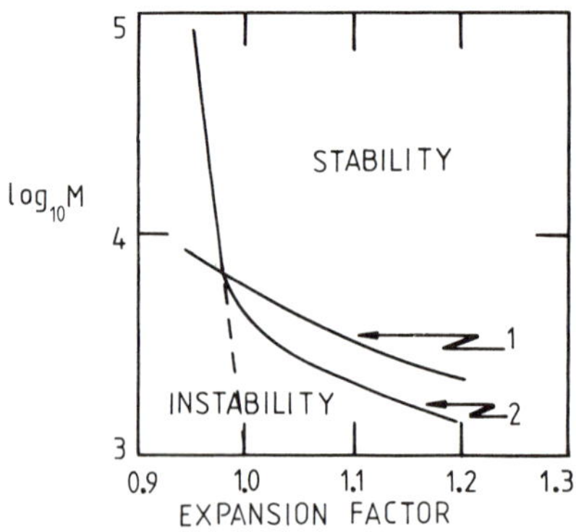

Fig. 11.3. The stability-flocculation domains for flat plates sterically stabilized by tails according to the theory of Hesselink, Vrij and Overbeek (1971). Curve 1: surface coverage = 0·2 mg m^{-2}, $A^* = 10^{-20}$ J; curve 2: surface coverage = 2 mg m^{-2}, $A = 3 \times 10^{-21}$ J (after Hesselink, 1977).

dominated by the elastic free energy which gives rise to stability in worse than θ-solvents. Of course, at low surface coverages, the assumption of uniform segment density parallel to the interface is unlikely to be correct. Moreover, the only experimentally observed instance where steric stabilization in worse than θ-solvents has been demonstrated is in the phenomenon of enhanced steric stabilization (see Chapter 9). This, however, is unlikely to correspond to the theoretical predictions of Hesselink *et al.* because, (i) enhanced steric stabilization occurs at high, not low, surface coverages, and (ii) enhanced steric stabilization is apparently only exhibited in conjunction with loopy adsorption whereas the theory predicts stability in worse than θ-solvents even for tails. As noted previously, whether an elastic contribution is significant in imparting enhanced steric stabilization is not yet known.

Subsequent to these initial calculations, Hesselink (1977) presented a typical stability/instability curve calculated at much higher polymer stabilizing densities and for a smaller Hamaker constant (curve 2 in Fig. 11.3). The experimentally observed incipient instability at the θ-point is now predicted theoretically, as is its insensitivity to the molecular weight of the stabilizing moieties at higher molecular weights. The weak coagulation that was discussed in Section 8.4.5, and which is evident in better than θ-solvents for lower molecular weight stabilizing moieties, is also predicted theoretically. The sharp upturn in curve 2 at ca $\alpha = 1{\cdot}0$ reflects a transformation in the mechanism of aggregation: coagulation, arising from the London attraction between the core particles, occurs if $\alpha \gtrsim 1{\cdot}0$ whereas flocculation resulting from the attraction between the segments of the steric layers is manifest if $\alpha < 1{\cdot}0$. This is illustrated in Fig. 11.4 by the dramatic increase in the depth of the potential energy well in worse than θ-solvents. Note that the steep

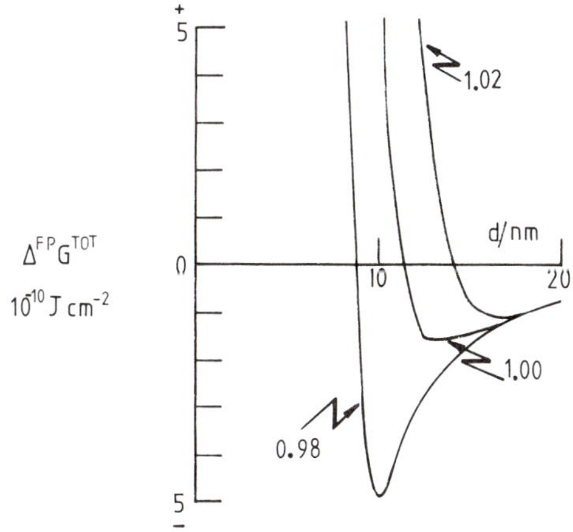

Fig. 11.4. The free energy of interaction of two flat plates sterically stabilized by tails of molecular weight 6 000 according to the theory of Hesselink, Vrij and Overbeek (1971). Three different intramolecular expansion factors are shown (after Hesselink, 1977).

repulsion predicted to occur on close approach in worse than θ-solvents arises from the elastic interaction. This is always repulsive, even though the mixing free energy change switches sign from positive to negative at the θ-point.

11.3.3. The effects of molecular weight

Tadros (1982) has presented theoretical potential energy diagrams for four different molecular weight stabilizing moieties. These were calculated using the formulae of Hesselink *et al.* (1971) and included the van der Waals attraction between the two latex particles. Poly(vinyl alcohol) was taken as the steric stabilizer. The results are shown in Fig. 11.5. As would be expected, any increase in the molecular weight moves the position of the pseudo-secondary minimum away from the surface, whilst at the same time decreasing its depth.

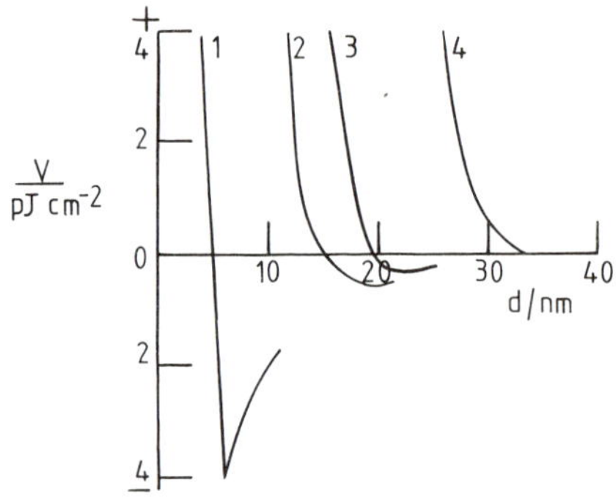

Fig. 11.5. The free energy of interaction of polystyrene latex particles stabilized by poly(vinyl alcohol) according to the theory of Hesselink, Vrij and Overbeek (1971); stabilizer molecular weight: 1, 8 000; 2, 17 000; 3, 28 000; 4, 43 000 (after Tadros, 1982).

11.3.4. The effects of polydispersity in molecular weight

Hesselink (1977) has shown that the introduction of polydispersity into the molecular weight of the stabilizing moieties leads to a rise in the repulsion that is less steep than that for the corresponding monodisperse polymer of the same mean molecular weight.

11.4. The third generation theories

11.4.1. The Edwards approach

The procedure adopted by Meier (1967) and Hesselink *et al.* (1971) assumes the superposition of the mixing and elastic contributions to the total steric interactional free energy. Their respective values are calculated separately, in the spirit of the original Flory calculation of the excluded volume. Dolan and Edwards (1975), Gerber and Moore (1977) and later Levine *et al.* (1978) have all endeavoured to by-pass this artifice and to evaluate the steric repulsion in its totality in one coordinated attack on the problem. This approach was pioneered by Dolan and Edwards (1975). They treated the interaction between segments as an excluded volume effect and estimated the whole of the free energy as a configurational entropy term.

The formulae derived by Dolan and Edwards demanded heavy numerical computations to estimate the magnitudes of the repulsion. Subsequently, Gerber and Moore (1976; 1977) and Levine *et al.* (1978) showed how the interactional free energy can be evaluated more readily. Note that the theory presented by Dolan and Edwards only applies to dispersion media that are θ-solvents or better than θ-solvents for the stabilizing moieties. No discussion of the predictions of their theory for worse than θ-solvents has yet been published. This is unfortunate in light of the flocculation behaviour observed for some sterically stabilized dispersions in marginally worse than θ-solvents.

11.4.1.1. The self-consistent field theory

An analogy with the Schrödinger wave equation

Edwards (1965) has noted a far-reaching analogy between the conformational properties of a polymer chain and the allowable trajectories of a nonrelativistic quantum particle. Recognition of this analogy is important for it permits the well-developed quantum mechanical techniques and concepts to be transmuted into a form capable of solving some of the problems of polymer science. de Gennes (1969) has given an especially lucid account of the concepts involved. His approach will be followed below.

To establish the analogy between a polymer chain and a particle trajectory, de Gennes started with a collection of n beads located at points $\mathbf{r}_1, \mathbf{r}_2, \ldots, \mathbf{r}_n$, the length l between the beads being fixed. Consider the link between the Nth and $(N+1)$th bead. The beads are subject to external constraints (e.g. thermal motion) so that the average elongation of this link is $\mathbf{u}$

$$\mathbf{u} = \mathbf{r}_{N+1} - \mathbf{r}_N$$

Note that $\mathbf{u} = 0$ corresponds to complete orientational disorder. A non-zero $\mathbf{u}$ implies some order in the link direction which corresponds to a reduction in entropy. This was shown by de Gennes to be given by

$$\Delta S \sim -3u^2/2l^2 \ (u \ll l) \tag{11.45}$$

Here, in the spirit of scaling theory, the Boltzmann constant has been taken to be unity. This entropy change is equivalent to a free energy change $\Delta G = -T\Delta S = (3T/2l^2)u^2$. Summing over all links, this formula yields

$$G = (3T/2l^2) \sum_N (\mathbf{r}_{N+1} - \mathbf{r}_N)^2 \tag{11.47}$$

Differentiation of this relationship predicts the *internal* force acting on the Nth bead to be

$$\frac{\partial G}{\partial \mathbf{r}_N} = -(3T/l^2)\,(\mathbf{r}_{N+1} - 2\mathbf{r}_N + \mathbf{r}_{N-1}) \sim -(3T/l^2)\left(\frac{\partial^2 \mathbf{r}}{\partial N^2}\right) \tag{11.48}$$

The latter result is readily derived from the preceding term through consideration of the difference between the differentials of $\mathbf{r}_N$ and $\mathbf{r}_{N-1}$ with respect to N, which can be equated to the second derivative.

Equation (11.48) can be readily generalized for a given set of external forces $F(\mathbf{r}_N)$ acting on each bead

$$\mathbf{F}(\mathbf{r}_N) = -(3T/l^2)\left(\frac{\partial^2 \mathbf{r}}{\partial N^2}\right) \tag{11.49}$$

This is analogous in form to the vector equation describing Newton's Second Law for the trajectory of a nonrelativistic particle if the mass is taken as being equivalent to $(3T/l^2)$ and N plays the role of an *imaginary* time (imaginary, because of the negative sign).

When $\mathbf{F}$ derives from an external potential, then by definition

$$\mathbf{F} = -\nabla U \tag{11.50}$$

where ∇ = the gradient ($\mathbf{i}\,\partial/\partial x + \mathbf{j}\,\partial/\partial y + \mathbf{k}\,\partial/\partial z$) and $U(\mathbf{r})$ is the potential acting on each bead. By analogy with the Newtonian equations, integration by parts of the combination of equations (11.49) and (11.50) yields the energy $U(\mathbf{r})$ (c.f. $V = \frac{1}{2}mv^2$)

$$U(\mathbf{r}) - (3T/2l^2)\left(\frac{\partial r}{\partial N}\right)^2 = \text{constant} \tag{11.51}$$

Edwards couched his theory in terms of $G_N(\mathbf{r},\mathbf{r}')$, which is the statistical weight of *all* conformations whereby a chain of N bonds starts at $\mathbf{r}$ and ends at $\mathbf{r}'$. To derive the differential equation for $G_N(\mathbf{r},\mathbf{r}')$, the statistical weight for a conformation where all beads have prescribed positions $\mathbf{r}_1(=\mathbf{r}), \mathbf{r}_2, \mathbf{r}_3, \ldots, \mathbf{r}_{N-1}, \mathbf{r}_N(=\mathbf{r}')$ is written down. This is simply the product of the appropriate normalized Boltzmann factors

$$f(r_{12})f(r_{23})\ldots f(r_{(N-1),N})\exp\{-\beta U(\mathbf{r}_1)\} \times \exp\{-\beta U(\mathbf{r}_2)\}\ldots\exp\{-\beta U(\mathbf{r}_N)\} \tag{11.52}$$

where $\beta = 1/T$. Here f_{ij} is the normalizing factor that ensures that $r_{ij} = l$

$$f(r_{ij}) = (1/4\pi l^2)\delta(r_{ij} - l) \tag{11.53}$$

where δ = Kronecker delta function ($\delta = 1$ if $r_{ij} = l$ but is zero otherwise).

Summation of equation (11.52) over all the intermediate points $\mathbf{r}_2, \mathbf{r}_3 \ldots, \mathbf{r}_{N-1}$ yields $G_N(\mathbf{r}, \mathbf{r}')$. The equation describing $G_N(\mathbf{r}, \mathbf{r}')$ is derived simply be adding one more bead

$$G_{N+1}(\mathbf{r}, \mathbf{r}') = \sum_{\mathbf{r}''} f(\mathbf{r} - \mathbf{r}'')\exp\left\{-\beta U(\mathbf{r})\right\} G_N(\mathbf{r}, \mathbf{r}'') \tag{11.54}$$

Equation (11.54) expresses the fact that any path of $N+1$ steps reaching r' must have reached one of its neighbours at the penultimate step.

Assuming that $G_N(\mathbf{r}, \mathbf{r}')$ is a slowly varying function of N and $\mathbf{r}'$ and that $U(\mathbf{r})/T$ is relatively small, we can expand the right hand side of equation (11.54) in a Taylor series to read

$$G(\mathbf{r}, \mathbf{r}') = f(\mathbf{r} - \mathbf{r}'')\ (1 - U(\mathbf{r})/T)\sum_{\mathbf{r}''}\left\{G_N(\mathbf{r}, \mathbf{r}')\right.$$

$$\left. + (\mathbf{r}' - \mathbf{r}'')\frac{\partial G_N}{\partial \mathbf{r}} + \tfrac{1}{2}(\mathbf{r}' - \mathbf{r}'')_\alpha(\mathbf{r}' - \mathbf{r}'')_\beta \frac{\partial^2 G}{\partial \mathbf{r}_\alpha \partial \mathbf{r}_\beta} + \ldots\right\} \tag{11.55}$$

If, for the purposes of illustration, the chain is placed on a Flory-Huggins centrosymmetric cubic lattice whose length parameter is l, then all terms that are linear in $(\mathbf{r}' - \mathbf{r}'')$ must vanish on summation since no direction is preferred. Moreover, normalization demands that

$$f(\mathbf{r} - \mathbf{r}'')\sum_{\mathbf{r}''}(\mathbf{r}' - \mathbf{r}'')_\alpha(\mathbf{r}' - \mathbf{r}'')_\beta = \delta(\alpha - \beta)\ (l^2/3) \tag{11.56}$$

where the factor of 3 enters because we are considering only one of three dimensions. These relationships allow equation (11.55) to be rearranged to become

$$G_{N+1}(\mathbf{r}, \mathbf{r}') - G_N(\mathbf{r}, \mathbf{r}') \sim \frac{\partial G_N(\mathbf{r}, \mathbf{r}')}{\partial N}$$

$$= -\frac{U(\mathbf{r})}{T}G_N(\mathbf{r}, \mathbf{r}') + (l^2/6)\ \nabla^2 G_N(\mathbf{r}, \mathbf{r}') + \ldots \tag{11.57}$$

This equation is usually written as

$$-\frac{l^2}{6}\nabla^2 G_N + \frac{U(\mathbf{r})}{T}G_N = -\frac{\partial G_N(\mathbf{r}, \mathbf{r}')}{\partial N} \tag{11.58}$$

because it is clearly analogous to the time dependent form of the Schrödinger wave equation

$$-\left(\frac{\hbar^2}{2m}\right)\nabla^2\psi + V(\mathbf{r})\psi = -\frac{\hbar}{i}\frac{\partial\psi}{\partial t} \tag{11.59}$$

N again plays the role of an imaginary time and $[-(l^2/6)\,\nabla^2 + U(\mathbf{r})/t]$ is the equivalent of the Hamiltonian operator.

In the *self-consistent mean field approximation*, $U(\mathbf{r})$ describes the average repulsion experienced by one bead situated at $\mathbf{r}$, due to all other beads. The Edwards approximation for $U(\mathbf{r})$ in the presence of an excluded volume is obtained by recognizing that in the statistical weight of a particular conformation $(\mathbf{r}_1, \mathbf{r}_2, \ldots, \mathbf{r}_N, \ldots)$, we must insert the factors

$$\{1-\tilde{v}\tilde{p}(\mathbf{r}_1)\}\,\{1-\tilde{v}\tilde{p}(\mathbf{r}_2)\}\ldots\{1-\tilde{v}\tilde{p}(\mathbf{r}_n)\}\ldots,$$

where $\tilde{p}(\mathbf{r}) = \sum_N p_N(\mathbf{r})$ is the total segment density at point $\mathbf{r}$ and $\tilde{v}$ is the excluded volume. In the more familiar Flory-Huggins notation, the latter is given by

$$\tilde{v} = (\tfrac{1}{2} - \chi_1)V_S{}^2/V_1 \tag{11.60}$$

To a good approximation, $\exp\{-\tilde{v}\tilde{p}(\mathbf{r})\} \sim \{1-\tilde{v}\tilde{p}(\mathbf{r})\}$. Thus $\{1-\tilde{v}\tilde{p}(\mathbf{r})\}$ is equivalent to a Boltzmann factor $\exp\{-\tilde{v}\tilde{p}(\mathbf{r})\} = \exp\{-U(\mathbf{r})/T\}$, where $U(\mathbf{r}) = T\tilde{v}\tilde{p}(\mathbf{r})$. This is the appropriate self-consistent mean field potential.

The reason for adopting the self-consistent field procedure resides in the circular nature of the problem: the conformation of the chain is determined by a random walk in the presence of a potential field that itself depends upon the solution of the random walk. A similar difficulty confronted Hartree in determining the wave functions of electrons in many-electron atoms. Hartree (1928), and later Fock (1930), circumvented this difficulty by resorting to the stratagem of modifying one-electron wave functions by the effect of the average field experienced by each electron arising from the presence of the other electrons. The appropriate set of orbitals can first be guessed at (e.g. the hydrogen-like wave functions might be used as a trial set) and the field calculated. This result can be used to generate a new set of wave functions which predict a new field potential. Iteration of these calculations leads to a set of wave functions that are consistent with the field potential that they predict. The electron problem is impossible to handle analytically but is susceptible to numerical solution with the aid of high speed digital computers. The corresponding polymer problem yields an asymptotic analytic solution in the limit of high molecular weight.

The Dolan and Edwards theory

Dolan and Edwards (1974) first solved equation (11.58) for steric stabilization in the complete absence of excluded volume effects (i.e. sterically stabilized dispersions in a melt or θ-solvent). Equation (11.58) then reduces to the simple diffusion equation

$$\left(\frac{l^2}{6}\right)\nabla^2 G_N=\frac{\partial G_N}{\partial N} \tag{11.61}$$

since $U(\mathbf{r})=0$ [cf. equation (11.4)]. This permitted analytic criteria to be established for the elastic steric stabilization of dispersions in θ-solvents or melts (see Section 14.2). Subsequently, the same authors extended the theory to encompass positive excluded volume effects (Dolan and Edwards, 1975).

Dolan and Edwards (1975) write the self-consistent field equation in a slightly modified form compared with equation (11.58)

$$(\partial/\partial s-[l/6]\nabla^2+[\tilde{v}/l^2]\tilde{p}(\mathbf{r}))G(\mathbf{r},\mathbf{r}',s,s')$$
$$=\delta(\mathbf{r}-\mathbf{r}')\delta(s-s') \tag{11.62}$$

Here δ is the delta function, $\mathbf{r}$ and $\mathbf{r}'$ are positions in space, s and s' are distances along the contour of the chain and $(1/l)\tilde{p}(\mathbf{r})$ is the segment number density. Note that $G(\mathbf{r},\mathbf{r}',s,s')$ is the unnormalized probability of a section of chain starting at $\mathbf{r}'$ in space (and s' along the contour of the chain) and ending at $\mathbf{r}$ in space (and s along the chain contour). Equation (11.62) is readily transformed into equation (11.58) by recalling that $\partial/\partial s=(1/l)\partial/\partial N$.

In the presence of the impenetrable surfaces, the segment number density is found from

$$(1/l)\tilde{p}(\mathbf{r})=(1/l)\int_0^L P(\mathbf{r},s)\mathrm{d}s \tag{11.63}$$

where $L=$ total contour length and $P(\mathbf{r},s)=$ the normalized probability that a segment at s along the chain is at $\mathbf{r}$. The latter quantity is given by

$$P(\mathbf{r},s)=\frac{G(\mathbf{r},\mathbf{r}',s)\int G(\mathbf{r}'',\mathbf{r},L-s)\mathrm{d}\mathbf{r}''}{\int G(\mathbf{r}'',\mathbf{r}',L)\mathrm{d}\mathbf{r}''} \tag{11.64}$$

At first sight, $P(\mathbf{r},s)$ might be considered simply equal to $G(\mathbf{r},\mathbf{r}',s)$, which is essentially what Meier (1967) assumed. $P(\mathbf{r},s)$, however, must include the probability $\int G(\mathbf{r}'',\mathbf{r},L-s)\mathrm{d}\mathbf{r}''$ which ensures that the rest of the chain lies within the confining surface, as first pointed out by Hesselink (1969). The denominator in the expression for $P(\mathbf{r},s)$ is just the normalizing factor.

Clearly, solution of equation (11.62) should, in principle, allow $P(\mathbf{r},s)$ to be calculated according to equation (11.64). Equation (11.63) then gives the segment density distribution function from which the steric free energy can be found. The difficulty resides in evaluating $G(\mathbf{r},\mathbf{r}',s)$ since it demands knowledge of the solution over the entire range of $\mathbf{r}$, $\mathbf{r}'$ and s. For this reason, Dolan and Edwards were forced to use an iterative numerical procedure. The steric free energy was evaluated entirely as a configurational free energy term

$$\Delta^{FP}G^{TOT} = kT[\ln\{\Omega(d)/\Omega(\infty)\}] \tag{11.65}$$

where $\Omega(d) = \int_0^d G(x,0,L,0,d)\mathrm{d}s$. Only the x-direction normal to the surface was considered, the segment density parallel to the surface being assumed to be reasonably large and constant. This high density is considered to overcome the criticism of the self-consistent field procedures that has been levelled by des Cloizeaux (1970). He pointed out that their application to a single polymer molecule was suspect because the fluctuations in segment density were quite large.

Predictions of the Dolan and Edwards theory. The dependence predicted by Dolan and Edwards for the free energy per tail as a function of the distance between the two parallel flat plates to which they are attached is shown in Fig. 11.6. The curves span three different values of the excluded volume parameter u, which is given in classical Flory-Huggins terms by

$$u = \tilde{\nu}\, i^2 \upsilon / (\langle r^2 \rangle / 3)^{\frac{1}{2}} \tag{11.66}$$

where $\tilde{\nu}$ = excluded volume, i = number of segments per chain and υ = number of stabilizing moieties per unit area. Also shown in Fig. 11.6 are the comparable predictions of the theory of Hesselink *et al.* (1971), whose equation (11.26) may be rewritten as

$$\Delta^{FP}G^{TOT}/kT = V(d) + (u/3^{\frac{1}{2}})M(d) \tag{11.67}$$

where G now refers to the Gibbs free energy per tail and some numerical values for $V(d)$ and $M(d)$ are listed in Table 11.1.

What is most striking about the curves presented in Fig. 11.6 is the

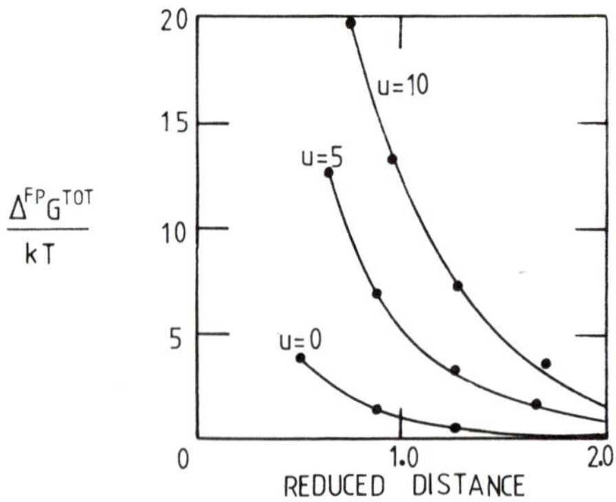

Fig. 11.6. Comparison of the predictions (●) of the theory of Dolan and Edwards (1975) with those of Hesselink, Vrij and Overbeek (1971) (full line) for the distance dependence of the increase in free energy per tail for three different values of the excluded volume parameter u (after Hesselink, 1977).

agreement between the second and third generation theories. The self-consistent mean field theory of Dolan and Edwards correctly allows for the change in the conformation of the stabilizing moieties as a result of their interaction with segments attached to the opposing surface. This conformational relaxation is ignored in the theory of Hesselink *et al.* It is apparent, as was anticipated previously, that the effect of any conformational change on the steric interactional energy is small.

The analytic solutions of Gerber and Moore

The results reported by Dolan and Edwards (1975) were obtained by a numerical solution of a complicated integro-differential equation. Gerber and Moore (1977) have pointed out that in the limit when the plate separation obeys the inequality $d < \sqrt{3} l n^{\frac{1}{2}}$, an approximate analytic solution to the Edwards equation is possible. The method adopted was the variational procedure well-known in quantum mechanics: the correct or best value of the wave function is that which minimizes the total energy of the system.

The energy integral that must be minimized can be written down on sight from equation (11.58) by analogy with that used to solve the Schrödinger wave equation (Landau and Lifshitz, 1977)

$$\Delta^{FP}G^{\text{TOT}}/kT = \upsilon \int_0^d \mathrm{d}x \left[(l^2/6)\left(\frac{\mathrm{d}\psi}{\mathrm{d}x}\right)^2 + \tfrac{1}{2}\upsilon\tilde{v}\psi^4 \right] \qquad (11.68)$$

The first term in this equation arises from the connectivity of the segments whereas the second is associated with their mutal repulsion. We note that the $\psi(x)$ that minimizes the energy integral is subject to the normalizing constraint

$$\int_0^d \psi^2(x)\mathrm{d}x = L/l \qquad (11.69)$$

The segment density of a macromolecule is analogous to the electron density. Therefore we may write

$$\rho_d(x) = \upsilon\psi^2(x) \qquad (11.70)$$

so that the evaluation of the ground state wave function determines the segment density distribution function.

The variational integral can be minimized whilst subject to the normalizing constraint (Arfken, 1970) by introducing a Lagrangian multiplier. The resulting Euler-Lagrange equation is

$$-\frac{l^2}{6}\frac{\mathrm{d}^2\psi}{\mathrm{d}x^2} + \upsilon\tilde{v}\psi = \lambda\psi(x) \qquad (11.71)$$

In the absence of excluded volume effects ($\tilde{v} = 0$), this equation reduces to the familiar Schrödinger equation for a particle-in-a-1D-box

$$-\frac{l^2}{6}\frac{d^2\psi}{dx^2}=\lambda\psi(x) \tag{11.72}$$

for which the solution is equally well-known

$$\psi_m(x)=(2L/ld)^{\frac{1}{2}}\ \sin(m\pi x/d) \tag{11.73}$$

where $m=1, 2, 3, \ldots$. Note that this solution satisfies the boundary conditions $\psi=0$ at $x=0$ and $x=d$. The corresponding energies (eigenvalues) are

$$\lambda_m=m^2\pi^2l^2/6d^2 \tag{11.74}$$

The lowest eigenvalue gives the desired smallest elastic free energy of repulsion so that

$$\Delta^{FP}G^{\text{TOT}}=\Delta^{FP}G^{\text{EL}}=\upsilon kT\pi^2lL/6d^2 \tag{11.75}$$

This approach is only valid when the energy is dominated by a single eigenvalue, which corresponds to the condition $d<\sqrt{3}ln^{\frac{1}{2}}$.

For a finite excluded volume ($\tilde{v}>0$), the solution to the Euler-Lagrange equation must be expressed in terms of the Jacobi elliptic functions

$$\psi(x)=\psi_1 sn((6\lambda-3\upsilon\tilde{v}\psi_1{}^2)^{\frac{1}{2}}x/l|m) \tag{11.76}$$

where $sn=$ the Jacobi sine-amplitude function and $m=\upsilon\tilde{v}\psi_1{}^2/(2\lambda-\upsilon\tilde{v}\psi_1{}^2)$. The unknown parameters $\psi_1(=\psi(x=d/2))$ and λ can be obtained by solving

$$(6\lambda-3\upsilon\tilde{v}\psi_1{}^2)^{\frac{1}{2}}d=2lK(m) \tag{11.77}$$

and

$$(6\lambda-3\upsilon\tilde{v}\psi_1{}^2)^{\frac{1}{2}}(K(m)-E(m))=3\upsilon\tilde{v}L/2l^2 \tag{11.78}$$

with $K(m)$ and $E(m)$ representing the complete elliptic integrals of the first and second kind respectively. Analytic solutions of these two simultaneous equations is only readily accomplished for small or large values of $z(=\upsilon\tilde{v}Ld/l^3)$. In the former case ($z$ small),

$$\Delta^{FP}G^{\text{TOT}}=\upsilon kT\frac{\pi^2lL}{6d^2}\left\{1+\frac{9\upsilon\tilde{v}Ld}{2\pi^2l^3}\right\} \tag{11.79}$$

whereas in the latter case (z large),

$$\Delta^{FP}G^{\text{TOT}}=(\tfrac{1}{2})kT\tilde{v}\left(\frac{\upsilon L}{ld^{\frac{1}{2}}}\right)^2 \tag{11.80}$$

Other extensions of the Dolan and Edwards approach

Levine *et al.* (1978) have developed an elegant matrix procedure to evaluate both the segment density distribution functions and the steric interaction for tails, loops and bridges that are subject to excluded volume considerations.

The chains were placed on a cubic lattice and parallel flat plate geometry was assumed.

The procedure adopted by Levine *et al.* is somewhat akin to the method developed for the rotational isomeric state theory in that a statistical weight was attached to each segment. The total statistical weight for the polymer chain was then obtained by serial multiplication of the individual statistical weights. These statistical weights were determined by the nearest neighbour contacts (a Flory-Huggins type of interaction coupled with an adsorption energy associated with the plate surface).

Statistical weight matrices allowed the sums of the statistical weights of all the conformations that start at the surface and end at a given layer to be found readily. These matrices are readily assembled from the segmental statistical weights. The matrix procedure allowed various modes of attachment of the chains to the surfaces of the plates to be handled conveniently.

Scheutjens and Fleer (1979; 1980; 1982) have recently elaborated a different matrix method for keeping track of the myriad of conformations accessible to an adsorbing polymer chain. Their theory leads to essentially the same results as those predicted by Levine *et al.* The starting point of the Scheutjens and Fleer approach is the partition function $Z\{n_2\}$ for a mixture of polymer molecules in specified conformations (n_c molecules with conformation c, etc.) and molecules of the dispersion medium in a lattice adjoining an adsorbing surface. The Bragg-Williams approximation of random mixing within each layer parallel to the surface was assumed. Differentiation of $\ln Z$ with respect to the number of chains in a given conformation yields the equilibrium properties of the system.*

Perhaps the most important conclusion to emerge from the Scheutjens and Fleer theory is the importance of tails, even for polymers of relatively high molecular weight (up to, say, 10^5), in the conformations of adsorbed homopolymers. Hesselink (1975) had previously suggested that this was likely to be the case for homopolymers. Tails may well be important for certain cases of steric stabilization (Lyklema and van Vliet, 1978; Israelachvili *et al.*, 1980; Sonntag *et al.*, 1982). The results of Scheutjens and Fleer will not be discussed further because they assumed that the stabilizing polymer was always in equilibrium with the bulk solution. This is rarely the case when robust steric stabilization is imparted. Proper anchoring is used to ensure that the stabilizing moieties do not desorb under the stress generated by a Brownian encounter.

11.5. Comparison of pseudo-'ab initio' theories with experiment

It should be stressed that in order to compare the predictions of the pseudo-*ab*

* Scheutjens and Fleer [*Adv. Colloid Interface Sci.* **18**, 309 (1983)] have recently recognized that their original method was erroneous and that some of their results require modification.

initio theories advanced to-date with the results of experiments, it is necessary to adopt some arbitrary scaling assumption. A moment's reflection leads to the inevitable conclusion that such pseudo-*ab initio* theories do not predict the barrier layer thickness in any absolute sense. Such theories therefore have little advantage over the pragmatic theories to be described in the next chapter. There it will be shown that the distance dependence of the steric repulsion in the interpenetrational domain is relatively insensitive to the precise form adopted for the segment density distribution function. Consequently, most experimental tests of the predictions of a pseudo-*ab initio* or pragmatic theory, which are scaled to the observed steric layer dimensions, are foredoomed to at least modest success. Perhaps a more sensitive comparison would be to scale to the dimensions of the polymer chains in free solution, although this really offers no major advance on other scaling methods.

Chapter 12

The Pragmatic Theories of Steric Stabilization

12.1. Introduction

In the preceding chapter, the pseudo-*ab initio* theories of steric stabilization were discussed. In order to facilitate the mathematical analysis, these theories, without exception, adopted primitive models for the polymer chains (e.g. the random flight model). To compare their predictions with experiment, it is necessary to introduce an arbitrary scaling assumption. Several such schemes immediately suggest themselves: e.g. the value of the rms end-to-end distance measured in free solution might be equated with that predicted theoretically for the model chain. Alternatively, the measured barrier layer thickness might be set equal to that predicted theoretically. The latter type of comparison is necessarily a very blunt instrument for discriminating between competing theories because it merely tests the general form of the predicted segment density distribution function. As we shall see, the distance dependence of the steric repulsion is relatively insensitive to the general form adopted by the segment density distribution function, provided that the thickness of the stabilizing sheath has been properly adjusted to the measured value.

The recent small angle neutron scattering and pulsed nmr studies reported by Cosgrove and coworkers (Barnett *et al.*, 1981; Barnett *et al.*, 1982; Cosgrove *et al.*, 1982) show that, as predicted by Clark and Lal (1978) and de Gennes (1981), the conformation of tails can be very extended, normal to the interface. This extension presumably arises from the high surface coverage in the systems studied. Apparently it is the crowding effect of the stabilizing chains at the interface that renders the theory for isolated chains inapplicable at high surface coverages. At the present time it does not seem possible to allow quantitatively for the effects of crowding and so the conformations of polymer chains at high surface coverages seem inaccessible theoretically. Yet high surface coverages are precisely what are found experimentally to be required to impart optimal steric stabilization and what are therefore commonly encountered in the practical context.

In these circumstances, a less ambitious attack on the problem seems justified. Such an approach could take as its starting point the measured segment density distribution functions. Alternatively, if these are not known, a mathematical form could be assumed for the segment density distribution function. The assumed form is then scaled to the measured steric layer thickness. This subterfuge circumvents in a pragmatic fashion the question of how crowding at the interface influences the conformation of the stabilizing moieties. Nonetheless, the arbitrary scaling procedure inherent in such an approach forfeits any claims to the absolute prediction of the magnitude of the steric repulsion.

12.2. The constant segment density theories

12.2.1. The theory of Fischer

The first theory to recognize clearly the prime importance of the solvency of the dispersion medium in steric stabilization was that published by Fischer (1958). Fischer considered the overlap of the steric layers attached to two spheres (see Fig. 10.4). The mixing free energy change $\delta(\Delta^S G^M)$ in the small volume element δV for one of the steric layers is given by

$$\delta(\Delta^S G^M) = \left(\frac{\partial \Delta^S G^M}{\partial N_1} \right) \delta N_1 \tag{12.1}$$

where N_1 = number of moles of solvent molecules. Note that $\Delta^S G^M$ is necessarily an excess Gibbs free energy. This can be grasped by recognizing that the overlap volume contains merely segments and not whole molecules whose centres of mass undergo independent translational motion. The partial differential in equation (12.1) can be written as

$$\frac{\partial \Delta^S G^M}{\partial N_1} = -(\mu_1 - \mu_1^{\ominus})^E = \pi_E \bar{V}_1 \tag{12.2}$$

where μ_1 = chemical potential of the solvent, π_E = excess osmotic pressure and $\bar{V}_1$ = molar volume of the molecules of the dispersion medium. The negative sign introduced into equation (12.2) is necessary to allow for the fact that overlap can be considered to be the reverse of dilution. We also invoke the relationship $\delta N_1 = \delta V / \bar{V}_1$. This permits equation (12.2) to be recast as

$$\delta(\Delta^S G^M) = \pi_E \bar{V}_1 \delta N_1 = \pi_E \delta V \tag{12.3}$$

It follows that for two layers, the total mixing free energy is predicted to be

$$\Delta^S G^M = 2\int_V \pi_E \mathrm{d}V \tag{12.4}$$

where the integration is performed over the entire overlap volume.

The excess osmotic pressure can be expressed by the virial expansion *viz*

$$\pi_E = RTB_2\, c_2^{\,2} + \ldots \tag{12.5}$$

where B_2 = second virial coefficient. The first virial coefficient is irrelevant in this context since it refers to the motions of the centres of mass of whole molecules. Fischer ignored the virial coefficients higher than the second. In that approximation, equation (12.4) becomes

$$\Delta^S G^M = 2\int_V RTB_2\, c_2^{\,2} \mathrm{d}V \tag{12.6}$$

If it is now assumed that the segment density is constant within the steric layer, the integration is readily performed

$$\Delta^S G^M = 2RT\, B_2\, c_2^{\,2} \int_V \mathrm{d}V = 2RTB_2\, c_2^{\,2}(\Delta V) \tag{12.7}$$

where ΔV = overlap volume.

The lens-shaped overlap volume can be calculated from

$$\Delta V = (2/3)\pi(\delta - H_0/2)^2(3a + 2\delta + H_0/2) \tag{12.8}$$

where δ = steric layer thickness, H_0 = minimum distance between the surfaces of the two spheres and a = particle radius. On substituting this value for ΔV into equation (12.7) Fischer obtained

$$\Delta^S G^M = (4\pi/3)RTB_2\, c_2^{\,2}(\delta - H_0/2)^2(3a + 2\delta + H_0/2) \tag{12.9}$$

The second virial coefficient B_2 can be related to the interaction parameter χ_1 via the Flory-Huggins relationship

$$B_2 = (\tfrac{1}{2} - \chi_1)/\, \overline{V}_1 \rho_2^{\,2}$$

where ρ_2 = density of the stabilizing moieties. This allows equation (12.9) to be transformed into

$$\Delta^S G^M = (4\pi c_2^{\,2}/3V_1\rho_2^{\,2})kT(\tfrac{1}{2} - \chi_1)\,(\delta - H_0/2)^2(3a + 2\delta + H_0/2) \tag{12.10}$$

where V_1 = volume of a molecule of the dispersion medium. An alternative expression for the free energy of interaction is (Ottewill and Walker, 1974)

$$\Delta^S G^M = (\pi c_2^{\,2} z^2/3V_1\rho_2^{\,2})kT(\tfrac{1}{2} - \chi_1)[3(a + \delta) - z/2] \tag{12.11}$$

where $z(=2\delta - H_0)$ = distance of interpenetration of the steric layers. Note that these expressions are only valid in the interpenetrational domain for steric layers of uniform segment density. The latter, as de Gennes (1981) has shown, is approximately valid only for highly extended stabilizing moieties. It might conceivably also be applicable to steric sheaths containing highly branched chains.

The equation derived by Fischer can be recast into a more convenient form as follows. Let L_S be the span of the attached polymer chains, which to a modest approximation can be equated to the layer thickness δ. Then if ω is the weight of stabilizing moieties attached to unit area of the surface, it follows that $\omega = L_S c_2$. Furthermore, the density of the attached polymer chains ρ_2 can be set equal to the reciprocal of their partial specific volume $\bar{v}_2$. These relationships on substitution into equation (12.10) yield

$$\Delta^S G^M = 4\pi a \omega^2 N_A (\bar{v}_2{}^2 / \bar{V}_1)\,(\tfrac{1}{2} - \chi_1) kT (1 - H_0/2L_S)^2 \qquad (12.12)$$

12.2.2. Extensions of the Fischer theory

Ottewill and Walker (1968) have used the relationship between the interaction parameter χ_1 and the entropy and enthalpy dilution parameters (ψ_1 and κ_1 respectively), *viz* $\frac{1}{2} - \chi_1 = (\psi_1 - \kappa_1)$, to write equation (12.10) as

$$\Delta^S G^M = (4\pi c_2{}^2 / 3 V_1 \rho_2{}^2) kT (\psi_1 - \kappa_1)\,(\delta - H_0/2)^2 (3a + 2\delta + H_0/2) \qquad (12.13)$$

One disadvantage of the approach of Fischer (1958) and Ottewill and Walker (1968) is that, as noted above, the formulae proposed by these authors are only valid for the mixing of constant segment density steric layers in the interpenetrational domain. Once the interpenetrational-plus-compressional domain is entered, no allowance is made for the elastic contribution to the free energy. The elastic interactions can become important, even paramount, when the minimum distance of separation between the surfaces of the particles (H_0) is less than the barrier layer thickness. An additional defect of the Fischer approach is apparent in this domain: the overlap volume is decreased below that given by equation (12.8) because part of it is occupied by the cores of the particles. Both the Fischer and the Ottewill and Walker theories disregard this decrease in volume. Implicit in their formulae is the notion that the solid cores become equivalent to the steric barriers. This is, of course, quite unphysical.

Doroszkowski and Lambourne (1971) endeavoured to improve upon the Fischer approach by assuming that when the polymer sheaths overlap, the segments distribute themselves uniformly over an arbitrarily chosen available volume. As this volume is larger than that assumed by both Fischer and Ottewill and Walker, the redistribution method necessarily predicts repulsions that are significantly weaker than those calculated by the Fischer-Ottewill-Walker approach. It must be stressed that the redistribution method adopted by Doroszkowski and Lambourne is entirely arbitrary and no real attempt has been made to justify the assumptions involved.

Doroszkowski and Lambourne calculated the free energy of steric interaction by numerical integration of the force of interaction ($\mathscr{F}$) between the particles. This force was given by

$$\mathscr{F} = (kTc_2{}^2/V_1\rho_2{}^2)\,(\tfrac{1}{2}-\chi_1)f(x,a,\delta) \tag{12.14}$$

where

$$f(x,a,\delta) = \frac{\pi x^2 R_4(3R_4 - x)\,(2R_4 - x)\,[x(R_4{}^3 - r^3) + xR_4(x - 3R_4)]}{[2(R_4{}^3 - r^3) + xR_4(x - 3R_4)]^2}$$

and $x = \delta - H_0/2$ and $R_4 = a + \delta$. Note that in common with the Fischer-Ottewill-Walker approach, this formula ignores any elastic contribution to the interaction.

12.3. The variable segment density models

The constant segment density model is, of course, only an approximation at best. It would be expected that in general the segment density would be a function of the distance from the surface of the particle. The precise form adopted by the segment density distribution function should depend upon the steric layer properties. These properties will be determined by such factors as the chemical nature of the surface and the polymer, the quality of the solvency of the dispersion medium, the surface coverage, and the mechanism of attachment of the polymer chains to the surface. Some of these expectations have been confirmed by the recent experimental determinations of the segment density distribution functions for several different systems.

12.3.1. Steric barrier layer properties

Vincent (1982) has pointed out that there are three different types of parameters that can be measured for steric layers:

(i) *thermodynamic parameters*, which include the excess amount of the stabilizer in the interfacial region and the corresponding excess enthalpies and excess entropies;

(ii) *structural parameters*, which comprise the barrier layer thickness and the molecular conformation as characterized by the segment density distribution functions, both parallel to and normal to the interface, and the separate tail, loop and train distribution functions.

(iii) *dynamic parameters*, by which is meant the normal modes and frequencies of both the molecular and segmental motions in the steric layers.

The thickness of the steric layers attached to colloidal particles can be measured by a variety of techniques: viscometry (Doroszkowski and Lambourne, 1968; Fleer *et al.*, 1972), sedimentation coefficient measurements (Garvey *et al.*, 1974), photon correlation spectroscopy (Garvey *et al.*, 1976), surface balance measurements (Doroszkowski and Lambourne, 1971; 1973), compression cell measurements (Cairns *et al.*, 1976) and even electrophoretic measurements (Homola and Robertson, 1975).

The fraction of stabilizing segments that is bound to the surface can be measured by microcalorimetry (Eisenlauer *et al.*, 1977), IR (Peyser *et al.*, 1967), ESR (Robb and Smith, 1977) and pulsed nmr (Cosgrove and Barnett, 1981) spectroscopy. The only method to-date that has allowed the segment density distribution function to be determined is small angle neutron scattering. Because of its likely importance in the future, this method will now be considered in detail.

12.3.1.1. The determination of the segment density distribution function

Cosgrove and coworkers (Barnett *et al.*, 1981a; Barnett *et al.*, 1982; Cosgrove *et al.*, 1982) have recently developed an elegant small angle neutron scattering (SANS) method for determining the segment density distribution function. These pioneering experiments provide for the first time experimental data against which the predictions of the various theories of the conformations of attached and adsorbed polymers may be tested.

The principles of neutron scattering

Ottewill (1982) has given a thorough survey of the basic principles of neutron scattering so that only the general concepts need be sketched in here. Neutrons are not, of course, electromagnetic radiation but, none the less, they do display wave character in accordance with the Uncertainty Principle. This is formally given by the de Broglie relationship: $\lambda = h/mv = h/(2mE)^{\frac{1}{2}} \simeq h/(2mkT)^{\frac{1}{2}}$, where m = mass of the neutron ($1{\cdot}67 \times 10^{-24}$ g) and E = energy. The thermal neutrons produced by a nuclear pile have a wavelength of order 0·1 nm. This is rather too small for many colloidal scattering studies but by cooling the neutrons (by passage through liquid deuterium at ca 20 K) their wavelength, which is inversely dependent upon the temperature, can be suitably increased (to, say, a value in the range 0·5–5 nm).

Neutrons are scattered by the nuclei of atoms in a fashion somewhat analogous to the scattering of electromagnetic radiation by the electrons in matter. In furtherance of this analogy, we recall that both neutrons and electromagnetic radiation can give rise to diffraction phenomena but this analogy will not be pursued further. The scattering of neutrons differs from one type of atomic nucleus to another, the scattering being characterized by a *coherent scattering length* (b). Of special importance in this context is the difference in scattering between the nucleus of a hydrogen atom (consisting simply of one proton) and that of a deuterium atom (composed of a proton and a neutron): the coherent scattering length for hydrogen is $-0{\cdot}37\times10^{-12}$ cm whereas that for deuterium is $+0{\cdot}67\times10^{-12}$ cm, the difference in sign reflecting a difference in the phase associated with the scattering.

The scattering of a molecular material is characterized by the neutron *scattering length density*, $\mathfrak{L}$, defined by

$$\mathfrak{L}=\sum_i b_i/V \tag{12.15}$$

where V = molecular volume. This merely sums the coherent scattering length of each nucleus in the molecule, with due regard to phase, and averages the sum over the entire scattering volume. The negative coherent scattering length of hydrogen drastically reduces the neutron scattering length densities for protonated, compared with deuterated, molecules of the same basic chemical structure. This is exemplified by protopolystyrene, with a scattering length density of $1{\cdot}4\times10^{10}$ cm^{-2}, and deuteropolystyrene with a corresponding value of $6{\cdot}5\times10^{10}$ cm^{-2}, almost five times larger.

The increase in the scattering power of a molecule on deuteration provides a powerful tool whereby various molecular regions of interest in a colloidal dispersion can be viewed against a weakly scattering background. Thus, for example, as shown by Barnett *et al.* (1982), it is possible in neutron scattering to contrast match sterically stabilized polystyrene core particles (ca 98% deuterated) with a suitable (2:98) H_2O/D_2O mixture as the dispersion medium. This is analogous to refractive index matching in light scattering. When contrast matching is achieved, the only scattering observed is that from the adsorbed layer, allowing in principle its structure to be elucidated.

The most powerful method devised by Cosgrove *et al.* (1982) is that whereby the core particles are not completely contrast matched. What is measured experimentally is the form factor $F(Q)$ of the adsorbed layer

$$I(Q)=F(Q)^2 \tag{12.16}$$

where I = scattered neutron intensity and Q = scattering vector. The latter parameter is given by $Q=(4\pi/\lambda)\sin(\theta/2)$, where θ = scattering angle. Wavelengths (λ) between 1–5 nm are usually employed. The form factor allows for the relative spatial distribution of the scattering centres in the sample and is given by

$$F(Q)^2=(K/Q^2)|\tilde{\rho}(Q)| \tag{12.17}$$

In this equation

$$\tilde{\rho}(Q)=\int\rho(x)e^{iQx}\mathrm{d}x \tag{12.18}$$

where $\rho(x)$ = segment density distribution function normal to the surface. To determine $\rho(x)$, $\tilde{\rho}(Q)$ must be found. This is related to the experimental quantity $|\tilde{\rho}(Q)|$ by

$$\tilde{\rho}(Q)=|\tilde{\rho}(Q)|e^{i\phi(Q)} \tag{12.19}$$

where $\phi(Q)$ = a (to be determined) phase factor. Cosgrove *et al.* (1982) have used the fact that $\rho(x)=0$ for $x<0$ (i.e. the polymer does not penetrate the surface of the particle) to determine the phase factor. This allows $\rho(x)$ to be calculated with only minor reservations with regard to its uniqueness.

Results of the neutron scattering studies

Two representative types of steric stabilizer have been studied by Cosgrove *et al.* (1982). The first was a partially hydrolyzed (88%) sample of poly(vinyl alcohol). It was, of course, a copolymer containing short blocks (e.g. of average length 6 repeat units) of poly(vinyl acetate). The second sample was poly(oxyethylene) that was terminally anchored to the surface of the latex particles. The normalized segment density distribution functions for these two samples are shown in Fig. 12.1a and 12.1b.

The very different shapes of the segment density distribution functions for these two stabilizers are readily discerned from these diagrams. Whereas the segment density distribution function for the poly(vinyl alcohol) showed an almost monotonic decrease with the distance from the surface, that for the poly(oxyethylene) exhibited a definite maximum at a distance of ca 4 nm from the surface. These qualitatively different shapes are precisely what was predicted by the theories of Silberberg (1962a,b; 1967; 1968), Meier (1967) and Hesselink *et al.* (1971). From its chemical structure, poly(vinyl alcohol) would be expected to adsorb in a loopy configuration, with the nominally insoluble

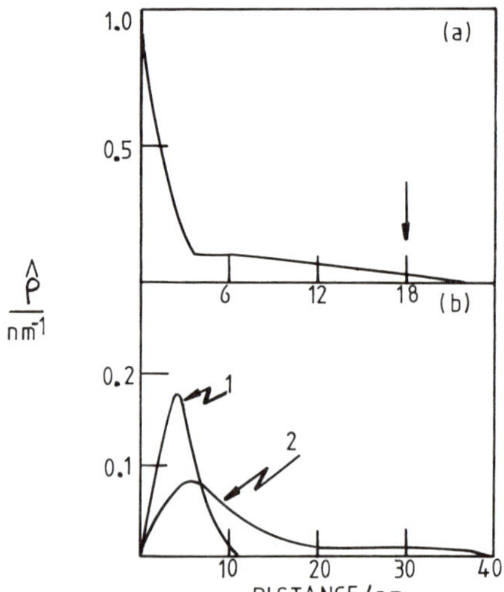

Fig. 12.1. The normalized segment density distribution functions determined by neutron scattering; (a) poly(vinyl alcohol) with the arrow showing the hydrodynamic thickness; (b) poly(oxyethylene) where curves 1 and 2 give the theoretical (according to Hesselink, 1969) and experimental results respectively (after Cosgrove *et al.*, 1982).

poly(vinyl acetate) blocks acting as the primary anchoring segments and the soluble poly(vinyl alcohol) blocks projecting into the dispersion medium. Notice, however, that the segment density curve appears to show a shallow maximum far from the surface, followed by a slow decay that extends up to a distance of 18 nm. The existence of this shallow maximum suggests that the observed segment density may be composite in character: it may well be comprised of a fast monotonic decay due to loops and trains coupled with a second segment density function arising from the tails, the latter displaying a maximum density somewhat removed from the surface. The importance of tails in loopy-type adsorption, as predicted by Hesselink (1975) and Scheutjens and Fleer (1979; 1980; 1982) appears to be confirmed in this instance.

The hydrodynamic thickness

One of the newer methods for determining the hydrodynamic thickness of a polymeric steric layer is photon correlation spectroscopy (PCS). The invention of lasers in 1961 led to the availability of intense well-collimated sources of coherent light. A dispersion of particles when irradiated with laser light behaves as a random 3-D diffracting array, the randomness arising from the Brownian motion of the particles (Pusey, 1982). As a result, a "speckle" pattern consisting of bright spots (where constructive interference predominates) and dark regions (where destructive interference occurs) is formed. Interference effects arise with a coherent source because phase relationships are maintained in the scattering process. The characteristic fluctuation time τ_c of the speckle pattern, which can be determined by photon correlation techniques, is related to the size of the colloidal particles, as can be seen from what follows.

An alternative way of considering dynamic light scattering is to recognize that the Brownian motion of the colloidal particles imparts Doppler shifts to the scattered light. The colloidal particles undergo perpetual Brownian motion with a Gaussian distribution of velocities. It follows that an incident monochromatic beam gives rise to a modulated scattered beam, whose distribution of frequencies can be characterized by a line width Γ. This width is inversely related to the correlation time τ_c. The spectral line width is also related to the diffusion coefficient of the particles: specifically, $\Gamma = K^2 D$, where K is the scattering vector ($[4\pi/\lambda]\sin[\theta/2]$) at a given scattering angle θ. The diffusion coefficient for spheres is related to the hydrodynamic radius of the particle a by the Stokes-Einstein equation $D = kT/6\pi\eta a$ (where η = viscosity of the dispersion medium). In this way the value of a can in principle be determined by measurement of the spectral line width, although some caution needs to be exercised with highly polydisperse systems. At the practical level, the spectral line width is not in fact measured. Rather, what is determined is the autocorrelation of the light intensity (i.e. the correlation time τ_c, which usually lies in the microsecond to millisecond range, rather than its reciprocal Γ). The barrier layer thickness δ can be estimated *inter alia* by determining the hydrodynamic radius of the particles both in the presence and absence of the adsorbed layer.

The first experiments in which it proved possible to measure both the PCS hydrodynamic layer thickness and the segment density distribution function were devised by Cosgrove *et al.* (1982). A fractionated poly(vinyl alcohol), of molecular weight 37 000, was found to give a hydrodynamic layer thickness of $\delta = 18 \pm 2$ nm. Surprisingly, it was found that this value of δ encompassed over 95% of the segments present in the layer. The value of δ in this instance was apparently determined by the longest tails. For comparison, we note that the radius of gyration of this sample of poly(vinyl alcohol) in free solution was found to be 7·2 nm and that the contour length of the chain was calculated to be 220 nm. The hydrodynamic thickness is thus *ca* 2·5 times the radius of gyration in solution. It is possible that the presence of a significant number of long tails in this system was a consequence of the high surface coverage.

The fraction of adsorbed segments

Nmr relaxation measurements allow the molecular motions of the segments of a polymer molecule to be investigated. The spectral line width, which is proportional to $1/T_2$ (i.e. the reciprocal of the spin-spin relaxation time of proton nmr) is a direct measure of segment mobility. It would be anticipated that the anchor segments in the trains attached to the surface of the adsorbent particle would display low mobility. On the other hand, those segments in loops and tails that project into the continuous phase should possess much higher mobility. Two distinct linewidths would therefore be expected for the two different types of segments.

The major difficulty confronting the determination of the fraction of adsorbed segments by nmr has always been the separation of the two different types of spectra, which unfortunately cannot be satisfactorily accomplished by simple curve fitting routines. Cosgrove and Barnett (1981) have overcome this difficulty by developing a pulsed nmr method. The high and low mobility segments respond in a dissimilar fashion to excitation pulses of different lengths because of their different relaxation times. This allows the responses of the two different types of segments to be separated.

The fraction of bound segments $\langle p \rangle$ was found to be ca 0·1 for the poly(vinyl alcohol) and zero (within experimental error) for the poly(oxyethylene) tails. The value of $\langle p \rangle \simeq 0{\cdot}1$ for poly(vinyl alcohol) is consistent with the assumption that most of the poly(vinyl acetate) segments, (present in the overall molecule at a mole fraction of ca 0·12) are in trains at the latex particle surface.

Quantitative comparison with theories

We also show in Fig. 12.1b the prediction by Hesselink (1969; 1971) of the segment density distribution function for isolated poly(oxyethylene) tails, whose radius of gyration is identical with that measured (2·8 nm). It is apparent that the predicted distance dependence for isolated tails is in poor agreement with the measured segment density distribution. This presumably is a consequence of the neglect of the effects of lateral interactions that arise from the crowding of the tails onto the surface. The effect of this crowding is to cause the chains to be more extended than isolated chains normal to the interface. The segment density distribution function is thus more extended (and perforce flatter) than that predicted for isolated chains.

12.3.1.2. Arbitrarily chosen segment density functions

Another indirect piece of evidence that suggests that the segment density distribution functions for isolated chains are inappropriate for many sterically stabilized dispersions with high surface coverage is presented in Fig. 12.2. This compares the experimental steric free energy of repulsion, measured by Doroszkowski and Lambourne (1973), as a function of the distance of particle separation with that predicted by combining the Fischer theory for the interpenetrational domain with the Hesselink segment density distribution function for isolated polymer chains (Smitham *et al.*, 1975). The poly(acrylonitrile) particles used in these 2-D compression studies were stabilized by polystyrene, of average molecular weight 6 000, and dispersed in toluene at 297 K.

It is apparent that the experimentally measured repulsion extends much further out from the surface than would be expected for isolated chains. The theoretical curve presented in Fig. 12.2 ignores the elastic repulsion that must

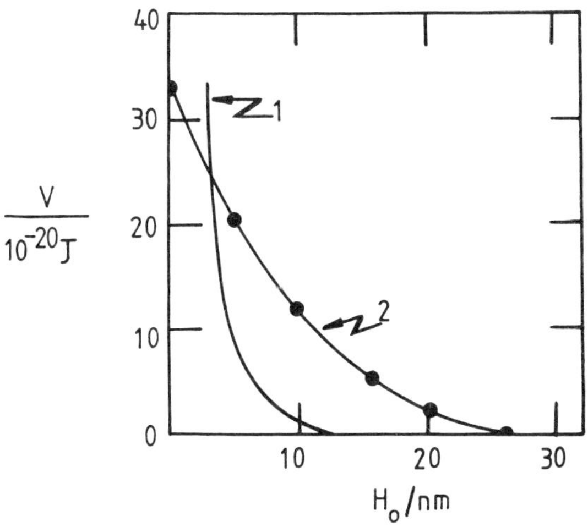

Fig. 12.2. The distance dependence of the repulsive potential energy for spheres sterically stabilized by polystyrene in toluene; curves 1 and 2 represent the theoretical (according to Hesselink *et al.*, 1971) and experimental (Doroszkowski and Lambourne, 1973) results respectively.

be included in the interpenetrational-plus-compressional domain. Inclusion of this, however, would not significantly alter the broad conclusion that isolated chains do not appear to model adequately such sterically stabilizing moieties.

Flat plates

It was shown in Section 11.2.1.2 that the mixing steric free energy per unit area for two parallel flat plates is given by

$$\Delta^{FP}G^M = 2kT(V_S{}^2/V_1)\,(\tfrac{1}{2}-\chi_1)\upsilon^2 i^2 P^M \tag{12.21}$$

where V_S = volume of a segment, V_1 = volume of a molecule of the dispersion medium, χ_1 = interaction parameter, υ = number of stabilizing moieties, i = number of segments per chain. The distance dependent parameter P^M is found from

$$P^M = (\tfrac{1}{2})\left[\int_0^h (\hat{\rho}_h + \hat{\rho}_h')^2 \mathrm{d}x - \int_0^\infty (\hat{\rho}_\infty + \hat{\rho}_\infty')^2 \mathrm{d}x\right] \tag{12.22}$$

where h = distance of separation. Note that for actual calculations, the quantities $(\upsilon i V_S)^2$ in equation (12.21) can be equated to $(\omega \bar{v}_2)^2$, where ω = weight of polymer attached to unit area of the surface and $\bar{v}_2$ = partial specific volume of the polymer. This allows equation (12.21) to be recast as

$$\Delta^{FP}G^M = 2kT\omega^2(\bar{v}_2{}^2/\bar{V}_1)N_A(\tfrac{1}{2}-\chi_1)P^M \tag{12.23}$$

where $\bar{V}_1$ = partial molar volume of the dispersion medium and N_A = Avogadro's constant.

In the interpenetrational domain it was shown that

$$P_I^M = \int_0^h \hat{\rho}_h\,\hat{\rho}_h'\mathrm{d}x$$

and in the interpenetrational-plus-compressional domain

$$\mathrm{P}_{I+C}^M = \int_0^h \hat{\rho}_h{}^2\mathrm{d}x - \int_0^\infty \hat{\rho}_\infty{}^2\mathrm{d}x + \int_0^h \hat{\rho}_h\,\hat{\rho}_h'\mathrm{d}x \tag{12.24}$$

Constant segment density models. We begin this discussion by reconsideration of the constant segment density model so as to be able to demonstrate later that this approach leads to the same result as is predicted by the Fischer theory.

The normalized segment density distribution functions in the interpenetrational domain ($L_S \le h \le 2L_S$, where L_S = the thickness of the steric layer) for the constant segment density model are $\hat{\rho}_h = \hat{\rho}_\infty = \hat{\rho}_h' = 1/L_S$ since $\int_0^{L_S}\hat{\rho}_h\mathrm{d}x = 1$. This leads to

$$P_I^M = \int_{h-L_S}^{L_S} \hat{\rho}_h\,\hat{\rho}_h'\mathrm{d}x = (1/L_S{}^2)\int_{h-L_S}^{L_S} \mathrm{d}x = (2/L_S)\,(1-h/2L_S) \tag{12.25}$$

In the interpenetrational-plus-compressional domain, $\hat{\rho}_h = \hat{\rho}'_h = (1/h)$ and $\hat{\rho}_\infty = 1/L_S$. These relationships give

$$P^M_{I+C} = (3/h - 2/L_S) \tag{12.26}$$

Denting. Bagchi (1974a,b; 1975; 1976) has proposed that when sterically stabilized particles interact, the polymer chains crumple rather than undergoing interpenetration (the *denting* mechanism). For the mutual denting of two surfaces there is no interpenetration term. Therefore

$$P^M_D = 2\left[\int_0^{h/2} \hat{\rho}_h{}^2 \mathrm{d}x - \int_0^{L_S} \hat{\rho}_\infty{}^2 \mathrm{d}x\right] \tag{12.27}$$

where $\hat{\rho}_h = (2/h)$ and $\hat{\rho}_\infty = (1/L_S)$ for an assumed constant density over the entire accessible space. Integration yields

$$P^M_D = 4(1/h - 1/2L_S) \tag{12.28}$$

The radial Gaussian distribution function. The Gaussian distribution function is symmetrically displaced about its maximum. Whilst this symmetry may be manifest approximately by chains sandwiched between two plates, it is unlikely to be correct for tails attached to a single plate. One advantage of the radial Gaussian distribution function is that it biases the segments away from the surfaces of the particles, as is indicated to be appropriate for tails by RIS-Monte Carlo simulations. Solutions to the diffusion equation lead to qualitatively similar conclusions. The segment density distribution function in the direction normal to the surface can be written for the radial Gaussian distribution function

$$\hat{\rho}_h = 4\pi(3/2\pi F^2 h^2)^{3/2} x^2 \exp\{-3x^2/2F^2h^2\} \tag{12.29}$$

where $F = \langle r^2 \rangle^{\frac{1}{2}}/h$ and $\langle r^2 \rangle^{\frac{1}{2}} =$ the rms end-to-end distance of the chain. The difficulty with this expression, which is of a form that can be easily dealt with in the calculation of the steric repulsion, lies in its unphysical prediction of the presence of segments at distances from the surface in excess of the contour length of the chain. Such a prediction is also at odds with the distinction made between the various domains of close approach.

Smitham (1976) has provided one way of rectifying this inconsistency by truncating the distribution function at the outer edge of the steric layer and normalizing the distribution function over the truncated space. This is, of course, equivalent to redistributing the truncated segments over the available space.

A revised version of the radial Gaussian function that is truncated at a distance L_S from the plate and renormalized is

$$\hat{\rho}_h = E\{(4\pi x^2)/(2\pi\sigma^2)^{3/2}\}\exp\{-\tfrac{1}{2}([x - m]/\sigma)^2\} \tag{12.30}$$

where m = mean of the distribution, σ = a modified standard deviation and the normalizing factor E is given by

$$E=[\mathrm{erf}\{L_S/2^{\frac{1}{2}}\sigma\}-(L_S/\sigma)\,(2/\pi)^{\frac{1}{2}}\exp\{-L_S^{\,2}/2\sigma^2\}]$$

Note that the standard deviation σ of the distribution determines its width, which would be expected to be closely related to the thickness L_S of the steric layer. The value of σ is readily determined by trial and error once the fraction F of segments redistributed after truncation is (arbitrarily) specified since

$$\int_0^{\infty}\hat{\rho}_h-\int_0^{L_S}\hat{\rho}_h=F \tag{12.31}$$

Some typical values of σ/L_S for different fractions of the segments distributed after truncation are shown in Table 12.1.

The expression for the distance dependence of the steric repulsion obtained in this way in the interpenetrational domain is

$$P_I^M=E^{-2}(1/\pi\sigma^5)\exp\{-h^2/2\sigma^2\}[\sigma(2L_S-h)\exp\{\mathrm{h}L_S-L_S^{\,2}/\sigma^2)\}$$
$$\times(h^2/4+hL_S-L_S^{\,2}-3\sigma^2/2)+$$
$$(\pi^{\frac{1}{2}}/2)\exp\{h^2/4\sigma^2\}\mathrm{erf}\{(2L_S-h)/2\sigma\}\{h^4/4-h^2\sigma^2+36^4\}] \tag{12.32}$$

The corresponding expression for the interpenetrational-plus-compressional domain will be omitted because in that domain the chains would be expected to adopt a more symmetrical Gaussian-type distribution.

The Gaussian distribution function after truncation at L_S and re-normalization reads

$$\hat{\rho}_h=(1/\sigma[2\pi]^{\frac{1}{2}}\mathrm{erf}\{L_S/2^{3/2}\sigma\})\exp\{-\tfrac{1}{2}([x-m]/\sigma)^2\} \tag{12.33}$$

The values of σ/L_S for different fractions of segments renormalized can be determined by the integration procedure described above. Some typical values are given in Table 12.2. In the interpenetrational domain, this form of the distribution function yields

$$P_I^M=\exp\{-(h-L_S/2\sigma)^2\}\mathrm{erf}\{2L_S-h/2\sigma\}/[2\pi^{\frac{1}{2}}\sigma(\mathrm{erf}\{L_S/2^{3/2}\sigma\})^2] \tag{12.34}$$

TABLE 12.1

Values of σ/L_S for different fractions of segments redistributed in the radial Gaussian function

Fraction of segments redistributed	σ/L_S
0·05	0·364
0·01	0·299
0·001	0·249

TABLE 12.2
Some values of σ/L_S for different fractions of segments redistributed for the Gaussian distribution function

Fraction of segments redistributed	σ/L_S
0·02	0·215
0·01	0·194
0·002	0·162
0·001	0·152

In the interpenetrational-plus-compressional domain, the Gaussian distribution yields

$$P_{I+C}^{M} = \mathrm{erf}\{h/2\sigma^*\}/[2\pi^{\frac{1}{2}}\sigma^*(\mathrm{erf}\{h/2^{3/2}\sigma^*\})^2]$$
$$+(1/2\pi^{\frac{1}{2}})[(1/\mathrm{erf}\{h/2^{3/2}\sigma^*\})^2\mathrm{erf}\{h/2\sigma\}/\sigma^* -$$
$$(1/\mathrm{erf}\{L_S/2^{3/2}\sigma\})^2\mathrm{erf}\{L_S/2\sigma\}/\sigma] \qquad (12.35)$$

where $\sigma^* = d\sigma/L_S$.

The exponential function. The mathematical forms chosen to this point were those considered possible for tails. We now consider the exponential distribution function that may be relevant to loopy adsorption, at least for high molecular weight polymers. The exponential distribution function must be artificially truncated at the physical limit of the steric layer. This leads to a normalized segment density distribution of the form

$$\hat{\rho} = be^{-bx}/(1-e^{-bL_S}) \qquad (12.36)$$

where b = an arbitrary scaling constant. The value of b is related to the fraction (F) of segments e^{-bL_S} that would normally reside beyond $x = L_S$ but here have been arbitrarily redistributed over the available space. Tails are ignored in this distribution and would have to be treated separately by one of the other distribution functions considered previously.

The distance dependent functions derived by Smitham and Napper (1979) for the exponential function were

$$P_I^M = b^2(2L_S - h)e^{-bh}/(1-F)^2 \qquad (12.37)$$

and

$$P_{I+C}^{M} = \{B(1-\mathrm{e}^{-2Bh}) + 2B^2he^{-Bh}$$
$$-b(1-\mathrm{e}^{-2bL_S})\}/2(1-F^*)^2 \qquad (12.38)$$

where $F = e^{-bh}$ and $F^* = e^{-Bh}$.

Spheres

The Derjaguin integration procedure. Derjaguin (1934) has shown how flat plate potentials can be converted into sphere potentials in those systems where the thickness of the stabilizing sheath is considerably less than the radius of the particles. The Derjaguin method calculates the sphere potentials by summing the contributions of infinitesimally thick planar rings of radius q parallel to each other at a distance h (see Fig. 12.3). The area of each ring is simply $2\pi q\, dq$. Therefore the sphere potential can be given by

$$\Delta^S G = \int_0^\infty 2\pi q(\Delta^{FP} G) dq \tag{12.39}$$

The upper limit is really the value of q at which the interaction cases.

For equal spheres of radius a, simple geometrical considerations give

$$(h - H_0)/2 = a - (a^2 - q^2)^{\frac{1}{2}} \tag{12.40}$$

where H_0 = minimum distance of surface separation of the spheres. Differentiation of this relationship gives

$$2q dq = a(1 - (q^2/a^2))^{\frac{1}{2}} \mathrm{d}h \sim a \mathrm{d}h \tag{12.41}$$

since $a \gg q$ in the Derjaguin approximation. Consequently another rendition of the Derjaguin integration for equal spheres is

$$\Delta^S G = \pi a \int_{H_0}^{2L_S} \Delta^{FP} G \mathrm{d}h \tag{12.42}$$

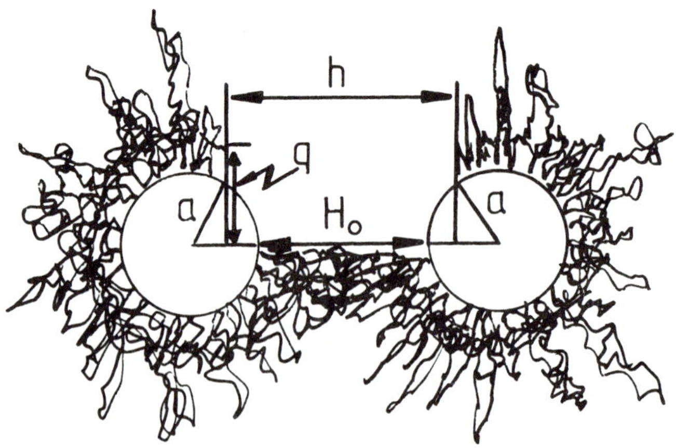

Fig. 12.3. Diagrammatic representation of the Derjaguin integration procedure for sterically stabilized spheres.

where the upper limit is the value of h where the interaction ceases. Geometries other than that of two equal spheres can be encompassed by the Derjaguin approximation. Some common relationships are summarized in Table 12.3 (Evans and Napper, 1973b).

The mixing interaction. The Derjaguin integration allows the flat plate potentials to be converted into sphere potentials *viz*

$$\Delta^S G^M = 2\pi a k T \omega^2 (\bar{v}_2{}^2/\bar{V}_1) N_A (\tfrac{1}{2} - \chi_1) S^M \tag{12.43}$$

where

$$S^M = \int_{H_0}^{2L_S} P^M \mathrm{d}h$$

TABLE 12.3
The relationship between the repulsive potentials for different geometries in the Derjaguin approximation

System	Repulsive energy
Equal spheres (radii a)	$\Delta^S G(a,a)$
Unequal spheres (radii a, b)	$(2b/[a+b])\Delta^S G(a,a)$
Flat plate/sphere (radius a)	$2\Delta^S G(a,a)$

Constant segment density model. The way by which the Derjaguin integration is performed for spheres is amply illustrated by the constant segment density model. In the interpenetration domain, equation (12.43) given above leads to

$$S_I^M = \int_{H_0}^{2L_S} (2/L_S)\ (1 - h/2L_S)\mathrm{d}h = 2(1 - H_0/2L_S)^2 \tag{12.44}$$

Substitution of this value into equation (12.43) leads to

$$\Delta^S G^M = 4\pi a k T \omega^2 (\bar{v}_2{}^2/\bar{V}_1) N_A (\tfrac{1}{2} - \chi_1)\ (1 - H_0/2L_S)^2 \tag{12.45}$$

This is identical with equation (12.12) derived by Fischer using an entirely different approach (Section 12.2.1).

In the interpenetrational-plus-compressional domain, for which the Fischer approach is incorrect, Smitham *et al.* (1975) showed that

$$S_{I+C}^M = 3\ln(L_S/H_0) + 2(H_0/L_S) - 1{\cdot}5 \tag{12.46}$$

The radial Gaussian distribution function. The radial Gaussian distribution function in the interpenetrational domain gives for spheres

$$S_I^M = (1/E^2\pi\sigma^5)\{\sigma\exp\{-L_S{}^2/\sigma^2\}[\sigma^2L^2/8 - \sigma^4/8 + \exp\{[2L_SH_0 - H_0{}^2]/2\sigma^2\}[-\sigma^2H_0{}^2/4 + \sigma^4 - 3L_S\sigma^2H_0/4 + 9L_S{}^2\sigma^2/4] + (\pi^{\frac{1}{2}}\sigma^2)/[32]^{\frac{1}{2}}\exp\{L_S{}^2/2\sigma^2\}\mathrm{erf}\{L_S/2^{\frac{1}{2}}\sigma\} - \mathrm{erf}\{[H_0 - L_S]/2^{\frac{1}{2}}\sigma\}(L_S{}^3 - 11L_S\sigma^2)] + (\pi^{\frac{1}{2}}/2)\times [\exp\{-L_S{}^2/\sigma^2\}(-4\sigma^2L_S{}^3 - 2L_S\sigma^4) + \exp\{-H_0{}^2/4\sigma^2\}\times (\sigma^4H_0 + \sigma^2H_0{}^3/2) + 4\pi^{\frac{1}{2}}\sigma^5(\mathrm{erf}\{L_S/2\sigma\} - \mathrm{erf}\{H_0/2\sigma\})]\} \quad (12.47)$$

where E and σ adopt values as explained in Section 12.3.1.2.

The Gaussian distribution function provides a somewhat simpler form for S:

$$S_I^M = 0{\cdot}5\{1/[\mathrm{erf}\{L_S/2^{3/2}\sigma\}]^2[1 - \mathrm{erf}\{(H_0 - L_S)/2\sigma\}] \quad (12.48)$$

In the interpenetrational-plus-compressional domain

$$S_{I+C}^M = 0{\cdot}5 + (1/K)\mathrm{erf}\{L_S/2\sigma\}[2\ln\{L_S/H_0\} - \{L_S - H_0\}/L_S] \quad (12.49)$$

where

$$K = (2\pi^{\frac{1}{2}}\sigma/L_S)(\mathrm{erf}\{L_S/2^{3/2}\sigma\})^2.$$

Exponential distribution function. Smitham and Napper (1979) showed that for an exponential segment density distribution function

$$S_I^M = \{e^{-2bL_S} + e^{-bH_0}[2bL_S - bH_0 - 1]\}/(1 - F^2) \quad (12.50)$$

and

$$S_{I+C}^M = bL_S\{e^{-2bL_S} + 2e^{-bL_S} - 1 + 2e^{-bL_S}[e^{-bL_S} - 1]/bL_S + [e^{-2bL_S} + 2e^{-bL_S} - 1]\ln(H_0/L_S) + [1 - e^{-2bL_S}](H_0/L_S)\}/2(1 - F)^2 \quad (12.51)$$

The elastic repulsion. Once the interpenetrational-plus-compressional domain is entered, not only must the mixing contribution to the steric repulsion be considered but so too must the elastic contribution. As the second particle approaches closer than the span of the stabilizing chains, the chains are compressed and so must lose configurational entropy. This is the origin of the elastic repulsion. The elastic repulsion is relatively insensitive to the solvency of the dispersion medium, being influenced by its nature only insofar as the solvency affects the chain conformation.

The loss of entropy of a single polymer chain attached to a colloidal particle, assumed to be a plate, on compression is found from the Boltzmann relationship to be

$$\Delta S = k\ln(\Omega_h/\Omega_\infty) \qquad (12.52)$$

where Ω_h = number of accessible configurational microstates at a separation of h. Consideration of υ such chains per unit area for two parallel flat plates leads to a free energy change from elastic effects of

$$\Delta^{FP}G_{EL}^{I+C} = -2\upsilon kT\ P_{EL} \qquad (12.53)$$

where $P_{EL} = \ln\ (\Omega_h/\Omega_\infty)$.

Evans *et al.* (1977) have shown how the loss of configurational entropy can be calculated assuming, somewhat unrealistically, that the segments behave as disconnected gas molecules with the appropriate number distribution function over the available space. Results obtained in this way for several distributions are summarized in Table 12.4. Also included in this table are the corresponding expressions for spheres obtained by a Derjaguin integration. These allow the elastic repulsion for spheres to be calculated from

$$\Delta^{S}G_{EL}^{I+C} = 2\pi a\upsilon kT\ S_{EL} \qquad (12.54)$$

Here a = particle radius and $\upsilon = \omega N_A/M$ where ω = weight of stabilizing chains for unit area of surface and M = molecular weight of the stabilizing chains.

The elastic repulsion calculated in this fashion is a rather sensitive function of the assumed form of the segment density distribution function. It is likely to be in error if only because of the unphysical nature of the model employed. Since the elastic term rises so steeply, however, its precise functional form is rarely of great moment.

Comparison of the different assumed forms. Figure 12.4 compares the predicted repulsions between two polystyrene latex particles stabilized by poly(vinyl alcohol) in water, assuming a range of mathematical forms (Smitham and

TABLE 12.4
The distance dependent functions for the elastic repulsion of spheres and flat plates

Segment density distribution function	P_{EL}	S_{EL}
constant	$^*-\ln\ \delta$	$^*L_S[1-\delta_0+\delta_0\ \ln\ \delta_0]$
Gaussian	$(\delta^2-1)/2-\ln\ \delta$	$L_S[2/3-\delta_0{}^3/6-\delta_0{}^2/2+\delta_0\ \ln\ \delta_0]$
radial Gaussian	$3[(\delta^2-1)/2-\ln\ \delta]$	$L_S[2-\delta_0{}^3/2-3\delta_0{}^2/2+3\delta_0\ \ln\ \delta_0]$
exponential	$(\delta-1)-\ln\ \delta$	$L_S[\frac{1}{2}-\delta_0{}^2/2+\delta_0\ \ln\ \delta_0]$

* Note that $\delta = h/L_S$ and $\delta_0 = H_0/L_S$, where L_S = barrier layer thickness, h = distance between the plates and H_0 = minimum distance between surfaces of the spheres.

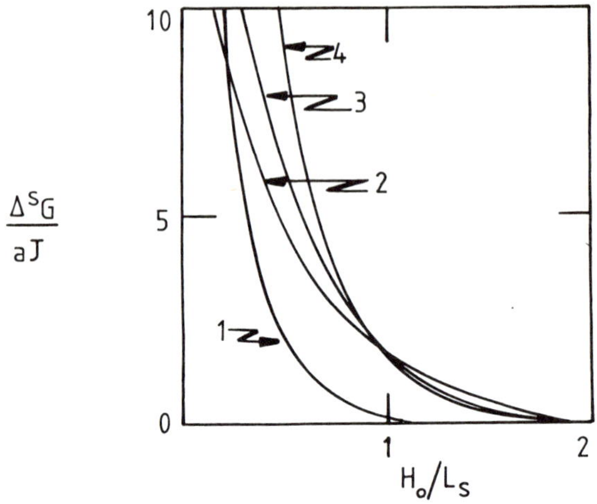

Fig. 12.4. The distance dependence of the steric interaction energy for two spheres of radius stabilized by polymer layers with different segment density distribution functions: 1, exponential; 2, constant; 3, Gaussian; 4, radial Gaussian (after Smitham and Napper, 1979).

Napper, 1979). The steric layer thickness was taken as being constant at 9·8 nm. It is apparent that the constant segment density, the radial Gaussian function and the Gaussian function all predict very similar distance relationships, showing that the distance dependence is relatively insensitive to the form of the distribution function chosen. Only the exponential distribution function gives a radically different result for the repulsion. The repulsion in the interpenetrational domain in that case is reduced by almost an order of magnitude compared with the other forms. This is a consequence of the removal of the stabilizing segments from the peripheral regions of the stabilizing sheath and an enhancement of their concentration in the regions close to the surface, as is appropriate for this type of distribution.

Comparison of theory with experiment. It will be shown in Section 13.3.2.1 that the flat plate potentials can be used to calculate the osmotic disjoining pressures in concentrated monodisperse sterically stabilized dispersions. Evans and Napper (1977) have compared the theoretical predictions using the above equations with those measured by Homola and Robertson (1976) for polystyrene latex particles stabilized by poly(oxyethylene) of molecular weight ca 2 000 in aqueous dispersion media. The elastic repulsion in the interpenetrational-plus-compressional domain was estimated from the following expression for the constant segment density model

$$\Delta^{FP} G_{EL}^{I+C} = (2RT\, L_S \bar{v}_2 / \bar{V}_2) \ln(1/\delta_0) \tag{12.55}$$

where $\overline{V}_2$ = molar volume of the polymer. The values of $a(=170$ nm), L_S ($=9{\cdot}0$ nm) and $\overline{V}_2$ ($=1\,950$ cm^3 mol^{-1}) were known from experiment. The value of χ_1 was the sole adjustable parameter. Its value was chosen as 0·497 to provide reasonable agreement between theory and experiment at $H_0 = L_S$. This value, although close to the θ-solvent value, appears to be reasonable in light of the observation that water behaves rather like a θ-solvent for very low molecular poly(oxyethylene) (Rossi *et al.*, 1963).

It is apparent from Fig. 12.5 that the measured distance dependence of the osmotic pressure is well described by the theoretical equations in the interpenetrational domain ($H_0 \gtrsim 9$ nm). On closer approach, however, the predicted repulsion rose too steeply, due presumably to an overestimation of the elastic repulsion. Better agreement between theory and experiment could be achieved by arbitrarily softening the elastic repulsion: this was accomplished by replacing δ_0 in equation (12.55) by $\delta_0(3-\delta_0)^2/4$.

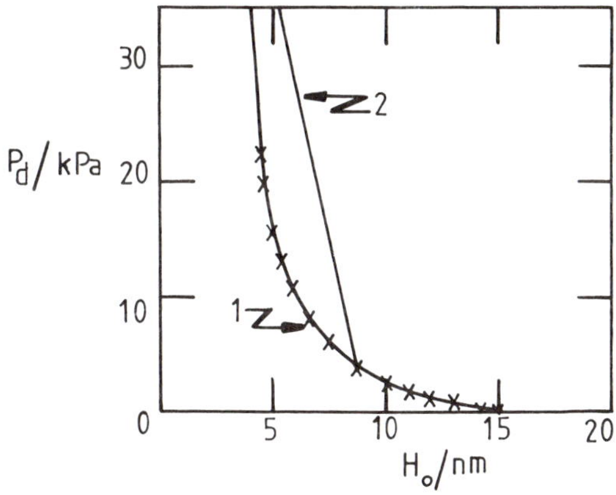

Fig. 12.5. The disjoining pressure p_d as a function of the interparticle distance of separation for spheres sterically stabilized by poly(oxyethylene); curve 1, the experimental results of Homola and Robertson (1976); curve 2, constant segment density model. The crosses (×) show the theoretical results for a softened elastic potential (after Evans and Napper, 1977).

12.3.2. Incorporation of a concentration dependent interaction parameter

Most theories of steric stabilization, whether pseudo-*ab initio* or pragmatic, are based upon a concentration independent interaction parameter. Yet it is known that the measured values of the interaction parameter are usually strongly concentration dependent. Commonly, the interaction parameter increases in magnitude as the volume fraction of the polymer chains increases, although such behaviour is by no means universal. Evans and Napper (1977) have shown how the concentration dependence of the interaction parameter can be incorporated into the pragmatic theories of steric stabilization.

It is important to recognize that the experimentally determined interaction parameter is usually derived from the measurement of the activity of the solvent in a polymer solution, coupled with the application of the Flory-Huggins expression for the excess (denoted by superscript E) chemical potential

$$RT \ln a_1 = \mu_1{}^E = RT\{\ln(1-v_2) + \chi v_2{}^2\} \tag{12.56}$$

where a = activity, μ = chemical potential and v = volume fraction. The subscripts 1 and 2 denote the solvent and polymer respectively.

To calculate steric repulsion, however, what is required is the parent excess free energy of mixing, rather than the corresponding chemical potential term. We showed in the previous chapter that for a volume element δV

$$\delta(\Delta G_M{}^E) = kT(\ln v_1 + \chi^* v_2)\delta n_1$$

where χ^* = interaction parameter appropriate to the free energy. Since $\delta n_1 = v_1\ \delta V/V_1$ and $v_1 = 1 - v_2$, we have

$$\delta(\Delta G_M{}^E) = (kT/V_1)\ (1-v_2)\{\ln(1-v_2) + \chi^* v_2\}\delta V \tag{12.57}$$

where V_1 = volume of a molecule of the dispersion medium.

Differentiation of equation (12.57) gives the excess chemical potential of the solvent

$$\mu_1{}^E = RT\left\{\ln(1-v_2) + v_2 + \left[\chi^* - (1-v_2)\frac{\partial \chi^*}{\partial v_2}\right]v_2{}^2\right\} \tag{12.58}$$

Comparison of equations (12.56) and (12.58) shows that only if χ^* is independent of the polymer concentration will χ^* be equal to the measured value χ. This is the exception rather than the rule.

12.3.2.1. The concentration dependence of χ

Both χ and χ^*, which are excess exchange free energies, may be expressed as power series in v_2 (Tompa, 1956; Orofino and Flory, 1957; Koningsveld and Staverman, 1968; Koningsveld *et al.*, 1970)

$$\chi^* = \chi_1{}^* + \chi_2{}^* v_2 + \chi_3{}^* v_2{}^2 + \ldots$$

$$= \sum_{i=1}^{\infty} \chi_i{}^* v_2{}^{i-1} \tag{12.59}$$

with an analogous expression for χ. Comparison of equations (12.56) and (12.58) gives

$$\chi = \chi^* - (1-v_2)\frac{\partial \chi^*}{\partial v_2} \tag{12.60}$$

Differentiation of equation (12.59) for χ^* yields

$$\frac{\partial \chi^*}{\partial v_2} = \sum_{i=1}^{\infty} i\chi_{i+1}^* v_2^{\,i-1} \tag{12.61}$$

so that

$$\chi = \sum_{i=1}^{\infty} i(\chi_i^* - \chi_{i+1}^*) v_2^{\,i-1} \tag{12.62}$$

$$= \sum_{i=1}^{\infty} \chi_i v_2^{\,i-1} \tag{12.63}$$

It can be inferred from equations (12.62) and (12.63) that

$$\chi_i = i(\chi_i^* - \chi_{i+1}^*) \tag{12.64}$$

for all integral i greater than zero. Thus

$$\sum_{i=j}^{\infty} \chi_i / i = \sum_{i=j}^{\infty} (\chi_i^* - \chi_{i+1}^*) = \chi_j^* \tag{12.65}$$

This result will be used below.

12.3.2.2. The mixing free energy

Equation (12.65) establishes the link required between the experimentally determined interaction parameter χ and the interaction parameter χ^* that appears in the free energy expression. Thus we can write

$$\chi^* v_2 = v_2 \sum_{i=1}^{\infty} \chi_i / i + v_2^{\,2} \sum_{i=2}^{\infty} \chi_i / i + \ldots$$

$$= \sum_{j=1}^{\infty} v_2^{\,j} \sum_{i=j}^{\infty} \chi_i / i \tag{12.66}$$

We may also expand the logarithmic term

$$\ln(1 - v_2) = -\sum_{j=1}^{\infty} v_2^{\,j} / j \tag{12.67}$$

Another rendition of equation (12.57) is therefore

$$\delta(\Delta G_M{}^E) = (kT/V_1)\,(1 - v_2)\{\sum_{j=1}^{\infty} v_2^{\,j}[\sum_{i=j}^{\infty} \chi_i / i - 1/j]\}\delta V$$

$$= (kT/V_1) \sum_{i=1}^{\infty} (v_2^{\,i+1}/i)\,[1/(i+1) - \chi_i]\delta V$$

$$-(kT/V_1) v_2 (1 - \sum_{i=1}^{\infty} \chi_i / i)\delta V \tag{12.68}$$

Flat plates

To proceed further, we shall consider the case of two parallel flat plates fully coated by attached polymers. The segment density parallel to the surface is assumed to be constant so that all the distance dependence of the repulsion will henceforth be taken to reside in the direction normal to the interface, which we denote as the x-direction. To calculate the total excess free energy of mixing, we must integrate over all volume elements in the attached layer:

$$\Delta^{FP}G_M{}^E=(kT/V_1)\sum_{i=1}^{\infty}(1/i)[1/(i+1)-\chi_i]\int_0^{\infty}v_2{}^{i+1}\mathrm{d}x-$$
$$(kT/V_1)\,(1-\sum_{i=1}^{\infty}\chi_i/\mathrm{i})\int_0^{\infty}v_2\mathrm{d}x \tag{12.69}$$

For the infinite separation of two flat plates, the excess free energy of mixing per unit surface area is

$$\Delta^{FP}G_M{}^E(\infty)=(2kT/V_1)\sum_{i=1}^{\infty}(\upsilon n_sV_s)^{i+1}(1/i)[1/(i+1)-\chi_i]\int_0^{\infty}\hat{v}_{\infty}^{i+1}\mathrm{d}x$$
$$-(2kT\upsilon n_sV_s/V_1)\,(1-\sum_{i=1}^{\infty}\chi_i/i) \tag{12.70}$$

where υ = number of attached chains per unit area of one surface, n_s = number of segments per chain, V_s = volume of a segment and $\hat{v}_d$ = normalized volume fraction distribution function in the x-direction for a plate separation of d. Note that the average volume fraction of segments ($\bar{v}_2$) in the entire barrier is given by $\bar{v}_2=\upsilon n_sV_s/L_S$, where L_S = span of the attached polymer molecules.

The interpenetrational domain. For parallel flat plates separated by a distance h, the interpenetrational domain is defined by $L_S \le h \le 2L_S$. The excess free energy of mixing at separation h may be expressed as

$$\Delta^{FP}G_M{}^E(d)=(kT/V_1)\sum_{i=1}^{\infty}[(\bar{v}_2L_S)^{i+1}/i][1/(i+1)-\chi_i]$$
$$\times\left\{\int_{h-L_S}^{L_S}(\hat{v}_{\infty}+\hat{v}'_{\infty})^{i+1}\mathrm{d}x+\int_0^{h}(\hat{v}_{\infty})^{i+1}\mathrm{d}x+\right.$$
$$\left.\int_L^{h}(\hat{v}'_{\infty})^{i+1}\mathrm{d}x\right\}-(2kT\bar{v}_2L_S/V_1)\,(1-\sum_{i=1}^{\infty}\chi_i/i) \tag{12.71}$$

The prime in this equation denotes the second chain, which is the mirror of the first chain at the second surface (i.e. $\hat{v}_x=\hat{v}'_{h-x}$), the origin being taken at the point of attachment of the first chain.

The change in the free energy of mixing on close approach is obtained by subtracting equation (12.70) from equation (12.71)

$$\Delta^{FP}G_M{}^I=(kT/V_1)\sum_{i=1}^{\infty}[(\bar{v}_2L_S)^{i+1}/i][1/(i+1)-\chi_i]$$
$$\times\left\{\int_{h-L_S}^{L_S}(\hat{\rho}_\infty+\hat{\rho}_\infty{}')^{i+1}\mathrm{d}x-\int_{h-L_S}^{L_S}[(\hat{\rho}_\infty)^{i+1}+(\hat{\rho}_\infty{}')^{i+1}]\mathrm{d}x\right\} \quad (12.72)$$

where $\hat{\rho}$ ($=\hat{v}$) = normalized segment density distribution function in the x-direction and the superscript I denotes the interpenetrational domain.

The first term ($i=1$) in equation (12.72) is simply

$$\Delta^{FP}G_M{}^I(i=1)=(2kT/V_1)\ (\bar{v}_2L_S)^2(\tfrac{1}{2}-\chi_1)\int_{h-L_S}^{L_S}\hat{\rho}_\infty\ \hat{\rho}'_\infty\mathrm{d}x \quad (12.73)$$

which is identical with the result obtained for a concentration independent interaction parameter. The next term ($i=2$) is the first order perturbation correction, which may be expanded as

$$\Delta^{FP}G_M{}^I(i=2)=(3kT/2V_1)\ (\bar{v}_2L_S)^3(1/3-\chi_2)$$
$$\times\int_{h-L_S}^{L_S}(\hat{\rho}_\infty{}^2\ \hat{\rho}'_\infty+\hat{\rho}_\infty\hat{\rho}'_\infty{}^2)\mathrm{d}x \quad (12.74)$$

Inspection of equations (12.73) and (12.74) suggests that the ratio of the perturbation term to the leading term is of order $\bar{v}_2$. Typically, $\bar{v}_2$ for uncompressed polymers lies in the range 0·001–0·20, where the upper bound is appropriate to low molecular weight stabilizers. This leads to the conclusion that the concentration dependence may be significant for low molecular weight stabilizers but is of minor consequence for high molecular weight species in the interpenetrational domain. The higher terms are of order of magnitude $\bar{v}_2{}^2$ or less and so do not contribute greatly to the interaction in most circumstances.

Note that, as mentioned previously, both χ_2 and χ_3 often tend to approach $\frac{1}{2}$ and $\frac{1}{3}$ respectively as the θ-point is approached. Accordingly, both free energy terms will approach zero at the θ-point and become negative in slightly worse than θ-solvents. Thus the concentration dependence of χ is unlikely to interfere with the expected correlation between the critical flocculation point and the θ-point. This accords with the results of several experiments.

The interpenetrational-plus-compressional domain. In the interpenetrational-plus-compressional domain, defined by $h<L_S$, the excess free energy of mixing is given by

$$\Delta^{FP}G_M{}^{I+C}=(kT/V_1)\sum_{i=1}^{\infty}[(\bar{v}_2L_S)^{i+1}/i][1/(i+1)-\chi_1]$$

$$\times\int_0^h(\hat{v}_h+\hat{v}'_h)^{i+1}\mathrm{d}x-(2kT\bar{v}_2L_S/V_1)\ (1-\sum_{i=1}^{\infty}\chi_i/i) \quad (12.75)$$

where the superscript $I+C$ specifies the interpenetrational-plus-compressional domain. Consequently the free energy of close approach of the flat plates in this domain becomes

$$\Delta^{FP}G_M{}^{I+C}=(kT/V_1)\sum_{i=1}^{\infty}[(\bar{v}_2L_S)^{i+1}/i][1/(i+1)-\chi_1] \times\left\{\int_0^h(\hat{\rho}_h+\hat{\rho}'_h)^{i+1}\mathrm{d}x-2\int_0^{\infty}\hat{\rho}_{\infty}{}^{i+1}\mathrm{d}x\right\} \quad (12.76)$$

The first term in this series is

$$\Delta^{FP}G_M{}^{I+C}(i=1)=(2kT/V_1)V_S{}^2n_S{}^2\upsilon^2(\tfrac{1}{2}-\chi_1) \times\left\{\int_0^h\hat{\rho}_h{}^2\mathrm{d}x+\int_0^h\hat{\rho}_h\,\hat{\rho}'_h\mathrm{d}x-\int_0^{\infty}\hat{\rho}_{\infty}{}^2\mathrm{d}x\right\}$$

which is again the result obtained for a concentration independent parameter. The first order perturbation term is

$$\Delta^{FP}G_M{}^{I+C}(i=2)=(kT/V_1)V_S{}^3n_S{}^3\upsilon^3(1/3-\chi_2) \times\left\{\int_0^h\hat{\rho}_h{}^3\mathrm{d}x+\frac{3}{2}\int_0^h\hat{\rho}_h{}^2\hat{\rho}'_h\mathrm{d}x+\frac{3}{2}\int_0^h\hat{\rho}_h\,\hat{\rho}'^2_h\mathrm{d}x -\int_0^{\infty}\hat{\rho}_{\infty}{}^3\mathrm{d}x\right\} \quad (12.77)$$

Again the perturbation term is of order $\bar{v}_2$ if compared with the leading term and so the remarks set forth in the preceding section on the importance of the concentration dependence of χ are applicable to this domain as well. Note, however, that the value of $\bar{v}_2$ in the interpenetrational-plus-compressional domain is significantly greater than that in the interpenetrational domain. It follows that the concentration effects, including the terms for $i \gtrsim 2$, are likely to be significant for most molecular weight stabilizing moieties.

The constant segment density model. Because the perturbation approach introduces considerable complexity into the mixing term, only the simplest model will be considered here. This assumes that the segment density in the steric layer is constant whereupon the following relationships hold: $\hat{\rho}_h=\hat{\rho}'_h=1/h$ *and* $\hat{\rho}_{\infty}=\hat{\rho}'_{\infty}=1/L_S$. Substitution of these values into equations (12.74) and (12.77) yields after integration

$$\Delta^{FP}G_M{}^I=(2RT\,L_S/\bar{V}_1)\sum_{i=1}^{\infty}[(\bar{v}_2)^{i+1}/i][1/(i+1)-\chi_i] \times(2^i-1)\,(2-\delta) \quad (12.78)$$

$$\Delta^{FP}G_M{}^{I+C} = (2RT\,L_S/\bar{V}_1)\sum_{i=1}^{\infty}[(\bar{v}_2)^{i+1}/i][1/(i+1)-\chi_i] \times \{(2/\delta)^i - 1\} \quad (12.79)$$

where $\delta = h/L_S$ and $\bar{V}_1$ $(=V_1N_A)$ = molar volume of the dispersion medium. Note that if $\delta = 1$, these two equations are identical so that the free energy of close approach is properly continuous at the boundary between the two domains of close approach.

The flat plate expressions for the free energy of steric interaction can be integrated using the Derjaguin (1934) method to yield the following sphere potentials

$$\Delta^{S}G_M{}^{I} = (4\pi a\,L_S{}^2RT/\bar{V}_1)\sum_{i=1}^{\infty}[(\bar{v}_2)^{i+1}/i][1/(i+1)-\chi_i] \times (2^i - 1)\,(1-\delta_0/2)^2 \quad (12.80)$$

$$\Delta^{S}G_M{}^{I+C} = (4\pi a\,L_S{}^2RT/\bar{V}_1)\Bigg\{\bar{v}_2{}^2(\tfrac{1}{2}-\chi_1)\,(\delta_0/2 - 1/4 - \ln\,\delta_0) + \sum_{i=2}^{\infty}[(\bar{v}_2)^{i+1}/i][1/(i+1)-\chi_i] \times \left[2^{i-2} + \delta_0/2 + (2/\delta_0)^{i-1}/(i-1) - 2^{i-1}/(i-1) - \frac{3}{4}\right]\Bigg\} \quad (12.81)$$

Here $\delta_0 = H_0/L_S$, where H_0 = minimum distance between the surfaces of the particles, and a $(\gg L,\ H_0)$ = the radius of the particles.

12.4. The denting hypothesis

As mentioned previously, Bagchi (1974a,b; 1975; 1976) has proposed that stabilizing moieties do not interpenetrate one another but rather undergo mutual compression, even in what has been termed above the interpenetrational domain. Such mutual compression has been referred to as *denting*. A diverse range of arguments can be marshalled against the occurrence of denting in many sterically stabilized systems.

12.4.1. A statistical mechanical argument against denting

The first thing that must be recognized about denting is that it would represent a decidely more severe constraint on the conformation of the stabilizing polymers. With denting, the polymer chains are excluded completely from the interactional zone. Interpenetration, on the other hand, represents only partial exclusion from the zone since the segmental volume fraction rarely exceeds 0·1.

Flory (1969) has stressed that the conformational partition function for a polymer molecule is extremely large (e.g. of order 2^{100} even for a short chain of 100 bonds). For this reason, a polymer chain has many ways of fulfilling its conformational proclivities in the presence of a second chain. Consequently, interpenetration is strongly favoured over denting.

12.4.2. Chemical evidence against the interpenetration of polymer chains

Considerable reluctance has been displayed, even by polymer chemists, in accepting the concept of free polymer interpenetration as proposed by Flory for θ-solvents, melts and higher polymer concentrations in good solvents. This reluctance has quite properly spawned attempts to devise experiments that disprove the interpenetration hypothesis using chemical means. As it happened, many of the early innovative studies of Vollmert, well summarized in his 1975 review (Vollmert, 1975), appeared to demonstrate that interpenetration did not occur significantly in more concentrated solutions.

Vollmert and coworkers attempted to study the phenomenon of interpenetration by exploiting the occurrence of crosslinking reactions that could proceed solely by an intermolecular, and not an intramolecular, mechanism. This is exemplified by the reaction of isocyanate groups attached to one set of chains with hydroxyl groups attached to a second set of chains to form carbamate (urethane) crosslinkages, as shown schematically in Fig. 12.6. Neither the isocyanate groups nor the hydroxyl groups undergo mutual self-reaction under the conditions of the experiment. The experiments were performed on mixtures of the statistical copolymers poly(*n*-butyl acrylate-*co*-2-isocyanatoethyl acrylate) and poly(*n*-butyl acrylate-*co*-4-hydroxybutyl acrylate).

Vollmert observed, as was expected, that the extent of crosslinking was a function of the polymer concentration. At concentrations below 5%, the

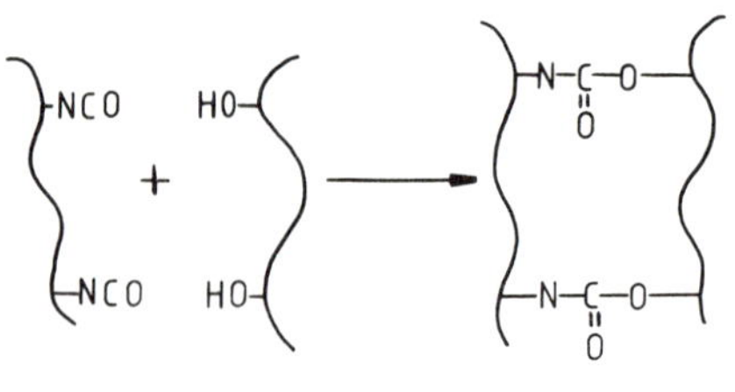

Fig. 12.6. Schematic representation of the intermolecular reaction of isocyanate and polyol reactions to yield urethane (carbamate) linkages.

extent of reaction was relatively small ($<20\%$). The gel produced in this manner appeared to consist merely of coils interlinked at their surfaces via urethane groups. The extent of the intermolecular reaction was found to increase steadily as the polymer concentration increased; but the extent of reaction never went to completion and, indeed, never exceeded ca 40%. Vollmert also studied *cis*-1,4-polybutadiene chains substituted with hydroxyl and isocyanate groups. These chains were looked at because, in contrast to the butyl acrylate polymers, their backbones were relatively bare. They lacked the pendant butyl side chains that could have shielded the substituents from chemical reaction in the study of the butyl acrylate polymers. A qualitatively similar pattern of behaviour, however, emerged with polybutadiene as was observed with the butyl acrylate polymers.

Braun (1976) has pointed out that there exist at least two objections that can be raised to this type of experiment: first, on mixing the predissolved polymers containing the two different functional groups, rapid chemical reaction of the functional groups on first contact could prevent the occurrence of complete interpenetration; second, the polymer chains may be incompatible due to the introduction of the two different types of functional groups.

Vollmert endeavoured to avert these criticisms by carrying out the reactions at various degrees of substitution by the functional groups. The results were then extrapolated to low values for the degree of substitution. Even under these conditions, the maximum attainable conversion of ca 20% in a 30% polymer solution suggested that interpenetration was minimal in these systems. Vollmert obtained similar results for the crosslinking of poly(methyl methacrylate) containing hydroxyl groups with an equivalent amount of polymer containing acyl chloride groups in a θ-solvent (a benzene/n-hexane mixture). Roughly the same degree of coil interpenetration (5–20%) was found in a θ-solvent as in good solvents. It was argued that steric hindrance must have been responsible for the observed noninterpenetration.

12.4.3. Evidence in favour of interpenetration

The cause of noninterpenetration in Vollmert's experiments was left up in the air because of the possibility of thermodynamic (and even steric) incompatibility. Braun (1976) endeavoured to circumvent this ambiguity by carrying out crosslinking reactions on identical polymer molecules. This was based upon the observation that 1,1-diphenylethylene in the presence of sodium metal forms radical anions that rapidly dimerize in quantitative yield. This reaction was applied to poly(1-phenyl-1-(4-vinylphenyl) ethylene-*co*-styrene), which can in principle undergo both intramolecular and intermolecular reactions. Intramolecular reactions cause coil contraction, which is manifest by a decrease in the intrinsic viscosity of the polymer solutions. Intermolecular reactions stiffen the polymer chains, their structure being analogous to ladder polymers, and so increase the intrinsic viscosity of the polymer solutions.

Braun found that at low polymer concentrations ($<2{\cdot}5\%$), the intrinsic viscosity was reduced after treatment of the polymer with sodium naphthalene. This pointed to the expected intramolecular crosslinking. Above 2·5% concentration, the intrinsic viscosity was found to increase on reaction, implying that intermolecular reactions had taken place. Quantitatively, it was found that for a 5% solution, more than 90% of the coils were linked intermolecularly. This is vastly in excess of that observed by Vollmert but in good accord with the Flory prediction of interpenetration at higher polymer concentrations. It suggests that the introduction of the functional groups, -NCO and -OH, in Vollmert's experiments prevented the chains from undergoing the expected interpenetration. This is scarcely surprising: the chains studied by Vollmert were as chemically different as would be copolymers of, say, *n*-butyl methacrylate with either vinyl chloride or styrene. If the latter comonomers were present, in significant amounts, incompatibility would be expected to prevent the mutual interpenetration of the resulting copolymers.

Neutron scattering studies have confirmed that polymer chains in the melt display the same unperturbed dimensions as are observed in θ-solvents (Benoit *et al.*, 1973a,b; Allen, 1976a,b). This would only be observed if the polymer chains underwent interpenetration. The absence of interpenetration in such concentrated systems would inexorably lead to highly compressed chains whose dimensions would be dramatically less than those observed in dilute θ-solvent solutions.

Further experimental proof of the occurrence of interpenetration resides in the verification of the scaling laws for concentrated polymers, reviewed in Section 4.6.3.5. Scaling law theories assume that polymer chains interpenetrate in the semi-dilute regime and their predictions appear to be confirmed experimentally.

12.5. Predictions of the free volume theory for nonaqueous dispersions

Free volume theory enables the relative magnitudes of the various components of steric stabilization to be calculated for nonaqueous dispersions. The basis of these theoretical predictions is the thermodynamic factor $(\frac{1}{2}-\chi_1)$, where χ_1 = interaction parameter. The way by which this factor enters the steric interaction is exemplified by the interpenetrational domain for spheres

$$\Delta^S G^M = 2kT(V_S{}^2/V_1)\,(\tfrac{1}{2}-\chi_1)S \qquad (12.82)$$

According to this equation, the steric free energy of interaction is proportional to $(\frac{1}{2}-\chi_1)$. In general terms, the steric free energy in the interpenetrational domain, which determines the incipient flocculation behaviour of colloidal dispersions, can be expressed as

$$\Delta G^M = \Gamma \mathcal{T} S \tag{12.83}$$

where Γ = geometrical factor, $\mathcal{T}$ = thermodynamic factor and S = segment density distribution function term. The thermodynamic term $\mathcal{T}$ incorporates the factor $(\frac{1}{2}-\chi_1)$.

According to the precepts of the original Flory-Huggins theory, the interaction parameter χ_1 should be independent of the volume fraction of polymer. The measured values of χ_1, however, usually display a strong dependence upon the volume fraction of polymer. To indicate that the interaction parameter is a variable, we will drop the subscript 1 and denote the variable by χ.

12.5.1. The temperature dependence of the interaction parameter

12.5.1.1. The macroscopic thermodynamic approach

Croucher and Hair (1978) have followed Eichinger (1970) in considering the temperature dependence of χ in macroscopic thermodynamic terms. In the classical Flory-Huggins theory, the enthalpy and entropy of dilution parameters κ_1 and ψ_1 are constants related to the θ-temperature by $\theta = (\kappa_1/\psi_1)T$ and to χ_1 by $(\frac{1}{2}-\chi_1) = \psi_1 - \kappa_1$. These relationships lead immediately to

$$\tfrac{1}{2}-\chi_1 = \psi_1(1-\theta/T) \tag{12.84}$$

which is the functional form of the excluded volume parameter in the vicinity of the θ-point. If ψ_1 and κ_1 were truly constant, then, as ψ_1 is usually positive in nonaqueous media at room temperature, the function $(\frac{1}{2}-\chi_1)$ would increase monotonically with temperature. This would imply that nonaqueous sterically stabilized dispersions should become more stable on heating whereas it is known that most undergo flocculation.

The underlying cause of this inconsistency with experiment is associated with the temperature dependence of ψ and κ (where the subscripts are also now dropped to indicate the possible variability of these parameters). On heating, the values of ψ and κ decrease and ultimately change sign. Negative values for ψ and κ, as Flory (1953) has pointed out, presage the occurrence of a θ_L-point on heating. Negative values for ψ and κ also forecast the onset of flocculation at an upper critical flocculation temperature (UCFT) for a dispersion of particles sterically stabilized by the polymer.

A change in sign of the parameters ψ and κ from positive to negative on increasing the temperature necessary implies that the first derivative of κ with respect to temperature must be negative. Recalling the classical thermodynamic relationship $\partial(\Delta H)/\partial T=\Delta C_P$, where H=the enthalpy and C_P=the heat capacity at constant pressure, we see that the sign change on heating implies that the excess (E) partial molar heat capacity of mixing (${}^E\bar{C}_P{}^M$) must have a negative sign and be of considerable magnitude. This heat capacity parameter is properly regarded as an excess function because it is not taken into account by the Flory-Huggins ideal expressions. Excess functions, as Flory (1953) has noted, are necessarily related to terms involving the square (and highest powers) of the volume fraction of the polymer v_2. This dependence reflects the importance of two-body (and higher) interactions. Accordingly, it is possible to write that

$$\zeta = {}^E\bar{C}_P{}^M/Rv_2{}^2 = T(\mathrm{d}\psi/\mathrm{d}T) = \mathrm{d}(T\kappa)/\mathrm{d}T \qquad (12.85)$$

The gas constant R is introduced here to ensure dimensional consistency. It compensates for the molar character of ${}^E\bar{C}_P{}^M$, remembering that both ψ and κ are dimensionless ratios.

The quantity ${}^E\bar{C}_P{}^M/Rv_2{}^2$ in equation (12.85) is fairly constant over the temperature range between θ_U and θ_L, provided that their temperature difference is not excessively large. Thus ζ in equation (12.85) can be assigned some mean value, $\bar{\zeta}$, in this temperature range. This allows the various relationships in equation (12.85) to be integrated readily

$$\psi(T) - \psi(\theta_U) \simeq \int_{\theta_U}^{T} (\bar{\zeta}/T)\mathrm{d}T$$

so

$$\psi(T) \simeq \psi(\theta_U) - \bar{\zeta}\ln(\theta_U/T) \qquad (12.86)$$

and, similarly

$$\kappa(T) \simeq \kappa(\theta_U)\ (\theta_U/T) + \bar{\zeta}(1-\theta_U/T) \qquad (12.87)$$

These results can be collected together in the temperature dependent thermodynamic factor that governs stability

$$\tfrac{1}{2} - \chi(T) = \psi(T) - \kappa(T)$$

$$\simeq \psi(\theta_U)\ (1-\theta_U/T) - \bar{\zeta}[(1-\theta_U/T) + \ln(\theta_U/T)] \qquad (12.88)$$

The interaction parameter

Croucher and Hair (1978) have presented a plot of temperature dependence of the interaction parameter $\chi(T)$ as a function of temperature for poly(α-methyl styrene) in *n*-butyl chloride. This is shown in Fig. 12.7. The curve was computed from equation (12.88) using the following numerical values for the parameters: $\psi(\theta_U)=0{\cdot}12$ (Cowie and McEwen, 1975), $\theta_U \simeq$ LCFT = 254 K and $\theta_L \simeq$ UCFT = 403 K (Croucher and Hair, 1978). These results allowed the value of $\bar{\zeta}$ to be calculated from equation (12.88) since $[\tfrac{1}{2}-\chi(T)]=0$ at $T=\theta_L$ ($\bar{\zeta}=-0{\cdot}48_3$).

The most striking feature of the plot of $\chi(T)$ *vs* temperature is its parabolic shape: $\chi = \frac{1}{2}$ at both the upper and lower θ-temperatures. The values of χ are less than 0·5 at intermediate temperatures, indicating solvency conditions for the polymer that are significantly better than those of a θ-solvent.

The parabolic shape of the curve can be readily understood from Fig. 12.8. Curve 1 presented therein displays the function $\psi(\theta_U)(1-\theta_U/T)$, which is a monotonically increasing function. This is the classical Flory-Huggins prediction: sterically stabilized particles should become progressively more stable on heating. Curve 2 displays the heat capacity term arising from the free volume contribution that decreases monotonically with increasing temperature. Curve 3 displays what happens on addition of the classical Flory-

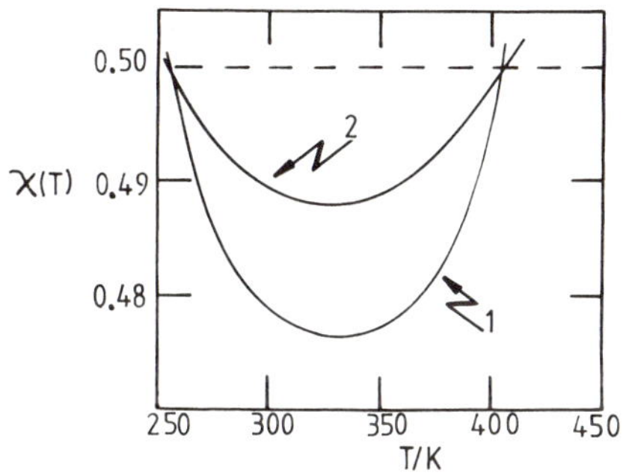

Fig. 12.7. The temperature dependence of the interaction parameter χ for poly(α-methylstyrene) in *n*-butyl chloride: 1, microscopic theory; 2, macroscopic theory (after Croucher and Hair, 1980b).

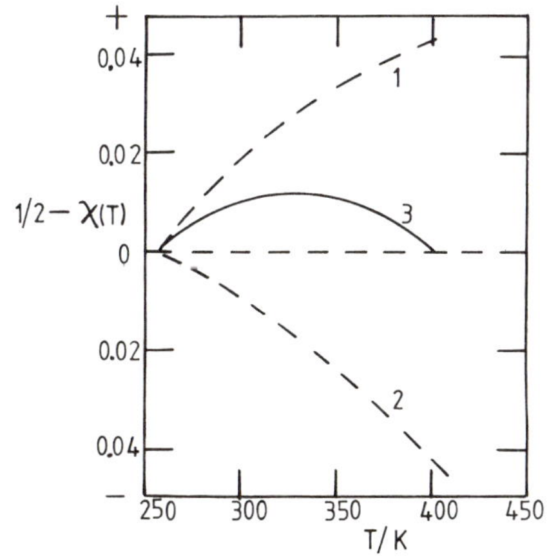

Fig. 12.8. The theoretical temperature dependence of the components of the interaction parameter for poly(α-methylstyrene) in *n*-butyl chloride: 1, Flory-Huggins contribution; 2, free volume contribution; 3, sum of 1 and 2 (after Croucher and Hair, 1978).

Huggins and the excess functions, whereupon the parabolic nature of the combined curve becomes apparent. This is also obvious from the plots of ψ(T) and κ(T) shown in Fig. 12.9. Both parameters are identical and positive at $T=\theta_U$ but both decrease monotonically as the temperature increases. They are again equal, when both are negative, at $T=\theta_L$.

If the athermal condition is taken to be that for which $\kappa=0$, then the temperature at which this condition occurs is readily found from equation (12.87) to be

$$\begin{aligned} T_M &= \theta_U\{1-[\kappa(\theta_U)/\bar{\zeta}]\} \\ &= \theta_U\{1-[\psi(\theta_U)/\bar{\zeta}]\} \end{aligned} \tag{12.89}$$

The subscript M is appended here because, as is shown in the next section, the temperature at which $\kappa=0$ also corresponds to the maximum in the thermodynamic function $[\frac{1}{2}-\chi(T)]$.

The temperature of maximum stability

Differentiation of equation (12.88) shows that the temperature dependent interaction parameter $\chi(T)$ has a minimum value at the temperature given by

$$T_M \simeq \theta_U\{1-[\psi(\theta_U)/\bar{\zeta}]\} \tag{12.90}$$

This temperature is important because the function $[\frac{1}{2}-\chi(T)]$ must reach its maximum value at T_M. Consequently, T_M corresponds, at least approximately, to the temperature at which the steric repulsion in the interpenetrational domain is maximal. This provides a relatively simple criterion by which the steric repulsion in the interpenetrational domain can be

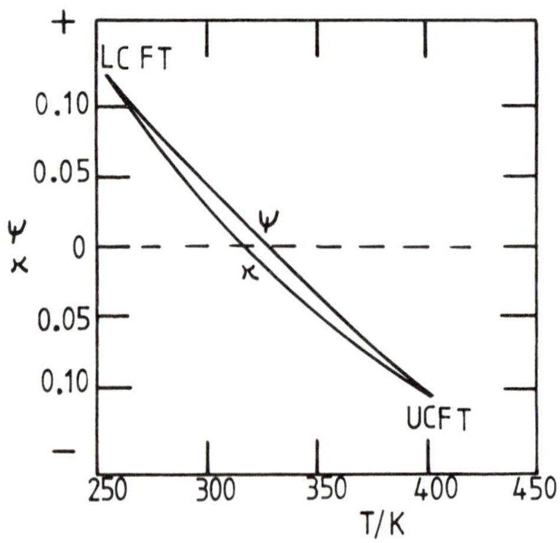

Fig. 12.9. The theoretical temperature dependence of the enthalpy and entropy of dilution parameters for poly(α-methylstyrene) in *n*-butyl chloride (after Croucher and Hair, 1978).

optimized. All that is required is for θ_U and θ_L to be known (and they often can be assumed without serious error to be equal to the corresponding measured LCFT and UCFT values) and for $\psi(\theta_U)$ (or equivalently, $\kappa(\theta_U)$) to be determined experimentally (e.g. by osmometry, viscometry, etc.). Substitution of these values into equation (12.88) yields a value for $\bar{\zeta}$ which is the only additional piece of information required to calculate T_M via equation (12.90)

Note that, as already mentioned, the temperature of maximum stability can be readily identified according to this model as the temperature at which $\kappa = 0$ (i.e. an athermal system).

The temperature dependence of the steric repulsion

The temperature dependence of the steric interaction between sterically stabilized particles mirrors the temperature dependence of $[\frac{1}{2} - \chi(T)]$. Croucher and Hair (1978) have calculated the temperature dependence of the steric repulsion between two sterically stabilized particles of poly(acrylonitrile) stabilized by poly(α-methylstyrene) in *n*-butyl chloride. The particle radius was taken as 100 nm and each steric layer, assumed to be of constant segment density, was taken to be 12 nm thick. The parabolic nature of the repulsion as a function of temperature for two different distances of particle separation is obvious from Fig. 12.10.

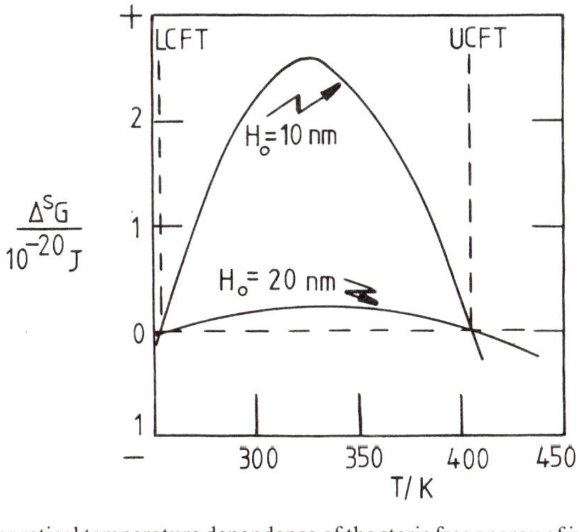

Fig. 12.10. The theoretical temperature dependence of the steric free energy of interaction for two spheres stabilized by poly(α-methylstyrene) in *n*-butyl chloride for a particle radius of 100 nm and a barrier layer thickness of 12 nm (after Croucher and Hair, 1978).

12.5.1.2. Microscopic free volume theory

The concentration dependence of the interaction parameter has been expressed by Orofino and Flory (1957) by the empirical power series

$$\chi = \chi_1 + \chi_2 v_2 + \chi_3 v_2^2 + \ldots \qquad (12.91)$$

Free volume theory allows the components of this expression to be evaluated. Thus Patterson (1968) has shown that

$$\chi_1 = -(U_1/RT)\upsilon^2 + (C_{P,1}/2R)\tau^2 \tag{12.92}$$

where

$$U_1 = p_1^* v_1^* \tilde{v}_1^{-1} \tag{12.93}$$

$$C_{P,1} = (p_1^* v_1^*/T_1^*)\ (4/3\tilde{v}^{-1/3} - 1) \tag{12.94}$$

and the subscript 1 refers to the dispersion medium. The starred quantities are the constant value reduction parameters and the tilde denotes the reduced parameter. The quantity τ in equation (12.92) is given by $(1 - T_1^*/T_2^*)$, where the subscript 2 denotes the stabilizing moieties. Moreover, $\upsilon^2 \simeq \delta^2/4 + 18\rho^2$. Here δ measures the difference in the self-attractive energy of the segments of the polymer and the dispersion medium and ρ measures the corresponding size difference. Exact expressions for δ and ρ are presented in the original publication.

The two terms in equation (12.92) are physically interpretable. The first term is the contact dissimilarity contribution arising from the difference in cohesive energy and size of the polymer and solvent segments. The second term is the free volume dissimilarity contribution.

The second interaction parameter χ_2 can also be derived from free volume theory as

$$\chi_2 = (C_{P,1}/R)\upsilon^2\tau - (T^2/3R)[\mathrm{d}(C_{P,1}/T)/\mathrm{d}T]\tau^3 \tag{12.95}$$

If only a rough idea of the temperature dependence is required, the series expansion in equation (12.91) can be truncated after the first (χ_1) term (i.e. any polymer volume fraction dependence is ignored). Under this assumption

$$\chi(T) \simeq \chi_1 = c_1\left[\frac{\tilde{v}_1^{\,1/3}}{\tilde{v}_1^{\,1/3} - 1}\upsilon^2 + \frac{\tilde{v}_1^{\,1/3}}{2(4/3 - \tilde{v}_1^{\,1/3})}\tau^2\right] \tag{12.96}$$

where $3c_1$ = number of external degrees of freedom of a molecule of the dispersion medium. The requisite values of the various parameters in equation (12.96) can be estimated as follows. The reduced volume $\tilde{v}_1$ and the connectivity parameter c_1 are readily found from the reduced equation of state. The values of υ^2 and τ^2 can be calculated by solving the two simultaneous equations obtained from equation (12.96) at the UCFT and LCFT where, to good approximation, $\chi = \frac{1}{2}$.

Croucher and Hair (1980b) have calculated the temperature dependence of $\chi(T)$ using this crude free volume approach for poly(α-methylstyrene) in butyl chloride. Their results are reproduced in Fig. 12.7 where they are compared with those calculated by using the crude macroscopic thermodynamic theory developed in the preceding section. It is clear that there exists general qualitative agreement between the predictions of these two methods, although their quantitative agreement leaves much to be desired. Neither method is truly quantitative and, without resort to experimental determination of $\chi(T)$, it is not possible to state which curve is a closer approximation for the correct temperature dependence of the interaction parameter.

Patterson (1968) has also derived expressions for the enthalpy and entropy of dilution parameters

$$\kappa = [(-U_1 + TC_{P,1})/RT]\upsilon^2 - (T/2R)\,(\mathrm{d}C_{P,1}/\mathrm{d}T)\tau^2 \tag{12.97}$$

and

$$\psi = \tfrac{1}{2} + (C_{P,1}/R)\upsilon^2 - [(C_{P,1} + T\mathrm{d}C_{P,1}/\mathrm{d}T)/2R]\tau^2 \tag{12.98}$$

The components of the steric repulsion

Patterson and Delmas (1969) have shown that the general expression for the interaction parameter is

$$\chi = c_1\{-(\tilde{U}_1/\tilde{T})\upsilon^2 + (\tilde{C}_{P,1}/2)\tau^2\} \tag{12.99}$$

where $\tilde{U}_1$ = reduced molar configurational energy and $\tilde{C}_{P,1}$ = reduced molar heat capacity and the subscript 1 refers to the dispersion medium. Note that $\tilde{U}_1 = -\tilde{V}_1^{\,-1}$ and $\tilde{C}_{P,1} = [(4/3)\tilde{V}_1^{\,-1/3} - 1]^{-1}$. The above relationship was derived for polymer solutions sufficiently dilute for the solvent to provide the major contributions to the respective quantities.

If the interaction parameter is written in terms of its respective entropy and enthalpy components χ_S and χ_H

$$\chi = \chi_H + \chi_S \tag{12.100}$$

then, as was shown in Section 3.2.8.2, the temperature dependence of χ gives χ_H and χ_S

$$\chi_H = -T(\partial\chi/\partial T) \tag{12.101}$$

and

$$\chi_S = \partial(T\chi)/\partial T \tag{12.102}$$

Note that division of both the numerator and denominator in each of these expressions by T^* leads to replacement of all the terms in T by the corresponding reduced temperature $\tilde{T}$.

Croucher and Hair (1981) applied these formulae to the stabilization of poly(acrylonitrile) particles by poly(isobutylene) in n-alkane dispersion media. Calorimetric studies suggested that for these systems $\upsilon^2 \sim 0$. Accordingly, the contact dissimilarity terms were disregarded and only the combinatorial and free volume terms were calculated. This is unlikely to be strictly correct, however. The combinatorial term is purely entropic in origin and is given by the classical Flory-Huggins expression. It is directly proportional to the kelvin temperature T. The enthalpic contribution to χ, which in this case arises solely from the free volume dissimilarity, can be estimated from

$$\chi_H = -\tilde{T}(\partial\chi/\partial\tilde{T}) = -(\tilde{T}/2)\,(\mathrm{d}\tilde{C}_{P,1}/\mathrm{d}\tilde{T})c_1\tau^2 \qquad (12.103)$$

The corresponding entropic contribution may be found from $\chi_S = \chi - \chi_H$. The results calculated by Croucher and Hair for particles of radius 150 nm and coated by poly(isobutylene) steric layers are displayed in Fig. 12.11. The steric layers were assumed to be of uniform segment density and of thickness 20 nm. In addition, the minimum distance between the surfaces of the colloidal particles was assumed to be 30 nm and the temperature was increased from above the freezing point of the dispersion medium to the UCFT.

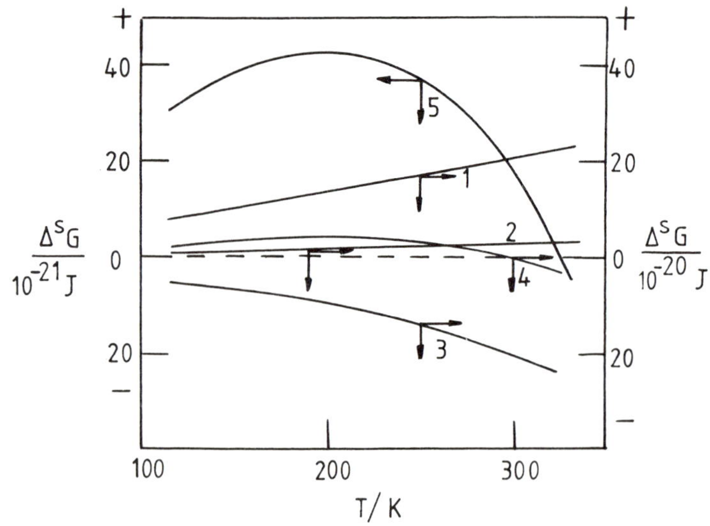

Fig. 12.11. The theoretical temperature dependence of the steric free energy of interaction of two spheres of radius 150 nm coated by steric layers of thickness 20 nm in 2-methylbutane: 1, combinatorial; 2, free volume enthalpy; 3, free volume entropy; 4, total entropy; 5, total free energy (after Croucher and Hair, 1981).

It is apparent from Fig. 12.11 that the enthalpic contribution from the free volume dissimilarity is relatively small, always positive and increases linearly with the temperature. The entropy terms are relatively large by comparison but of opposite signs: the combinatorial term is repulsive whereas the contact dissimilarity term is attractive. Both results are as expected physically. The increase in segment concentration in the stabilizing zone opposes flocculation whereas the "vaporization" of the excluded solvent molecules increases their configurational entropy and promotes flocculation. The summation of the three contributions yields the characteristic parabolic curve for the total interaction in the interpenetrational domain.

Croucher and Hair (1981) argued that because the combinatorial entropy term was contributing significant entropic repulsion, it was misleading to refer to these dispersions as being enthalpically stabilized near to the UCFT. This, however, is a misconception that is readily dispelled by reference to Fig. 12.11. There the net entropic contribution to the steric interaction free energy is plotted as a function of temperature. It is apparent that below a temperature of ca 300 K, both the enthalpic and total entropic contributions to the overall steric energy are positive, i.e. the dispersions display combined enthalpic-plus-entropic stabilization. Above a temperature of ca 300 K, only the enthalpic contribution is positive, the net entropy term having changed sign and become negative due to the dominance of the contact dissimilarity entropy. Hence above ca 300 K, the enthalpy term is indeed responsible for stability, the net entropy term opposing stability. The designation of enthalpic stabilization is therefore entirely justified since any such thermodynamic designation is necessarily concerned with the overall thermodynamic changes in entropy and enthalpy. The criticism articulated by Croucher and Hair does, however, lead to other possible schemes whereby sterically stabilized dispersions could be classified. Thus, for example, the primary phenomenological free energy that gives rise to stability might be specified, irrespective of whether one of its component entropy or enthalpy changes promotes flocculation (e.g. combinatorial stabilization). Such a classification scheme is not especially useful because it is quite inadequate in predicting, e.g., the effects of temperature changes.

12.5.2. The pressure dependence of the critical flocculation temperature

As Patterson and Delmas (1969) have implied, application of the Euler chain rule in the vicinity of a critical flocculation point (CFT) gives for its pressure dependence

$$(\mathrm{dCFT}/\mathrm{d}p) = (\mathrm{d}T/\mathrm{d}p) = -(\partial\chi/\partial p)_T/(\partial\chi/\partial T)_p \qquad (12.104)$$

The utility of this thermodynamic relationship resides in the fact that the pressure and temperature dependence of χ near to the UCFT can be discerned intuitively on phenomenological grounds. Consider, first, the temperature dependence of χ. As χ increases on heating near to the UCFT, $(\partial\chi/\partial T)_p$ must be positive. Increasing the applied pressure at constant T must decrease the free volume difference between the polymer and the dispersion medium. Therefore the solvency of the dispersion medium for the stabilizing moieties must increase, implying a reduction in χ on compression. It follows that near to the UCFT, $(\partial\chi/\partial p)_T$ is negative. Since $(\partial\chi/\partial T)_p$ is positive, equation (12.104) predicts that d(UCFT)/dp is unambiguously positive. Thus, increasing the pressure must raise the UCFT of a sterically stabilized nonaqueous dispersion.

The pressure dependence of the lower critical flocculation temperature (LCFT) is not established intuitively with comparable certainty. The temperature dependence of χ can be established unequivocally near to the UCFT. The value of χ must decrease because the solvency of the dispersion medium improves on heating. Hence $(\partial\chi/\partial T)_p$ must be negative. Unfortunately, because there is no dominant process, the sign of $(\partial\chi/\partial p)_T$ can be positive or negative which means that d(LCFT)/dT can also be positive or negative. What can be stated unequivocally is that the sign of d(LCFT)dT is determined directly by the sign of $(\partial\chi/\partial p)_T$.

It might be expected that even at the LCFT, there would be a significant contribution from free volume effects so that, just as at the UCFT, $(\partial\chi/\partial p)$ would be negative. In those circumstances, d(LCFT)/dT would also be negative. This was what Croucher and Hair (1979) predicted theoretically for poly(α-methylstyrene) in n-butyl chloride (see Fig. 12.12). Note that the absolute magnitudes of the pressure dependencies of the UCFT and LCFT are very different: d(UCFT)/d$T \simeq 4 \times 10^{-6}$ KPa^{-1} whereas d(LCFT)/d$T \simeq 5 \times 10^{-7}$ KPa^{-1} in this specific instance. The strong pressure dependence of the UCFT, which was ca an order of magnitude greater than that for the LCFT, is associated with the large compressibilities of liquids near to their critical temperatures.

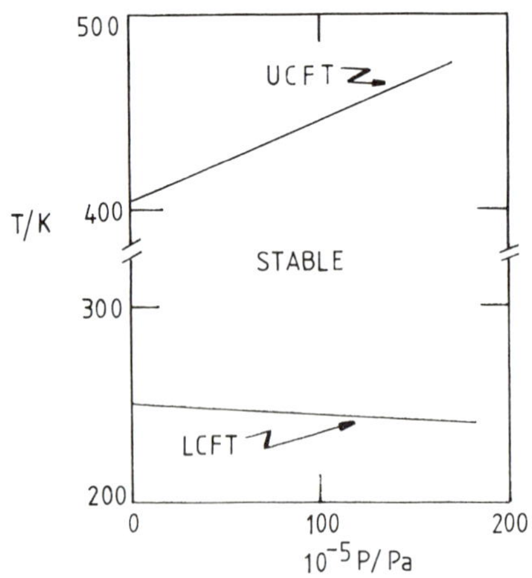

Fig. 12.12. The theoretical pressure dependence of the UCFT and LCFT of particles sterically stabilized by poly(α-methylstyrene) (after Croucher and Hair, 1979).

Finally, we stress that the free volume approach is only applicable to nonpolar systems. Aqueous dispersions fall outside its scope. This is vividly illustrated by the data of Evans *et al.* (1975), who determined experimentally that $d(\text{UCFT})/dT \simeq -1 \times 10^{-7}$ KPa^{-1} for latex particles sterically stabilized by poly(oxyethylene) in aqueous 0·43 molal magnesium sulphate solutions. Both the sign and magnitude of this quantity is different from that predicted by free volume theory for the UCFT of non aqueous dispersions. Paradoxically, it falls in line with the predictions, both in sign and magnitude, published by Croucher and Hair (1979) for the pressure dependence of the LCFT of poly(α-methylstyrene) in *n*-butyl chloride. This may be merely coincidental, but the very small pressure dependence exhibited by the UCFT of aqueous sterically stabilized dispersions emphasizes the major differences between the origins of flocculation at the UCFT for aqueous and nonaqueous dispersions. The small pressure dependence observed for aqueous systems is scarcely surprising since the UCFT of an aqueous dispersion occurs far from the critical point of water whereas that for nonaqueous dispersions is quite close to the critical point of the dispersion medium.

12.6. The configurational entropy of colloidal particles

12.6.1. The dispersion configurational entropy

Feigin *et al.* (1981) have calculated the order of magnitude of the configurational entropy of sterically stabilized particles. They employed a lattice approach, in the spirit of the Flory-Huggins theory, in which the total available volume is subdivided into cubes of equal size. The system is assumed to contain N_3 latex particles at a volume fraction v_3. The number of lattice sites, whose individual volumes are taken to be the effective volume of the particles when close packed, is simply given by $\mathfrak{N} = 0{\cdot}74\ N_3/v_3$.

The number of ways of distributing the N_3 indistinguishable particles among the available lattice sites is

$$\Omega_{\text{disp}} = \mathfrak{N}!/N_3!(\mathfrak{N} - N_3)! \tag{12.105}$$

so that the configurational entropy of the particles is

$$S_{\text{disp}} = k \ln \Omega_{\text{disp}} = k \ln\{\mathfrak{N}!/N_3!(\mathfrak{N} - N_3)!\} \tag{12.106}$$

Application of the Stirling approximation yields

$$S_{\text{disp}} \simeq k\ N_3\{\ln(\mathfrak{N}/N_3 - 1) - (\mathfrak{N}/N_3)\ln(1 - N_3/\mathfrak{N})\}$$

so that the configurational entropy per particle is

$$S_{\text{disp}}/N_3 \simeq k\{\ln(0{\cdot}74/v_3 - 1) - (0{\cdot}74/v_3)\ln(1 - v_3/0{\cdot}74)\} \tag{12.107}$$

At low volume fractions of particles ($v_3 \ll 1$), this reduces to

$$S_{\mathrm{disp}}/N_3 \simeq k\{\ln(0{\cdot}74/v_3 - 1) + 1\} \quad (12.108)$$

12.6.2. Configurational free energy change on flocculation

To calculate the change in configurational free energy on flocculation, it is necessary to specify the structure of the aggregates that are produced on flocculation. A simple, but extreme, assumption is that the single floc formed is close packed so that the individual particles contained therein have no freedom of movement. The configurational free energy change per particle

$$\Delta G^{\mathrm{CONF}}/N_3 = TS_{\mathrm{disp}}/N_3 = kT\{\ln(0{\cdot}74/v_3 - 1) + 1\} \quad (12.109)$$

Feigin *et al.* also calculated a second extreme case, that of doublet formation. They showed that the configurational free energy change per doublet is given by an expression closely analogous to the other limiting case

$$\Delta G^{\mathrm{CONF}}/N_3 \simeq kT\{\ln(0{\cdot}031/v_3) + 1\} \quad (12.110)$$

Values of ΔG^{CONF} calculated for each of these limiting cases is shown in Table 12.5. Although the basis of the comparison presented therein is slightly different (per particle *vs* per doublet), the results predicted by both extreme models are not too widely disparate. The configurational free energy change associated with flocculation can obviously be as large as 10 kT in systems of usual interest. Dilute dispersions are clearly more stable on this basis than more concentrated systems. What this means in terms of the effect on the critical flocculation point depends critically upon the nature of the particular system concerned.

TABLE 12.5
Change in configurational free energy on flocculation

	$\Delta G^{\mathrm{CONF}}/kT$	
v_3	Floc	Doublets
10^{-1}	3·0	—
10^{-2}	5·3	2·1
10^{-3}	7·6	4·4
10^{-4}	9·9	6·7
10^{-5}	12·2	9·0
10^{-6}	14·5	11·3

Chapter 13

Experimental Studies of the Distance Dependence of Steric Interactions

13.1. Introduction

In the last three chapters we dealt with the theoretical predictions of the magnitudes of steric interactions and their dependence upon distance. The incipient flocculation behaviour that was discussed in Chapters 5–9 is determined primarily by the magnitude and the sign of these steric interactions. Incipient flocculation studies, however, do not provide direct information about the distance dependence of the interaction. An additional drawback is that it is inevitable in incipient flocculation experiments that the dispersions be studied under conditions of pathological instability. This may change the very properties that we wish to study in order to determine how stability is imparted. It is therefore necessary to complement the incipient flocculation experiments with studies on dispersions that exhibit robust stability.

13.2. Classification of types of distance dependence studies

A diverse range of experiments has been devised to measure the distance dependence of the steric interaction. These may be classified under two headings:

(i) Compression of dispersions of close-packed sterically stabilized colloidal particles. This may be accomplished in

(a) 2-D experiments

or

(b) 3-D experiments, the latter using either

1. osmotic pressure methods,

or

2. a filtration method.

(ii) The interaction of polymer chains attached to macroscopic objects that undergo highly-controlled close approach. This has been performed in a variety of ways using

(a) crossed cylinders,
(b) crossed fibres,
(c) spherical caps, and
(d) thin films.

These different approaches will now be reviewed in turn and the results of the experiments compared with theory.

13.3. Studies of the compression of dispersions

13.3.1. 2-D compression studies

The 2-D studies of the compression of dispersions of sterically stabilized particles have all been carried out using a *surface balance*. This method was pioneered by Doroszkowski and coworkers (Doroszkowski and Monk, 1969; Doroszkowski and Lambourne, 1971; 1973) at the oil/water interface. Subsequently, it was extended by Garvey *et al.* (1979) to encompass the air/water (2 M NaCl) interface. An automatic surface balance suitable for such studies has been described by Doroszkowski and Monk (1969).

13.3.1.1. The results of Doroszkowski and Lambourne

The surface pressure (π) *vs* the area per particle (A) curves for monodisperse poly(acrylonitrile) particles stabilized by poly((methyl methacrylate-*co*-methacrylic acid)-*g*-styrene) have been reported by Doroszkowski and Lambourne (1973). In these experiments, the grafted polystyrene chains, of molecular weights 6 000 and 12 000, functioned as the stabilizing moieties and the statistical copolymer poly(methyl methacrylate-*co*-methacrylic acid) functioned as the anchor polymer. The quality of the solvency of the dispersion medium for the stabilizing moieties was varied by using toluene (a good solvent for polystyrene at the temperature of the experiment, 24 °C) and cyclohexane (a slightly worse than θ-solvent since $\theta = 34$ °C) as the supernatant oil phase in the surface balance.

It should be realized at the outset that although the experiments were conducted on monolayers of particles spread at a suitable interface, the interactions between the particles were of a 3-D, not a 2-D, character. It can be shown that the depth of immersion of a small particle in a liquid is dependent upon the contact angle. In the case of poly(acrylonitrile) coated by stabilizer, the contact angle was found to be 149° at the toluene/water interface. The zones of contact between compressed particles were therefore primarily in the nonaqueous phase, the particles being held at the oil/water interface by their partial immersion in the aqueous phase. This anchoring of the particles to the interface is crucial to the success of the method because of the tendency of the particles to escape into the bulk of the oil phase, both on spreading and on compression. Electron microscopy was used to examine compressed monolayers, which for particles at the toluene/water interface were reported to form hexagonally close packed 2-D arrays that displayed no holes. Holes were, however, apparent to a minor degree for particles compressed at the cyclohexane/water interface. These holes apparently arose because flocculation prevented the free movement of particles past one another. Flocculation was also inferred to be present in the systems because large hysteresis effects were apparent in the π *vs* A curves. The free energy of interaction per contact for pairs of steric layers was calculated by integrating the area under the π *vs* A curves (i.e. $\Delta G^{\mathrm{TOT}} = \int \pi dA$) and dividing by the total number of contacts.

Some results obtained by Doroszkowski and Lambourne for the distance dependence of the steric repulsion for polystyrene stabilizing moieties in toluene are shown in Fig. 13.1. Also shown are the predictions of the theory of Hesselink *et al.* (1971) including the individual osmotic and elastic components. These were obtained by a numerical Derjaguin-like integration procedure that transformed flat plate potentials into potentials for spheres. It

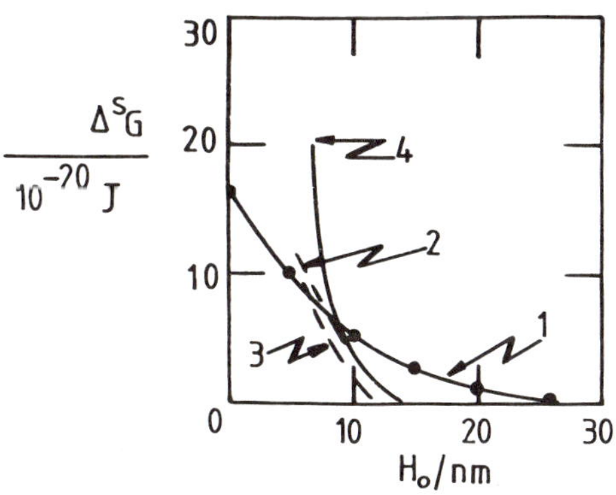

Fig. 13.1. The distance dependence of the steric interaction free energy for polystyrene stabilized latex particles in toluene; 1, experimental results; 2, theoretical mixing term; 3, theoretical elastic term; 4, total theoretical terms (after Doroszkowski and Lambourne, 1973).

would appear from the results of these experiments that the theory of Hesselink *et al.* predicts interactions that are too weak at large separations and too strong at close distances of approach. It should be realized, however, that any comparison such as this, involving as it does the theory of random flight chains, necessarily requires an arbitrary scaling assumption. A different scaling assumption could conceivably have improved the agreement between theory and experiment.

Figure 13.2 compares the results of experiment with the theoretical predictions of Meier for the slightly worse than θ-solvent case, cyclohexane at 24 °C. In that instance, Doroszkowski and Lambourne measured the repulsion that arose from the interactions that were generated on compression, even though the particles were flocculated. The comparison is made with the theoretical predictions of the Meier (1967) theory for tails. Under near θ-conditions, the osmotic term, which was slightly underestimated by Meier, is relatively small. The Meier treatment of the elastic repulsion appears to be correct for the model adopted and was that utilized by Hesselink *et al.* (1971). Meier's theory correctly predicts the occurrence of a repulsive interaction in worse than θ-solvents. Although Doroszkowski and Lambourne described the agreement between theory and experiment as satisfactory, close inspection of the curves in Fig. 13.2 shows that the predicted repulsion appears to be too large when compared with experiment. This result is in close accord with the trend observed on close approach of the particles with the good solvent toluene as the dispersion medium.

Note that Klein (1980, 1982) has made a much more detailed study of the

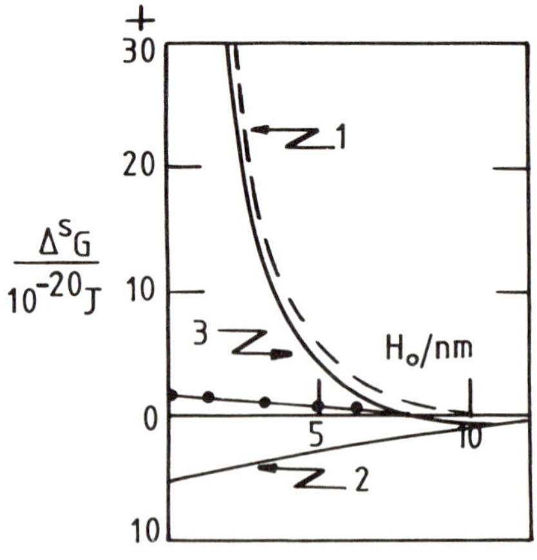

Fig. 13.2. The distance dependence of the steric interaction free energy for polystyrene stabilized particles in cyclohexane at 24 °C (a worse than θ-solvent); (●) experimental points; 1, theoretical elastic contribution; 2, theoretical mixing contribution; 3, total theoretical contribution (after Doroszkowski and Lambourne, 1973).

polystyrene/cyclohexane system at 24 °C. He has confirmed the repulsion observed by Doroszkowski and Lambourne but showed that it was preceded by an attractive interaction regime. Unfortunately, attraction cannot apparently be measured by the surface balance technique. A more comprehensive discussion of the origins of the repulsion observed by Doroszkowski and Lambourne will be deferred until after Klein's results have been set forth in detail later in this chapter.

Doroszkowski and Lambourne (1971) also performed similar experiments on poly(methyl methacrylate) particles stabilized by poly(12-hydroxystearic acid) of low molecular weight (1600). These were spread at the *n*-heptane/water interface. The results for this system are less amenable to comparison with theory than those described above for polystyrene stabilized latices because of the lower molecular weight of the stabilizing chains and their highly-branched character.

13.3.1.2. The results of Garvey, Mitchell and Smith

Garvey *et al.* (1979) extended the surface balance compression studies to encompass aqueous sterically stabilized dispersions. They studied the interactions of monodisperse polystyrene latex particles stabilized by nominally unanchored poly(vinyl pyrrolidone) of molecular weight ca 40 000. The particles were spread at the air/water interface, the water containing 2 M NaCl. The problem of escape of the latex particles into the bulk phase from the interface on spreading was overcome by ensuring that the density of the aqueous electrolyte solution was greater than that of the latex particles.

The typical compressional curves obtained by Garvey *et al.* displayed some hysteresis, perhaps because the polymer chains, which were nominally unanchored, underwent desorption as a result of the stress generated on compression. One example of an initial compressional curve is shown in Fig. 13.3. The thickness of the steric layer was determined by photon correlation

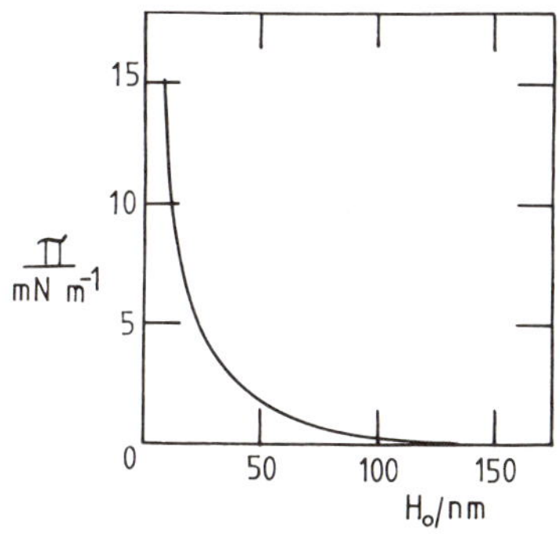

Fig. 13.3. The distance dependence of the surface pressure of a monolayer of latex particles sterically stabilized by poly(vinyl pyrrolidone) of molecular weight 40 000 in 2 M NaCl (after Garvey *et al.*, 1979).

spectroscopy to be 44 ± 10 nm. Yet the steric repulsion appeared to extend to 75 nm. This discrepancy may well be a consequence of the occurrence of tails predicted by the theories of Scheutjens and Fleer (1979; 1980; 1982). Such tails could well project into the continuous phase and behave in a free-draining fashion. They would not then contribute to the hydrodynamic radius of the particle, as measured by photon correlation spectroscopy, but could still contribute effectively to the steric repulsion. In this way, the apparently contradictory results obtained for the steric barrier layer thickness by dynamic light scattering and compressional measurements may perhaps be reconciled.

13.3.2. 3-D compression studies

13.3.2.1. Osmotic pressure methods

Experimental

The determination of the distance dependence of the steric repulsion between latex particles by measuring the osmotic pressure of dispersions at different volume fractions was first accomplished by Ottewill and coworkers (Barclay *et al.*, 1972; Cairns *et al.*, 1976; Ottewill, 1976; 1980). The apparatus used to make these measurements is shown in Fig. 13.4. The cell body was cylindrical in shape. Mercury was used as a piston in the cell. The dispersion was layered onto the top of the mercury and its level adjusted until it was just proud of the top of the cell. A millipore filter, of 0·2 μm mean diameter pore size, was placed centrally over the dispersion. A stainless steel top, fitted with a uniform bore capillary, completed the pressure cell.

The latices studied by Cairns *et al.* (1976) were composed of poly(methyl methacrylate) particles stabilized by low molecular weight poly(12-hydroxystearic acid). Dodecane was chosen as the dispersion medium. The volume fraction of latex particles could be varied by adjusting the pressure on the mercury column.

Some results obtained in this manner are shown in Fig. 13.5. At low volume fractions the resistance to compression was negligible. Equilibrium was attained very slowly, however, perhaps because the filter displayed a low permeability when in contact with nonaqueous dispersion media. At a volume fraction of ca 0·55, the resistance to compression increased considerably with even small changes of volume fraction. Extremely strong resistance to compression was experienced at a particle volume fraction of 0·566; indeed, the system virtually became incompressible ($\partial\pi^c/\partial v_3 \to \infty$ where π^c = osmotic pressure of the particles at volume fraction v_3). Interestingly, no hysteresis was observed in these systems, the compression and decompression cycles behaving quite reversibly. As noted previously, however, equilibrium was achieved only slowly.

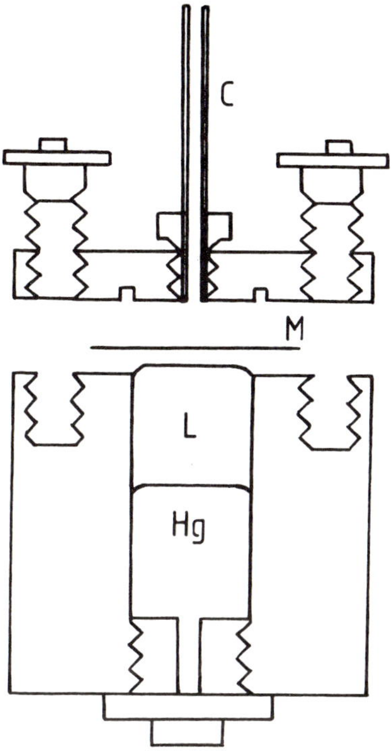

Fig. 13.4. Schematic representation of the osmometer used to measure the disjoining pressures of latices (after Cairns *et al.*, 1976). Note that C = capillary, L = latex and M = membrane.

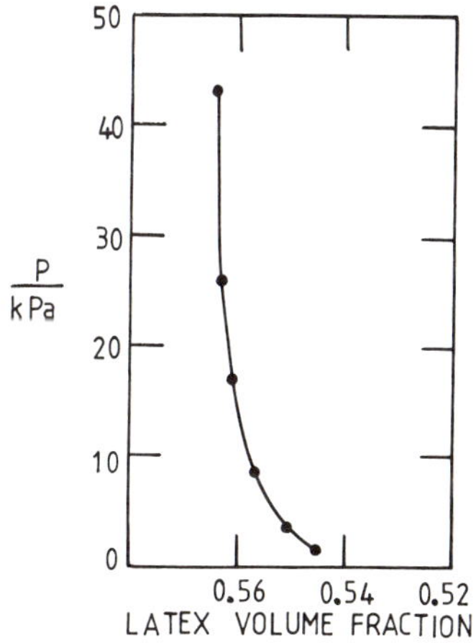

Fig. 13.5. The osmotic pressure of poly(methyl methacrylate) latex particles sterically stabilized by poly(12-hydroxystearic acid) in *n*-dodecane as a function of the volume fraction of latex (after Cairns *et al.*, 1976).

The experimental curve shown in Fig. 13.5 indicates that the general shape of the potential energy versus distance curves predicted theoretically for sterically stabilized particles dispersed in a good solvent for the stabilizing moieties is at least qualitatively correct. That is to say, the shapes of the curves differ drastically from those characteristic of electrostatically stabilized particles. One consequence of the characteristic shape of the potential energy curves for sterically stabilized particles, to which the stabilizing moieties are irreversibly attached, is that they can display *thermodynamic stability*. This contrasts with the thermodynamic *metastability* exhibited by electrostatically stabilized dispersions. Of course, to be thermodynamically stable, the dispersed state must be the thermodynamically preferred state (i.e. that which is lowest in free energy) and this is only true if the relevant van der Waals attraction between the core particles is negligibly small in comparison with the thermal kinetic energy of the particles. Note, however, that the total potential energy curve should strictly include the configurational entropy of the particles, which is rarely the case.

The shape of the curve determined by 3-D osmotic pressure measurements is similar to that determined by the 2-D compression studies. This congruence is strongly suggestive of the absolute validity of this shape for steric stabilization in better than θ-solvents.

Returning to the detailed discussion of the experimental data displayed in Fig. 13.5, we note that the core radius of the nearly-monodisperse particles used in these studies was 77·5 nm. At the onset of almost-hard sphere repulsion ($v_3 = 0{\cdot}566$), the distance of separation for face centred cubic (or hexagonal) arrays of such particles corresponds to 14·5 nm or a contribution of 7·25 nm for each shell. The fully extended length of poly(12-hydroxystearic acid) chains of the molecular weight studied was calculated to be ca 9 nm, which if attached to particles of this size would just touch at a volume fraction of 0·53. Therefore, if the chains are extended in the shell layers, which may not be correct, then either the chains must undergo mutual interpenetration or mutual denting.

Homola and Robertson (1976) have used a very similar device to that described by Ottewill and coworkers to measure the osmotic pressure of polystyrene latex particles stabilized by anchored poly(oxyethylene) of low molecular weight (n-alkyl poly(oxyetheylene) chains of the type C_nE_{20}, so that the stabilizing moieties were of ca 900 molecular weight). The nature of the anchor polymer, an alkyl chain that was varied in length from dodecyl (C_{12}) through hexadecyl (C_{16}) to octadecyl (C_{18}), was found not to influence the measured osmotic pressures at constant chain length of the stabilizing moieties. The one major difference in the experiments of Homola and Robertson and those of Ottewill and coworkers resides in the choice of the independent variable. Homola and Robertson controlled the volume fraction of particles and recorded the resulting pressures. Ottewill and coworkers, in contrast, varied the pressure by means of movement of the mercury piston and determined the volume fraction of colloidal particles and their osmotic pressure by measuring the column height of the expelled liquid in the capillary above the body of the cell. For sterically stabilized systems, these differences are essentially trivial, except if flocculation occurred.

The results obtained by Homola and Robertson are shown in Fig. 12.5 and have already been discussed in another context (Section 12.3.1.2.) so that they will not be considered further. The results confirm, however, the general shape established theoretically for steric repulsion curves.

Theoretical

The potential energy-osmotic pressure transformation formula. The measurement of the osmotic pressure of colloidal dispersions, in principle, provides a relatively straightforward method for determining the distance dependence of the steric repulsion. As Ottewill (1980) has pointed out, the relatively small osmotic pressures involved render the actual measurements difficult to accomplish. None the less, it is worth establishing how to transform osmotic pressures into potential energies of interaction between the particles involved and vice versa.

To achieve this end, it is necessary to assume some regular geometrical packing of the concentrated dispersion. The following analysis is therefore confined to monodisperse systems.

Let us assume that the particles pack as either an ordered hexagonal or face-centred cubic lattice. Both arrays yield identical results in the following analysis. There is definite evidence for the latter from light diffraction and small angle neutron scattering studies (Ottewill, 1980). The case of a hexagonally closed packed array is shown in Fig. 13.6. We will assume the complete absence of defects, which may well be very important in actual compression studies. We will calculate the pressure on one side of a hypothetical plane inserted into the lattice along a slip-plane.

The effective area, A, occupied per particle at this hypothetical plane is simply a regular hexagon constructed about an inscribed circle of radius $(a+H_0/2)$, where a=particle radius and H_0=minimum distance of surface separation of contiguous particles. Simple geometrical considerations yield

$$A=2(3)^{\frac{1}{2}}(a+H_0/2)^2 \tag{13.1}$$

The force on each particle at the wall is transmitted equally from three other particles in the adjacent layer in the array, the centres of the four particles forming the apices of a regular tetrahedron. The component of the force normal to the wall exerted by each of these three particles is just $F\cos\tau$, where F=the interparticle force along the line of centres and τ=the angle between the centre-to-centre direction and the normal to the wall ($\tau=\cos^{-1}(2/3)^{\frac{1}{2}}$). The total force on the wall ($\mathscr{F}_T$) is therefore given by

$$\mathscr{F}_T=3F\cos\tau=(6)^{\frac{1}{2}}F \tag{13.2}$$

Fig. 13.6. Schematic representation of the closepacking of a monodisperse latex.

The osmotic pressure of the particles, which ignores their configurational entropy, can be calculated according to this model from $\mathscr{F}_T/A = (6)^{\frac{1}{2}}F/A$, a result of encouraging simplicity.

Colloid stability theories usually predict the distance dependence of the total potential energy of interaction (${}^{S}V_T$) between pairs of spheres. The force along the line of centres is readily obtained by differentiation

$$F = -(\partial^{S}V_T/\partial H_0) \tag{13.3}$$

so that the osmotic pressure π^c becomes

$$\pi^c = -(\partial^{S}V_T/\partial H_0)/(2)^{\frac{1}{2}}(a + H_0/2)^2 \tag{13.4}$$

This equation allows theoretically predicted values of ${}^{S}V_T$ to be transformed into an excess osmotic pressure, or a *disjoining pressure* in the terminology of Derjaguin (1955, 1964).

The general equation (13.4), which was first derived by Evans and Napper (1978), can be simplified for those systems where the Derjaguin integration transforms a flat plate potential into a sphere potential. If ${}^{FP}V_T(h)$ denotes the total potential energy per unit area between two parallel flat plates separated by a distance h, then the Derjaguin integration can be approximated by

$$^{S}V_T \sim \pi a \int_{H_0}^{\infty} {}^{FP}V_T(h)\mathrm{d}h \tag{13.5}$$

It follows on differentiation that

$$F = -\frac{\partial}{\partial H_0}\{\pi a \int_{H_0}^{\infty} {}^{FP}V_T(h)\mathrm{d}h\}$$

$$= \pi a {}^{FP}V_T(H_0) \tag{13.6}$$

The disjoining pressure in this special case becomes

$$\pi^c \sim [\pi/(2^{\frac{1}{2}})a]^{FP}V_T(H_0)/(1 + H_0/2a)^2 \tag{13.7}$$

This shows that the excess osmotic pressure can be calculated from the theoretically predicted flat plate potential energy. It is restricted to those instances where the range of the potential is small compared with the particle size. Neglect of the configurational entropy of the particles should not be debilitating so far as the practical applications of equation (13.4) and (13.7) are concerned. At the high volume fractions required for close packing, the configurational entropy must necessarily be small (Feigin *et al.*, 1981). Evans and Napper (1978) have suggested an upper limit for the contribution from this source of 10 Pa on the basis of the Percus-Yevick equation (Percus and Yevick, 1958). One example of the application of equation (13.7) to steric stabilization has already been presented in Section 12.3.2.1.

The foregoing derivation of the transformation formula for converting potential energies into osmotic pressures was based upon the assumption of pair-wise additivity. Evans and Napper (1978) showed that if the barrier layer thickness was less than ca 15% of the particle radius, many-body interactions are distinctly unlikely and the assumption of pair-wise additivity introduces negligible error.

Statistical mechanical theories. The application of classical statistical mechanics to predicting the behaviour of ensembles of interacting colloidal particles has been vigorously pursued by Snook and van Megen (van Megen and Snook, 1975; Snook and van Megen, 1976; van Megan and Snook, 1976a,b; Homola *et al.*, 1977; Cairns *et al.*, 1981). Two general approaches have been adopted for investigating steric stabilization: a cell model theory and Monte Carlo calculations. Note that the first-order perturbation theory, which has been successfully applied to electrostatically stabilized dispersions (van Megen and Snook, 1976b), has not apparently been exploited for sterically stabilized systems. For sterically stabilized particles, the cell model approach is usually preferred because the Monte Carlo calculations tend to be more time consuming.

A. The cell model theory. The cell model theory was first introduced by Lennard-Jones and Devonshire to calculate the bulk properties of molecular fluids (Barker, 1963). In this approach, the array of particles is replaced by an array of hypothetical cells inside which the movement of each particle is confined.

Most cell model calculations to-date have been performed on electrostatically stabilized dispersions. The canonical partition function for such a system is given by

$$Z=(2\pi mkT/h^2)^{3N/2}\exp\{-U_0/kT\}v_f^{N_3} \tag{13.8}$$

where U_0 = the static energy of the lattice and v_f = the free volume of the particles. The first term of the partition function is the conventional momentum (translational) partition function. The second term is the Boltzmann term and the third term is the free volume term introduced in an endeavour to compensate for the fact that the cell model underestimates the communal entropy of the particles.

A spherical smeared-out (mean field) potential may be used to evaluate the free volume term for an electrostatically stabilized dispersion

$$v_f=4\pi\int_0^{r_c} r^2\exp\{-[\psi(r)-\psi(0)]/kT\}\mathrm{d}r \tag{13.9}$$

where r_c = cell radius appropriately truncated to allow for the particle size and $\psi(r)$ = the potential at distance r from the particle. Inclusion of the first three shells of neighbours is usually sufficient for the evaluation of this integral, the final result then being insensitive to the inclusion of additional shells. The calculation of the static lattice energy for electrostatically stabilized particles, however, may require the inclusion of significantly more shells (e.g. 20) before it becomes insensitive to the inclusion of more cells. In this way, the properties of a large collection of particles can be calculated by replacing the large system by a periodic one based upon a unit cell of a small number of particles. Of course, sterically stabilized particles are easier to handle by the cell model than electrostatically stabilized particles because the interactional potential is usually nowhere near as long range in character.

The osmotic pressure π^c of an isotropic system of spherical particles is given by

$$(\pi^c V/N_3kT)=1-(1/3N_3kT)\langle\sum_{i<j}\mathbf{r}_{ij}\cdot\nabla_{r_{ij}}U_{ij}\rangle \tag{13.10}$$

where N_3 = number of particles, V = volume of the system, r_{ij} = centre-to-centre distance between the ith and jth particles, U_{ij} is assumed to be the pair potential and the bracketed term is the average virial of the system. The first term in equation (13.10) is, of course, the ideal van't Hoff term. A reduced osmotic pressure $\hat{\pi}^c$ may be defined by

$$\hat{\pi}^c = \pi^c V/N_3 kT$$
$$= \pi^c(3kT/4\pi a_3^3)v_3 \quad (13.11)$$

where a = particle radius and v_3 = volume fraction of the particles given by $v_3 = (N_3/V)(4\pi a^3/3)$.

To compare the predictions of cell model calculations with the results of the experiments of Cairns *et al.* (1976) on poly(methyl methacrylate) latices stabilized by poly(12-hydroxystearic acid) in *n*-dodecane, Cairns *et al.* (1981) adopted a simple model for the pair potential U_{ij}. This was the constant segment density model of Fischer (1958), as elaborated by Ottewill and Walker (1968). They also considered the modification of that model by Doroszkowski and Lambourne (1971; 1973). The latter approach, which allows in a completely arbitrary way for a supposed redistribution of stabilizer segments on close approach of the opposing surface, yields the potential

$$\Delta^S G^M = U_{ij} = (2kTc_2{}^2/V_1)(\tfrac{1}{2} - \chi_1)(v_H + v_R)(v_R/v_H) \quad (13.12)$$

where c_2 = initial average concentration of polymer in the layer, V_1 = volume of a molecule of the dispersion medium, χ_1 = the interaction parameter. The volume terms v_H and v_R are given respectively for

$$v_H = (\pi/3)(\delta - H_0/2)[(a + H_0/2)(2a + \delta + H_0/2) -$$
$$(2a^3/\{a + \delta\})] \quad (13.13)$$

and

$$v_R = (\pi/3)(\delta - H_0/2)^2(3a + 2\delta + H_0/2) \quad (13.14)$$

where δ = the thickness of the steric layer and H_0 = minimum distance of surface separation. The corresponding expression according to the Fischer-Ottewill-Walker approach is

$$\Delta^S G^M = U_{ij} = (4\pi kTc_2{}^2/3V_1\rho_2{}^2)(\psi_1 - \kappa_1)(\delta - H_0/2)^2$$
$$\times (3a + 2\delta + H_0/2) \quad (13.15)$$

where ρ_2 = density of the stabilizing moieties, and ψ_1, κ_1 = entropy and enthalpy of dilution parameters respectively. The van der Waals attraction, which was assumed to be additive to the steric repulsion, was calculated using the conventional Hamaker equation.

Comparison of the predictions of equation (13.10) and (13.12) with the results of experiments showed that the Doroszkowski and Lambourne approach yielded repulsions that were too soft, presumably as a consequence of the arbitrary redistribution of the stabilizing segments. The results presented herein were accordingly obtained using equation (13.15).

The predictions of the cell model theory for various assumed values of the average polymer concentration are shown in Fig. 13.7 where the results of the experiments are also plotted. It can be seen that reasonable agreement with experiment is possible using χ_1 as an adjustable parameter (c_2 and δ are also arbitrarily chosen but are constrained by approximate experimental determinations). The value of $(\frac{1}{2}-\chi_1)$ that yielded the best fit with the results of experiment (0·35) gave $\chi_1 = 0{\cdot}15$. This seems to be an unexpectedly small value for χ_1 in light of the experimentally determined value of χ_1 for poly(12-hydroxystearic acid) in *n*-heptane ($\chi_1 \sim 0{\cdot}4$). Nevertheless, the volume fraction at which the steric repulsion first manifested itself was accurately predicted by the theory, as was the initial distance dependence of the steric repulsion. At higher volume fractions, the experimental pressure increased more rapidly than was predicted theoretically. This is scarcely surprising since the Fischer theory does not include any elastic repulsion arising from the loss of configurational entropy of the stabilizing moieties on close approach of the opposing surface. Note, too, that the Fischer-Ottewill-Walker model is unphysical in the interpenetrational-plus-compressional domain because it assumes that the core particles somehow become identical with the steric layers.

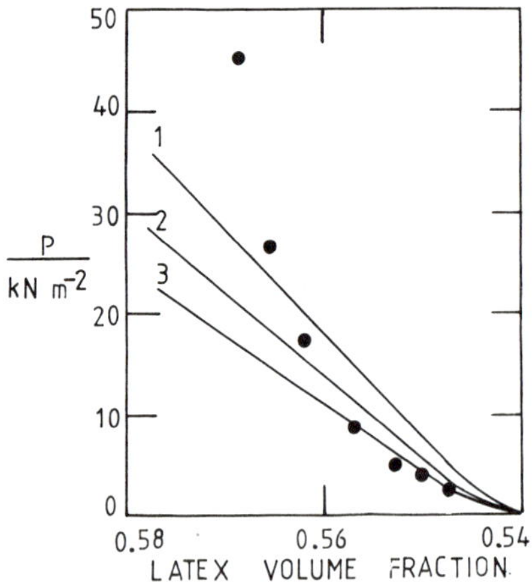

Fig. 13.7. Comparison of the measured disjoining pressures (●) of poly(12-hydroxystearic acid) stabilized latices in *n*-dodecane with the theoretical predictions of Cairns *et al.*, 1981; average segmental concentrations: curves 1, 0·25 g cm^{-3}; 2, 0·22 g cm^{-3}; 3, 0·20 g cm^{-3}.

B. Monte Carlo simulations. Monte Carlo methods designate simulation procedures whereby certain probabilistic features are introduced into the numerical computations. They contrast sharply with the completely deterministic algebraic expressions that are common in traditional theories of stochastic processes. It follows that some care must be exercised to ensure that the precision of Monte Carlo calculations is adequate for the job at hand.

The statistical mechanical starting point for the Monte Carlo calculations of the osmotic pressure of dispersions is again [see equation (13.10)]

$$(\pi^c V/N_3kT) = 1 - (1/3N_3kT)\langle \Psi_{N3} \rangle \tag{13.16}$$

where

$$\langle \Psi_{N3} \rangle = \langle \sum_{i<j} \mathbf{r} \,.\, \nabla_{rij} U_{ij} \rangle \tag{13.17}$$

The statistical mechanical value of $\langle \Psi_{N3} \rangle$ must be correctly weighted according to the properties of the radial distribution function $g(r_{12})$. The latter can be related to the two-body distribution function $n^{(2)}(\mathbf{r}_1,\mathbf{r}_2)$ through

$$g(r_{12}) = (V/N_3)^2 n^{(2)}(\mathbf{r}_1\mathbf{r}_2) \tag{13.18}$$

where

$$n^{(2)}(\mathbf{r}_1,\mathbf{r}_2) = \frac{N_3! \int \exp\{-\Phi_{N_3}/kT\} \mathbf{dr}_3 \ldots \mathbf{dr}_{N3}}{(N_3-2)! \int \exp\{-\Phi_{N_3}/kT\} \mathbf{dr}_1 \ldots \mathbf{dr}_{N_3}} \tag{13.19}$$

Physically, $g(r_{12})$ measures the probability of finding a particle at a distance r_{12} from a particle held at the origin. Note that for a pair-wise additive potential in a canonical ensemble

$$\Phi_{N3} = \sum_{i<j} U_{ij}$$

The Monte Carlo method evaluates the multidimensional integrals in equation (13.18) by generating particle configurations in a computer using the random displacement of particles. The configurations are selected or rejected according to a prescription that ensures that a particular configuration appears in the final distribution of configurations with its correct Boltzmann probability $\exp\{-\Phi_{N3}/kT\}$. This is accomplished as follows.

A particle is selected, either serially or at random, and subjected to a displacement from configuration i to configuration j. The change in configurational energy of the systems, Φ_{N3}, is then calculated from the pair potentials. If Φ_{N3} is negative, the new configuration is accepted since it is more probable than the old. If Φ_{N3} is positive, then the configuration is accepted with a probability of $\exp\{-\Phi_{N3}/kT\}$. This is done by the computer by generating a random number in the range 0–1 (hence the name *Monte Carlo*). This random number is then compared with $\exp\{-\Phi_{N_3}/kT\}$. The move is allowed if the exponential is larger than the random number but otherwise it is rejected and the previous configuration counted again. Snook and van Megen chose the maximum displacement of a particle so that approximately half of the moves were accepted.

To simulate an infinite system as closely as possible, a small number of particles N_3 was enclosed in a cell of volume V and *periodic boundary conditions* were applied to the system. These overcome the problem intrinsic to the use of a small number of particles which is the escape of particles from the cell. The periodic boundary conditions are implemented as follows. Each particle inside the cell has a "ghost" particle outside that mimics its position and motion exactly. As a result when one particle moves out of the cell, another compensatory particle enters from the other side (Fig. 13.8).

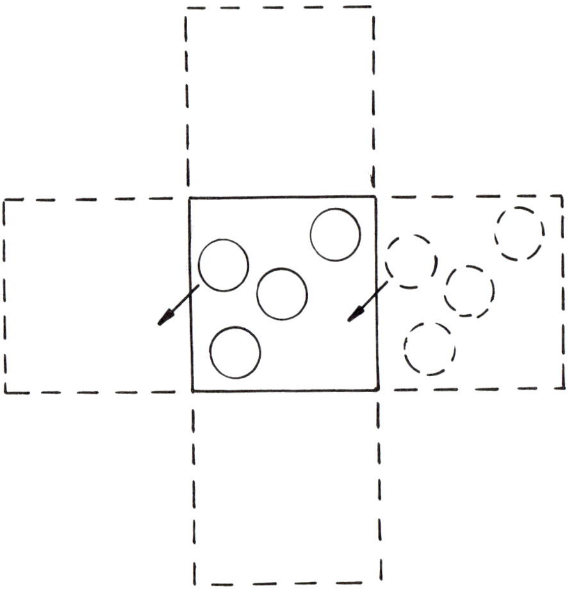

Fig. 13.8. Schematic representation of the periodic boundary conditions utilized in the Monte Carlo calculations.

The Monte Carlo calculations provide a yardstick against which the results of the cell model theory is compared. The agreement between the two methods is, in general, excellent.

13.3.2.2. A filtration method

Homola and Robertson (1975) have described a filtration method for determining the interaction between colloidal particles. This measures the minimum pressure required to initiate coagulation. It is therefore only applicable to dispersions of marginal stability, although it does represent a simple and convenient method of assessing qualitatively the stability of suspensions of larger particles.

The method consists essentially of repetitive filtration of a series of dispersions through a filter bed at different constant pressures. The minimum coagulation pressure can be estimated from the number of latex particles retained in the bed (measured either gravimetrically or by an optical difference technique). This shows a sharp increase at the pressure corresponding to the onset of coagulation.

13.4. Interaction of polymer chains attached to macroscopic objects

Probably the most direct and, in principle, the most powerful method for determining the distance dependence of steric interactions is the compression of polymer chains attached to macroscopic objects as they undergo close approach. The most successful method to-date has unquestionably been the crossed mica cylinders technique. This was originated by Tabor and coworkers (Tabor and Winterton, 1969; Israelachvili and Tabor, 1972) and improved by Israelachvili and Adams (1976; 1978). In the hands of Pashley and Israelachvili (1981), it has been successfully exploited to test the predictions of the Derjaguin-Landau-Verwey-Overbeek (DLVO) theory of electrostatic stabilization. It is also likely in the future to prove to be the most sensitive method for testing the predictions of theories for the distance dependence of steric interactions.

13.4.1. Crossed cylinders

13.4.1.1. Experimental apparatus

The experimental apparatus used to measure the interaction between crossed mica cylinders, of equal radii of curvature and coated with polymer chains, is shown schematically in the insert to Fig. 13.9. The force measured was that between two crossed cylindrical mica sheets that were glued to two optically polished glass discs. The cylindrical surfaces were mounted so that their axes were at right angles and positioned in the apparatus as shown in Fig. 13.11. Mica plates were used because muscovite mica cleaves to give surfaces that are smooth at the molecular level. Satisfying this stringent requirement of surface smoothness at the molecular level was essential for the measurement of forces at separations of order 1 nm or less. The thickness of the mica sheets was relatively large (ca 3 μm) compared with their separation and so they behaved like semi-infinite solids. The crossed cylinder geometry is equivalent in its interaction to that of a sphere (whose radius is equal to the radius of curvature of the plates) interacting with a flat plate.

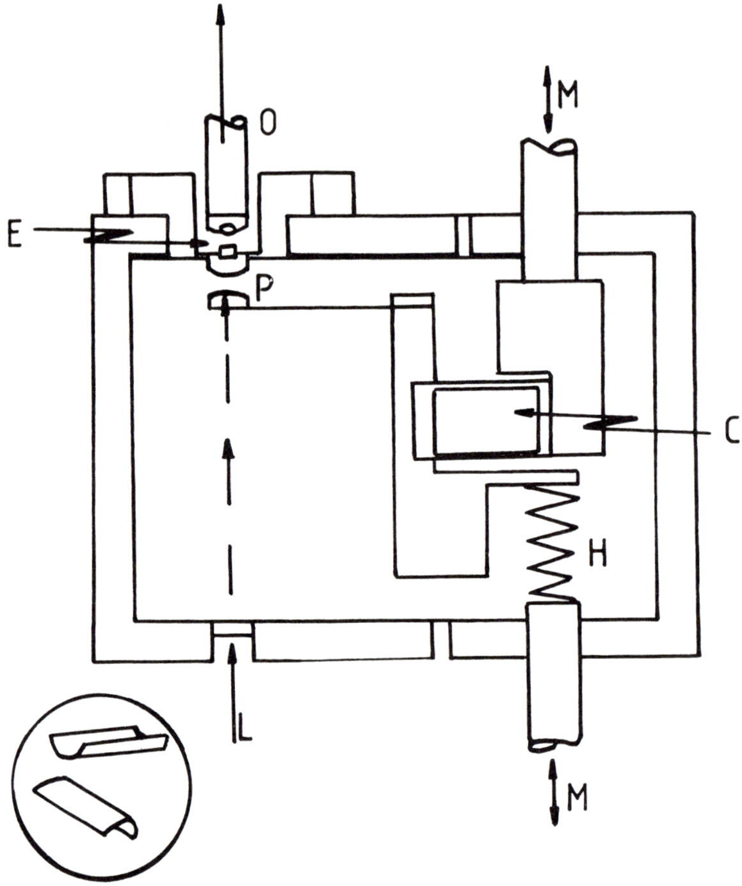

Fig. 13.9. The experimental apparatus used to measure the interaction between crossed mica cylinders (see inset) (after Israelachvili *et al.*, 1979; 1980). Note that M = micrometer drive, O = microscope objective, P = mica plates, E = piezoelectric crystal, H = helical spring, C = stiff double cantilever spring and L = ir light.

The upper glass disc was rigidly attached to a piezoelectric crystal tube. The lower glass disc was mounted on a weak cantilever spring whose force constant had been previously measured. The distance between the two surfaces could be controlled coarsely (to 1 μm) by an upper micrometer rod. Fine control (to 0·1 nm) was achieved through two movements: the piezoelectric crystal moved the upper mica surface whereas the lower mica surface was moved by a differential spring mechanism driven by a micrometer. These mechanisms were hypersensitive, allowing accurate measurement of weak forces down to a few nanonewtons (10^{-9} N)†.

† A newton is roughly the force exerted by gravity on a medium sized apple!

The distance between the mica plates was measured to ±0·1–0·2 nm (1–2 Å) by multiple beam interferometry. Both mica plates were silvered on their back (i.e. non-adjacent) surfaces. White light was passed through the interacting region. After multiple reflections, only selected wavelengths, corresponding to those undergoing constructive interference, passed through into the monitoring spectroscope. The positions of the interference fringes allowed both the separational distance and the average refractive index of the material between the plates to be calculated. The shape of the fringes also gave the mutual geometry of the crossed cylinders, allowing any surface deformations to be monitored.

13.4.1.2. Results

Nonaqueous systems in worse than θ-solvents
Klein (1980; 1982) has performed the most definitive experiments to date on the distance dependence of the steric interactions of polymer coated macroscopic surfaces. Monodisperse polystyrene, of molecular weight 1×10^5 or 6×10^5, was adsorbed onto the mica plates by incubation of the plates with a very dilute solution (ca 7×10^{-3} g dm^{-3}) of the polymer in cyclohexane at 24 °C. The polymer solution (save for ca 1 cm^3 between the plates) was then replaced by pure cyclohexane.

The force curves. The forces measured by Klein for the higher molecular weight polystyrene sample as a function of the distance of separation are shown in Fig. 13.10. These forces were stable, reproducible and reversible over

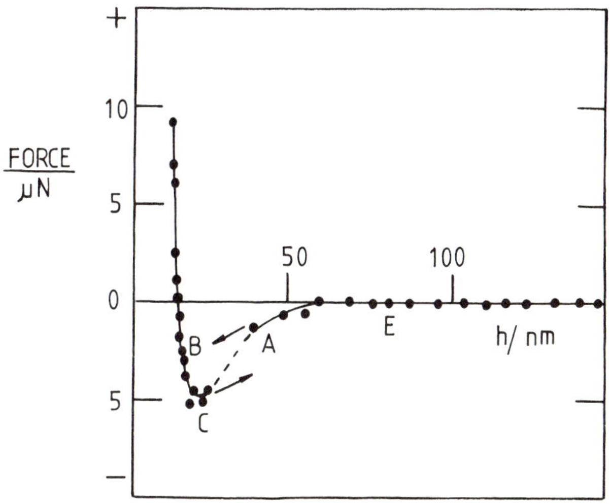

Fig. 13.10. The experimental results of Klein (1980) for the interaction of crossed mica plates coated by polystyrene of molecular weight 6×10^5 in cyclohexane at 24 °C.

several compression and decompression cycles. No forces were detectable down to ca 65 nm (about 3 R_g, where R_g = radius of gyration). Further compression resulted in a jump from point A to point B indicated in Fig. 13.10, with a concomitant increase in the attraction. Closer approach led to very strong repulsion. Decompression resulted in a decrease in the repulsion and the reappearance of a strong attraction, the potential energy showing a minimum around C, which is located near to a separation of R_g. Further separation caused the surfaces to jump apart from C to a new stable position at E. Note that both the jumps $A \rightarrow B$ and $C \rightarrow E$ were due to mechanical instability.

The origins of the steric interaction. The clue to the interpretation of the shapes of the curves measured by Klein lies in the recognition that cyclohexane at 24 °C is a worse than θ-solvent for polystyrene ($\theta = 34 \pm 1$ °C). The central features to be explained are the existence of an attractive region in the profile, followed ultimately by very strong repulsion nearer to the surface. In a worse than θ-solvent, the interpenetration of the attached segments is favoured, which generates an attractive interaction. This manifests itself in the potential energy curve by a decrease in potential energy in the domain $R_g \lesssim D \lesssim 3R_g$, where D is the minimum distance between the mica plates.

The simplest interpretation of the dramatic increase in the measured force when $D \lesssim R_g$ is that it is the elastic repulsion arising from the compression of the attached polystyrene chains by the close approach of the opposite plate. Since the first sign of steric interaction occurs at ca $3R_g$, this suggests that the steric layer thickness is $1{\cdot}5R_g$. Consequently, elastic compression of the chains would certainly be expected when $D \lesssim R_g$.

Whilst this elastic repulsion must assuredly contribute to the repulsion observed close in, Flory has suggested privately to Klein that another factor must contribute to the repulsion. This can be understood by reference to the polystyrene/cyclohexane phase diagram shown in Fig. 13.11, which displays an upper critical solution temperature (UCST). The solutions in which the mica plates were immersed were extremely dilute and corresponded to the one phase region. It is represented by point S_1 on the phase diagram. Bringing the plates closer together, effectively from infinite separation, results in interpenetration of the polymer chains so that the segmental volume fraction (v_2^{int}) in the interaction zone may lie in the two phase region ($v_2^d < v_2^{\mathrm{int}} < v_c^c$).

Klein claimed that a slight repulsion should be observed prior to entry into the two phase domain. No such repulsion was observed experimentally, perhaps because the chains were apparently attached irreversibly to the surfaces. It is the configurational entropy of the centres of mass of the particles that counteracts the unfavourable enthalpic segmental interactions at low free polymer concentrations and leads to the downturn in the coexistence curve. Attachment of the chains to the mica surfaces eliminates this configurational entropy term. It is as if the chains were of infinite molecular weight, which would render the coexistence curve much flatter in this region than that for free chains. This would reduce the repulsion expected from consideration of the phase diagram for free chains.

In the two phase region (signified by the letter T in Fig. 13.11), the overlap of polymer segments to form a more concentrated phase is spontaneous and so overlap is thermodynamically favoured. This would contribute to the observed attraction.

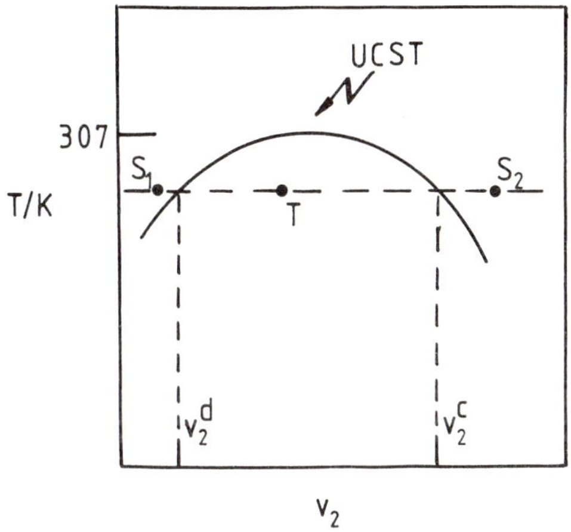

Fig. 13.11. Schematic representation of the phase diagram for high molecular weight polystyrene in cyclohexane. Note that S denotes a single phase whereas T denotes two phases.

Even closer approach can result in the volume fraction of segments in the interactional zone exceeding v_2^c. The volume fraction then passes back into the single phase region (denoted by S_2 in Fig. 13.11). In the single phase region, the formation of a more concentrated phase is obviously thermodynamically disfavoured. Thus, even though the dispersion medium is a worse than θ-solvent for the polymer segments, a mixing term can contribute to the repulsion. In this high segmental volume domain, it is the increase in the configurational entropy of the molecules of the dispersion medium on mixing that causes the downturn in the coexistence curve, outweighing the unfavourable enthalpic interaction. Crudely, the segment/solvent mixture should be envisaged as a solution of the dispersion medium in a polymeric solvent rather than vice versa. It is the first virial coefficient term of the dispersion medium that generates the repulsion and as such the magnitude of the repulsion depends upon its molecular weight. This contribution to the repulsion does not appear to be incorporated into any theory of steric stabilization that has been published to-date. It is effectively a very high virial coefficient term of the polymer interaction terms. It is known that the third and higher virial coefficients do not vanish even in θ-solvents and so it should come as no surprise that repulsive higher order coefficient terms persist into the worse than θ-solvent domain.

Scheutjens and Fleer (1982) have argued that Klein's explanation for the origins of the principal features of the potential energy curve is erroneous. They claim that the theory of phase separation, upon which it is based, is only valid for solutions that are homogeneous over distances large in comparison with the size of the polymer molecules. Between the plates, the segment density is far from homogeneous. Simply taking average segmental concentrations between the plates, as Klein has done, is therefore invalid. Scheutjens and Fleer correctly point out that the segment concentration in the steric layers is, in a small region, of order close to the critical segmental concentration, both at infinite separations and on close approach. Yet phase separation cannot occur whenever the bulk solution is composed of only one phase since the interplate region is in thermodynamic equilibrium with the bulk continuous phase.

Notwithstanding these criticisms, it remains an experimental fact that the theories of steric stabilization which consider the sole origin of repulsion in worse than θ-solvents to be the elastic repulsion, predict a rise in the repulsion that is seemingly nowhere near sufficiently steep to explain Klein's results. This strongly suggests the operation of another source of repulsive interaction, of which that suggested by Flory is the most soundly based.

The effect of polymer molecular weight. The effect of the polymer molecular weight on the measured force is shown in Fig. 13.12. Two molecular weights

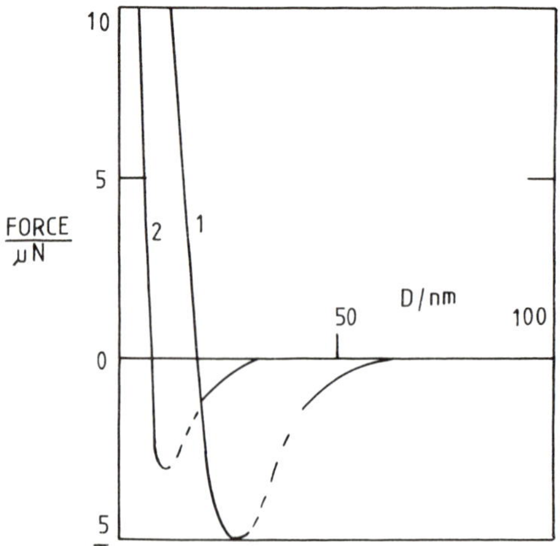

Fig. 13.12. The distance dependence of the force of interaction between crossed mica plates coated by polystyrene in cyclohexane at 24 °C; polystyrene molecular weight: 1, 6 × 10^5; 2, 1 × 10^5 (after Klein, 1982).

were studied: 1×10^5 and 6×10^5. It is apparent that the shapes of the two curves are very similar. For both molecular weights, attraction was apparent in the distance regime $R_g < D \lesssim 3R_g$ and a steep repulsion set in on closer approach ($D \lesssim R_g$). Obviously, an increase in the molecular weight pushed the whole interactional profile away from the surfaces towards the continuous phase, as would be expected intuitively. Note that the depth of the attractive potential energy well increased with increasing molecular weight. It should also be pointed out that the measured van der Waals attraction between the mica plates was negligible over the distance regime where the steric interaction could be measured.

It is apparent from these measurements that the mutual interactions of any long tails, as predicted by the theory of Scheutjens and Fleer (1980, 1982), is not seemingly detectable in these experiments. This suggests that, in these systems at least, the tails are unimportant insofar as the steric interaction is concerned.

The effects of solvency. Figure 13.13 displays Klein's results for the interactional energy per unit area, in the Derjaguin approximation, as a function of the reduced distance (D/R_g) for two polystyrene samples at two temperatures below the θ-temperature (26·5 °C and 24 °C). It is apparent that, as expected intuitively, the depth of the potential energy well increases as the solvency of the dispersion medium for the stabilizing segments becomes worse.

Aqueous systems in better than θ-solvents

The forces of repulsion between crossed mica cylinders coated by poly(oxyethylene) in 0·04 M magnesium sulphate solution have been meas-

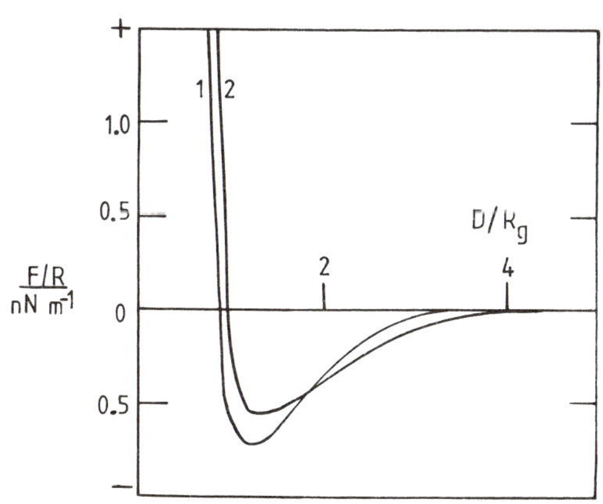

Fig. 13.13. The distance dependence of the forces of interaction between crossed mica plates coated by polystyrene in cyclohexane at two temperatures: 1, polystyrene molecular weight 6×10^5 at 24 °C; 2, molecular weight 1×10^5 at 26·5 °C (after Klein, 1982).

ured by Israelachvili *et al.* (1980). Unlike the cyclohexane in Klein's experiments, the continuous phase in these experiments was a better than θ-solvent for the stabilizing chains. The most striking feature of the results of Israelachvili *et al.* were the large irreversibilities and time-dependent effects exhibited by the forces. For this reason, their results do not yield the same fundamental information as those of Klein. Note that Klein did not observe these time-dependent, irreversible phenomena. Presumably, these curious, and haplessly debilitating, phenomena spring from at least two sources: first, the poor anchoring of the poly(oxyethylene) chains to the mica surfaces and, second, possible molecular association phenomena between the free and adsorbed poly(oxyethylene) chains.

The results reported by Israelachvili *et al.*, which are really only of qualitative interest, suggested the existence of dilute polymer layers at the mica surfaces that probably possessed gel-like properties. Even though the forces between the plates were monotonically repulsive, once significant overlap occurred there existed adhesion between the plates, probably through bridging. This observation is diametrically opposed to the conclusion of Klein (1982), who could find no evidence for polymer bridging in the polystyrene/cyclohexane system. Israelachvili *et al.* also claimed that as the rate of segmental rearrangement may be slow, the effective forces between colloidal particles stabilized by high polymers may often be determined by nonequilibrium relaxation mechanisms. Again, this is in sharp conflict with the conclusions of Klein who argued that whilst the process of segmental adsorption/desorption at the mica surfaces was relatively slow, the equilibrium between the segments in the interactional zone away from the surfaces is relatively rapid.

It is apparent that the crossed cylinder method provides direct access to the measurement of the distance dependence of the steric forces. To-date, only preliminary results, some of which are apparently conflicting, have been reported. Rapid advances in this area can be confidently anticipated in the immediate future.

13.4.2. Crossed filaments

Sonntag and coworkers (1982) have developed a crossed filaments technique for measuring the distance dependence of steric repulsion. One of the filaments, both of which were made from quartz, was connected to both a micrometer screw and a piezoelectric column. The latter allowed the distance between the two filaments to be changed, the limit of change being ± 1 nm. The other filament, together with a small mirror, was attached to a coil that was suspended in a magnetic field. A light beam was reflected from the mirror and, after passing through a slit, illuminated a photocell. Interaction between the filaments caused a change in the photocurrent. Forces as small as 5 nN could be measured by this technique.

Sonntag *et al.* (1982) have measured the distance dependence of the steric interaction between filaments of quartz coated by partially (88%) hydrolysed poly(vinyl alcohol). Some of their results are presented in Fig. 13.14. They interpreted their results in terms of the theory of Hesselink *et al.* (1971) and concluded that the average number of loops and tails was relatively small. These authors inferred that a small number of longer tails (<0·1%) was responsible for the observed stabilization. However, at smaller distances of

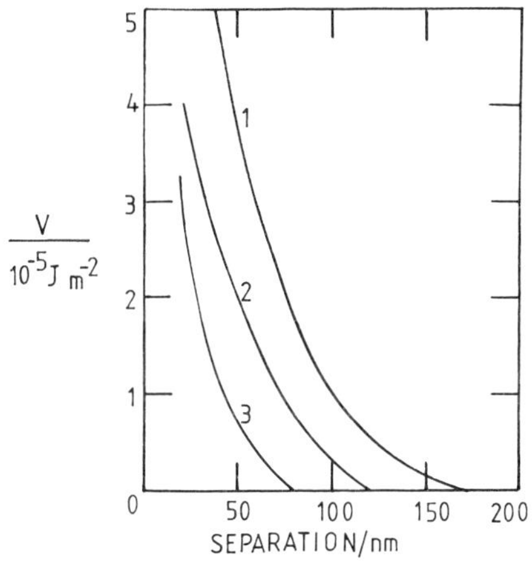

Fig. 13.14. The distance dependence of the potential energy of interaction for crossed filaments of quartz coated by poly(vinyl alcohol) in water (after Sonntag *et al.*, 1982).

close approach, repulsion due to loops, and shorter tails, became important. They concluded, in agreement with the theoretical predictions of Scheutjens and Fleer (1979; 1980; 1982), that there was less than one tail per molecule protruding into solution.

13.4.3. Spherical caps

13.4.3.1. Apparatus

Ottewill and coworkers (Cain *et al.*, 1978) have used a different geometry from those cited above to measure the distance dependence of steric repulsion. This was an extension of the apparatus first devised by Roberts and Tabor (1971) to measure the interactions between electrostatically stabilized macroscopic surfaces immersed in a liquid. Their technique consisted of pressing an optically smooth spherical cap of transparent rubber against a flat glass surface so as to produce a thin liquid film between the glass and the rubber. The rubber deformed readily over local protrusions so that in the compression zone, essentially parallel surfaces were formed. The thickness of the liquid film between these surfaces could be monitored interferometrically. One problem associated with this technique is that the two surfaces are intrinsically different. Hence the interaction between different types of surfaces is what is commonly measured by this technique.

Cain *et al.* (1978) have extended this method to allow the homostabilization between identical surfaces to be studied. The apparatus developed by them is shown schematically in Fig. 13.15. The central elements of the apparatus were two hemispherically capped silicone rubber cylinders which had been silvered and coated with a thin layer (ca 5 μm) of poly(methyl methacrylate) (see insert

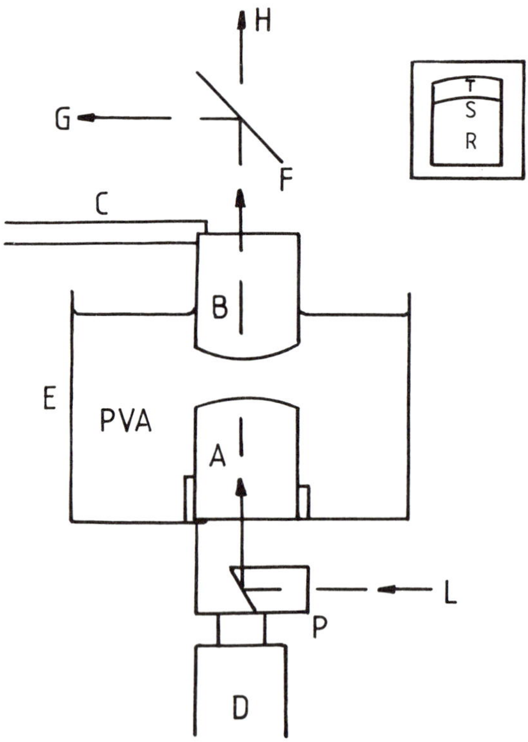

Fig. 13.15. Schematic representation of the apparatus used to measure the interactions between two silicone rubber cylinders (see inset) coated by polymer (after Cain *et al.*, 1978). Note that A, B = coated silicone rubber, C = clamp, D = micrometer drive for A, E = bath for polymer solution, F = beam splitter, G = eye-piece, H = prism of spectroscope, P = prism, R = silicone rubber, S = silver layer and T = poly(methyl methacrylate).

Fig. 13.15), so as to resemble the surface of a latex particle. The two rubber cylinders, A and B, were mounted vertically, one above the other. The upper cylinder was held fixed during the experiment whereas the position of the lower cylinder could be moved up and down by a micrometer drive unit (D). Moving the cylinder A up towards B, after polymer had been adsorbed onto the poly(methyl methacrylate) surfaces, distorted the soft, elastic surfaces in the central region, resulting in flat discs in the zone of interaction. The distance of separation between the surfaces was measured by multiple beam interferometry. The applied pressure caused an equilibrium contact area to be formed, the magnitude of which could also be determined interferometrically. The relationship between contact area and the equilibrium pressure was established by direct calibration. In this fashion, the equilibrium pressure could be measured for a given distance of separation.

13.4.3.2. Experimental results

Cain *et al.* (1978) studied the distance dependence of the disjoining pressure generated by a partially (88%) hydrolysed, polydisperse sample of poly(vinyl alcohol) of mean molecular weight 45 000. Some results obtained by them are displayed in Fig. 13.16, using both water and 0·1 M sodium chloride. The curve shown is qualitatively similar to that exhibited by sterically stabilized particles in a good solvent. Extrapolation of the observed curve suggests that the initial interactions occurred at a separation of 160 ± 20 nm. If the interaction occurs when the peripheral regions of the steric layers just touch, then the thickness of the adsorbed layer must be ca 80 nm. This was

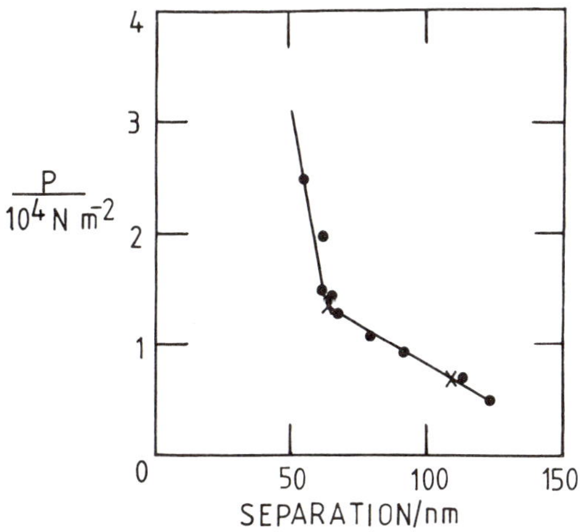

Fig. 13.16. The distance dependence of the disjoining pressure between flat plates coated by partially hydrolysed poly(vinyl alcohol) in both water (●) and (×) 0·1 M NaCl (after Cain *et al.*, 1978).

comparable to the layer thickness observed by Garvey *et al.* (1976) for the higher molecular weight components of the sample.

The compressibility of the steric layers, as expressed by $-(\partial P/\partial \ln h)$ where P = applied pressure and h = distance of separation of the flat discs, was found to be ca 1×10^4 N m^{-2} at $h = 100$ nm. If interpreted as an elastic modulus, the figure suggests that poly(vinyl alcohol) forms a weak viscoelastic gel. Whether this is due to the overlap of the adsorbed polymer layers or the entrapment of molecules in the layer, preventing proper draining, is not yet clear. These results indicate, in agreement with those of Israelachvili *et al.* (1980) for poly(oxyethylene), that the rheological properties of the steric layers may well influence their steric interaction.

It should not pass unnoticed, however, that poly(vinyl alcohol), despite its immense industrial importance, is not a homopolymer, and is scarcely an ideal polymer for carrying out fundamental studies of steric stabilization. Indeed, its solution properties suggest that it may even be relatively atypical in its behaviour.

13.4.4. Thin liquid films

13.4.4.1. Oil/water films

The first exploitation of thin liquid films to measure the distance dependence of the steric interaction of two parallel surfaces was reported by Andrews *et al.* (1970). They formed optically black films in aqueous media from solutions of glycerol mono-oleate in aliphatic hydrocarbons (*n*-heptane up to *n*-hexadecane and cyclohexane). The thickness of the hydrocarbon cores of the films was estimated from electrical capacitance measurements. When an electrical potential was applied across the liquid films, they were subjected to large compressive forces. The films as a result usually became significantly thinner. The steric interactions could be calculated from the thickness measured at a given applied potential.

One example of the form of the distance dependence of the steric repulsion constructed by Andrews *et al.* is given in Fig. 13.17. The steric repulsive force for different separations was determined by experiment whereas the London–van der Waals attraction was calculated from the Hamaker formula for flat plates. It is clear that the steric repulsion showed a sharp increase as the thickness of the film decreased. Because the films proved to be relatively compressible, Andrews *et al.* concluded that the volume fraction of the stabilizing chains in the overlapping regions of the film must be small compared with the average volume fraction in the entire film, which was usually found to lie between 0·5–1·0. Yet, paradoxically, the thickness of the films was close to twice the extended chain length of the stabilizing tails. From this they inferred that only a small fraction of the stabilizing chains were fully extended at any given instant. Note that the very thin lipid films formed

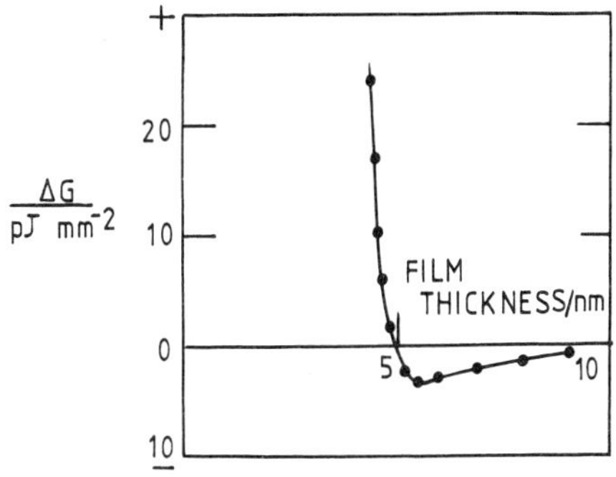

Fig. 13.17. The distance of the steric repulsion between two thin liquid films of glycerol mono-oleate in *n*-decane (after Andrews *et al.*, 1970).

from solutions of the higher alkanes (e.g. *n*-hexadecane) contained so little *n*-alkane that it was analytically undetectable. The stability of such films may then have an important contribution from genuine steric hindrance effects.

13.4.4.2. Air/water films

The stability of aqueous films in air, sterically stabilized by poly(vinyl alcohol), have been investigated by Lyklema and van Vliet (1978). Small horizontal films were formed inside a glass ring. Hydrostatic suction could be applied to the films by lowering (or raising) a container with polymer solutions, which was connected to the film.

The steric repulsive force measured by Lyklema and van Vliet for two different samples of poly(vinyl alcohol), one of molecular weight 143 000 and the other 42 500, are shown in Fig. 13.18. The general shape of the curves is qualitatively consistent with those presented previously for steric stabilization

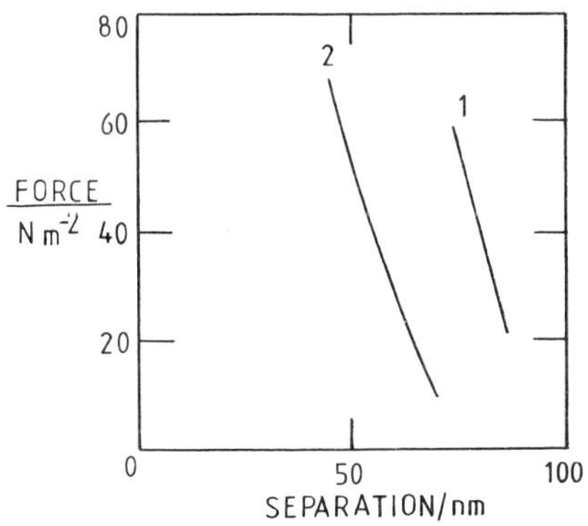

Fig. 13.18. The steric repulsive force measured by Lyklema and van Vliet (1978) for thin films stabilized by poly(vinyl alcohol) in water; poly(vinyl alcohol) molecular weight: 1, 143 000; 2, 42 500.

in a good solvent. The appearance of the repulsive force at larger film thicknesses for the higher molecular weight polymer is apparent.

The equilibrium thickness of the film greatly exceeded twice the ellipsometric or hydrodynamic thickness of one adsorbed layer. This was attributed to the steric repulsion associated with isolated tails, in agreement with subsequent predictions of Scheutjens and Fleer (1979; 1980; 1982). Such isolated tails could be essentially free draining, contributing little to the hydrodynamic radius and being essentially invisible to ellipsometry.

13.5. General conclusions

The foregoing experimental studies present a pattern of agreement that leaves little doubt that the overall shapes of the potential energy curves for sterically stabilized systems are in qualitative accord with the theoretical predictions. The general shape is determined *inter alia* by the quality of the solvency of the dispersion medium for the stabilizing moieties, the molecular weight of the stabilizing moieties and the particle size.

Figure 13.19 summarizes schematically the three potential energy profiles predicted theoretically and observed experimentally for a range of sterically stabilized systems. Curves 1 and 2 refer to dispersions for which the dispersion medium is a good solvent for the stabilizing moieties. The molecular weight of the stabilizing moieties is sufficiently large in the case of curve 1 that the van der Waals attraction is negligible, at least in comparison with the thermal energy of the particles and their configurational entropy. Only repulsion is evident in the potential energy diagram. This implies a thermodynamically stable system, provided that the stabilizing chains are attached irreversibly (e.g. by grafting) to the particle surface.

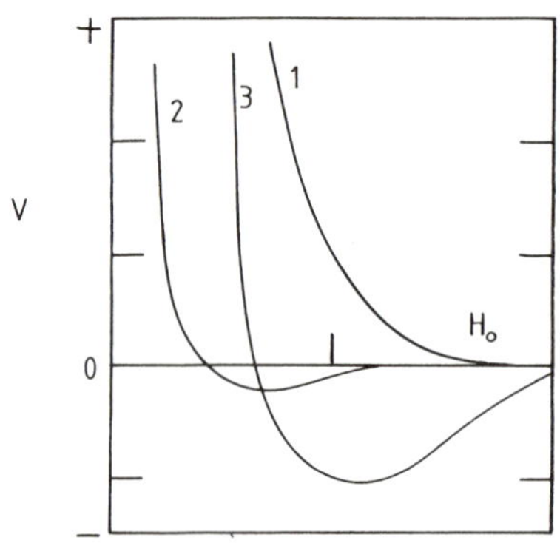

Fig. 13.19. The potential energy profiles for sterically stabilized systems: curves 1 and 2 refer to a good solvent whereas curve 3 applies to a worse than θ-solvent.

The pseudo-secondary minimum evident in curve 2 arises from the appreciable van der Waals attraction between the core particles. This can be a consequence of large particle sizes and/or thin steric layers, usually resulting from the presence of low molecular weight stabilizing moieties. Whether stability is observed or not depends upon whether the depth of the well is sufficiently large to exceed the thermal kinetic energy of the particles, coupled with their configurational entropy. It would typically be associated with weak coagulation, usually in better than θ-solvents.

The case where the dispersion medium is a worse than a θ-solvent for the stabilizing moieties is illustrated by curve 3. This shows a relatively deep potential energy well (again a pseudo-secondary minimum) that arises not from the core attraction, but rather from the mutual attraction of the segments of the stabilizing moieties. On closer approach, a steep repulsion is observed which is generated by the elastic compression of the chains, allied with a mixing repulsion that is manifest even though the solvent is a worse than θ-solvent. Again, whether or not stability is observed depends critically on the depth of the pseudo-secondary minimum: usually deviations of only a few degrees from θ are sufficient to ensure a well-depth of many kT, adequate to induce flocculation.

In conclusion it may be stated that there is now abundant experimental evidence to justify the assertion that the potential energy diagram for sterically stabilized systems is qualitatively different from that for electrostatically stabilized systems. The primary maximum that is so important in determining whether or not stability is observed in electrostatically stabilized systems appears to be absent for many sterically stabilized dispersions.

Chapter 14

Heterosteric Stabilization and Elastic Stabilization

14.1. Heterosteric stabilization

14.1.1. Introduction

In the preceding chapters, we have been concerned primarily with the stabilization of colloidal particles by steric layers that are chemically identical. We now turn our attention to the case where the particles are coated by polymer layers that are chemically different. For this reason, the stability so imparted is designated as *heterosteric stabilization.* By way of specific example, we cite the mixing of two aqueous, homosterically stabilized polystyrene latices, one stabilized by poly(oxyethylene) and the other by poly(acrylamide). Conditions can be chosen such that mixing occurs without the onset of flocculation. This phenomenon appears to be quite widespread: particles homostabilized by pairs of water soluble polymers chosen from the set poly(vinyl alcohol), poly(acrylic acid), poly(oxyethylene) and poly(acrylamide) may exhibit heterostabilization on mixing.

Instability can be observed, however, on mixing under other conditions. This shall be referred to as *heteroflocculation.*

14.1.2. Theory

The method for calculating the Gibbs free energy of mixing in the interpenetrational domain is a relatively straight forward extension of that elaborated above for *homosteric stabilization.* As before, the dispersion medium will be denoted by subscript 1. The two polymers will be designated by the subscripts 2 and 3.

The Gibbs free energy of mixing of polymer segments and dispersion medium in a volume element δV is given by

$$\delta(\Delta G) = \delta(\Delta G_M{}^*) + \delta(\Delta G_{SP}) \tag{14.1}$$

where $\Delta G_M{}^*$ = free energy of mixing calculated solely from external considerations of mixing segments with solvent and ΔG_{SP} = free energy changes due to all other interactions. The latter term encompasses all interactions between different species i and j so that

$$\delta(\Delta G_{SP}) = \sum_{i \neq j} \delta(G_{SP,i-j}) \tag{14.2}$$

To evaluate $\Delta G_{SP,i-j}$, it is necessary to introduce the segmental interaction parameter χ_{ij}. This is defined such that $kT\chi_{ij}$ represents the excess free energy associated with the transfer of one (solvent sized) segment of type i (or one solvent molecule if $i=1$) from pure i to pure j. If there are δn_i segments of i in δV, and the volume fraction of j therein is v_j, then it follows that

$$\delta(\Delta G_{SP,i-j}) = kT\chi_{ij}\delta n_i v_j \tag{14.3}$$

Flory has shown that the ideal mixing of solvent molecules with polymer segments is given by

$$\delta(\Delta G_M{}^*) = kT\delta n_1 \ln v_1 \tag{14.4}$$

For heterosteric stabilization, we obtain for the mixing free energy in the volume element:

$$\delta(\Delta G^M) = kT\{\delta n_1 \ln v_1 + \chi_{12}\delta n_1 v_2 + \chi_{13}\delta n_1 v_3 + \chi_{23}\delta n_2 v_3\} \tag{14.5}$$

If the segment density of species i in δV is denoted by ρ_i then $v_i = \rho_i V_1$, where V_1 = volume of a solvent molecule, and $\delta n_2 = \rho_2 \delta V$. Recalling that $v_1 = 1-(v_2+v_3)$, we obtain an alternative version of equation (14.5)

$$\delta(\Delta G^M) = -kT\{\delta n_1(\ln[1-V_1(\rho_2+\rho_3)] + \chi_{12}\rho_2 V_1 + \chi_{13}\rho_3 V_1) + \chi_{23}\rho_3 \delta V\} \tag{14.6}$$

To eliminate δn_1 from this equation, we note that $v_1 = 1-(v_2+v_3) = \delta n_1 V_1/\delta V$. Accordingly, $\delta n_1 = [1 - V_1(\rho_2+\rho_3)]\delta V/V_1$. This can be substituted into equation (14.6) to give on expansion

$$\delta(\Delta G) = kT\{\rho_2(\chi_{12}-1) + \rho_3(\chi_{13}-1) + V_1(\tfrac{1}{2}-\chi_{12})\rho_2{}^2 + V_1(\tfrac{1}{2}-\chi_{13})\rho_3{}^2 + V_1(1-\chi_{12}-\chi_{13}+\chi_{23})\rho_2\rho_3\}\delta V \tag{14.7}$$

where powers higher than the second are neglected.

14.1.2.1. Parallel flat plates

The simplest geometry to consider initially is parallel flat plates from which the sphere potentials can be calculated by the Derjaguin integration procedure. Again the segment densities parallel to the interfaces will be considered constant. The segment densities ρ_{ih} at separation h are given by $\nu_i \bar{p}_i \hat{\rho}_{ih}$, where ν_i=number of polymer chains of type i per unit area, $\bar{p}_i$=number of solvent sized segments per molecule of polymer i and the normalized segment density $\hat{\rho}_{ih}$ is such that

$$\int_0^h \hat{\rho}_{ih} \mathrm{d}x = 1,$$

the x-direction being normal to the interface.

The general expression for $\Delta^{FP}G^M(h)$ for unit area of flat plates is

$$\Delta^{FP}G^M(h) = kT\Bigg\{ \upsilon_2 \bar{p}_2 (\chi_{12} - 1) \int_0^h \hat{\rho}_{2h} \mathrm{d}x + \upsilon_3 \bar{p}_3 (\chi_{13} - 1)$$

$$\times \int_0^h \hat{\rho}_{3h} \mathrm{d}x + V_1 \Bigg[\upsilon_2^2 p_2^2 (\tfrac{1}{2} - \chi_{12}) \int_0^h \hat{\rho}_{2h}^2 \mathrm{d}x$$

$$+ \upsilon_3^2 \bar{p}_3^2 (\tfrac{1}{2} - \chi_{13}) \int_0^h \hat{\rho}_{3h}^2 \mathrm{d}x + \upsilon_2 \upsilon_3 \bar{p}_2 \bar{p}_3$$

$$\times (1 - \chi_{12} - \chi_{13} + \chi_{23}) \int_0^h \hat{\rho}_{2h} \hat{\rho}_{3h} \mathrm{d}x \Bigg] \Bigg\} \qquad (14.8)$$

The corresponding expression for infinite separation is readily derived from this equation by replacing $\hat{\rho}_{ih}$ by $\hat{\rho}_{i\infty}$ and recognizing that $\hat{\rho}_{2\infty}\hat{\rho}_{3\infty} = 0$. The free energy change on bringing the plates from infinite separation to a separation h is given by

$$\Delta^{FP}G^M = \Delta^{FP}G^M(h) - \Delta^{FP}G^M(\infty)$$

$$= \omega_2^2 (\bar{v}_2^2 / \bar{V}_1) N_A (\tfrac{1}{2} - \chi_{12}) kT \left[\int_0^h \hat{\rho}_{2h}^2 \mathrm{d}x - \int_0^\infty \hat{\rho}_{2\infty}^2 \mathrm{d}x \right]$$

$$+ \omega_3^2 (\bar{v}_3^2 / \bar{V}_1) N_A (\tfrac{1}{2} - \chi_{13}) kT \left[\int_0^h \hat{\rho}_{3h}^2 \mathrm{d}x - \int_0^\infty \hat{\rho}_{3\infty}^2 \mathrm{d}x \right]$$

$$+ \omega_2 \omega_3 (\bar{v}_2 \bar{v}_3 / \bar{V}_1) N_A (1 - \chi_{12} - \chi_{13} + \chi_{23}) kT \int_0^h \hat{\rho}_{2h} \hat{\rho}_{3h} \mathrm{d}x \qquad (14.9)$$

where ω_i=mass of polymer i attached to unit surface area, $\bar{v}_i$=partial specific volume i, $\bar{V}_1$=partial molar volume of the dispersion medium. In deriving equation (14.9), we have used the relationship $V_i p_i \upsilon_i = \omega_i \bar{v}_i$ since $\bar{p}_i = (V_i / V_1) p_i$, where p_i=number of real segments.

The interpenetrational domain

For heterosteric stabilization, the interpenetrational domain is defined by $L_2 \le h \le L_2 + L_3$, where L_2, L_3=span of the stabilizing moieties and $L_2 > L_3$. Equation (14.9) then reduces to

$$\Delta^{FP}G_I{}^M = \omega_2\omega_3(\bar{v}_2\bar{v}_3/\bar{V}_1)N_A(1-\chi_{12}-\chi_{13}+\chi_{23})kT\int_0^h \hat{\rho}_{2\infty}\hat{\rho}_{3\infty}\,dx \qquad (14.10)$$

The sign of $\Delta^{FP}G_I{}^M$ is clearly determined by $(1-\chi_{12}-\chi_{13}+\chi_{23})$. The significance of this term is more easily comprehended if it is recast into the form $[(\frac{1}{2}-\chi_{12})+(\frac{1}{2}-\chi_{13})+\chi_{23}]$. Interpenetration results in the replacement of some 1–2 and 1–3 contacts by 2–3 contacts. The first two terms specify the contributions to the free energy change resulting from the loss of 1–2 and 1–3 contacts respectively, whereas the third term indicates the contribution arising from 2–3 contacts. For example, if 1 is a good solvent for 2 ($\chi_{12}<\frac{1}{2}$), loss of 1–2 contacts is disfavoured and there is a positive contribution to the free energy from interpenetration. If, say, polymers 2 and 3 are compatible ($\chi_{23}<0$), the third term promotes interpenetration.

14.1.2.2. Qualitative predictions of the theory

Provided that L_3 is not excessively small, the interpenetrational domain will determine the flocculation behaviour in heterosterically stabilized systems, just as it does in homosteric stabilization. In the limit of small L_3, the interpenetrational-plus-compressional domain may well become important in predicting incipient instability. The elaboration of the general principles that govern heterosteric stabilization is then quite different.

The foregoing predictions for flat plates can be converted into sphere potentials by the Derjaguin integration. This will only change the term involving the segment density distribution functions, leaving intact the premultiplying thermodynamic factor $[(\frac{1}{2}-\chi_{12})+(\frac{1}{2}-\chi_{13})+\chi_{23}]$. The sign of this thermodynamic term determines the sign of the steric free energy of interaction, i.e. whether the interaction is attractive or repulsive. It proves convenient in discussing the sign of this term to differentiate between positive and negative values of χ_{23}.

Positive values for χ_{23}

A positive value of χ_{23} is the rule, rather than exception, for binary mixtures of nonpolar polymers. As Flory (1953) has pointed out, the critical value of χ_{23} required for the phase separation in a 50 : 50 melt of two different polymers of the same molar volume is $2/\bar{p}_2$, where $\bar{p}_2$ = number of solvent sized segments per polymer. This is readily seen as follows.

Consider, first, the mixing of two chemically different polymers of identical molecular weight (i.e. $x_2 = x_3 = x = 1$) in the absence of solvent. According to the Flory-Huggins theory, the chemical potentials of the polymers are given by

$$\mu_i - \mu_i^{\ominus} = RT[\ln v_i + \chi_{ij} v_j^2]$$

where we have set the *molecular* interaction parameters such that $\chi_{ij} = \chi_{ji}$. The critical condition for phase separation requires that both terms of the type $(\partial\mu_i/\partial v_j)$ equal zero and further that both of the corresponding second derivatives $(\partial^2\mu_i/\partial v_j^2)$ also be zero. The symmetry of these relationships for polymers of equal molar volume obviously demands that $v_2 = v_3 = \frac{1}{2}$, which only satisfies the critical relationship $\chi_{23}=1/2(1-v_2)^2$ if $\chi_{23}=2$. The result is identical, as would be expected, with the classical Bragg-Williams result for the critical mixing parameter of minimolecules (Moore, 1972). The interactional free energy *per segment* is therefore just $2/\bar{p}_2$. Since $\bar{p}_2$ is large for high polymers, almost any positive value for χ_{23} suffices to render the polymers incompatible. Whilst the foregoing discussion has been concerned solely with binary molten systems, Flory (1953) has shown that the presence of solvent does not alter the conclusion just reached for melts because the magnitude of the solvent-polymer interaction parameters does not enter into the critical condition. The latter is determined solely by χ_{23}. These considerations explain why polymer incompatibility is so widespread.

If χ_{23} is positive and the dispersion medium is a good solvent for both polymers (χ_{12}, $\chi_{13}<\frac{1}{2}$) then the thermodynamic term $[(\frac{1}{2}-\chi_{12})+(\frac{1}{2}-\chi_{13})+\chi_{23}]$ must be positive. Heterosteric stabilization will be evident on mixing homosterically stabilized particles.

Selective flocculation. Suppose χ_{23} is positive but now the dispersion medium is adjusted to be a worse than θ-solvent for, say, polymer 2 ($\chi_{12}>\frac{1}{2}$ but $\chi_{13}<\frac{1}{2}$). Homoflocculation of the particles coated by polymer 2 will ensue. Provided that χ_{12} is not excessively large, heteroflocculation would be expected to be absent. This implies that it should be possible under such conditions to flocculate *selectively* one species of sterically stabilized particles in a mixed dispersion. Apparently polymer coated particles are able to discriminate between "self" and "not-self" when undergoing Brownian collisions in such dispersion media, coagulation only occurring between particles of the same type. Of course, if the dispersion medium were made a worse than θ-solvent for both polymers (χ_{12}, $\chi_{13}>\frac{1}{2}$), both 2–2 and 3–3 homoflocculation would occur. Heteroflocculation would only be possible if $\chi_{23}<\chi_{12}+\chi_{13}-1$.

Experimental observation of selective flocculation. Croucher and Hair (1980a) have demonstrated the phenomenon of selective flocculation in mixtures of heterosterically stabilized dispersions. They prepared particles of poly(vinyl acetate) stabilized by polystyrene and polyacrylonitrile particles stabilized by polyisobutylene, both in cyclopentane. The value of χ_{23} for polystyrene and polyisobutylene is known to be positive (Hyde and Tanner, 1968), corresponding to mutually incompatible polymers. The poly(vinyl acetate) particles stabilized by polystyrene exhibited both upper and lower critical flocculation temperatures whereas the polyisobutylene stabilized particles only flocculated on heating.

When a mixture of the two different types of particles dispersed in cyclopentane was cooled to the lower critical flocculation temperature (280 K) of the poly(vinyl acetate) particles stabilized by polystyrene, those particles underwent *selective homoflocculation*. This was demonstrated by the complete dissolution of the floc phase in toluene, polyacrylonitrile being insoluble in toluene. Selective homoflocculation of the poly(vinyl acetate) particles was also achieved by heating to their upper critical flocculation temperature (410 K). Raising the temperature even higher (to 455 K) resulted in the unflocculated particles undergoing aggregation at the upper critical flocculation temperature of the polyacrylonitrile particles stabilized by polyisobutylene, thus confirming the existence of the phenomenon of selective flocculation.

Summary. The foregoing analysis shows that for positive values of χ_{23} corresponding to incompatible polymers, stability will normally be observed on mixing homosterically stabilized dispersions. Heteroflocculation is not possible without accompanying 2–2 and/or 3–3 homoflocculation. The selective flocculation of one type of particle in a mixture of different particles is predicted to be possible under suitable conditions, and has been demonstrated experimentally.

Negative values for χ_{23}

Negative values for χ_{23} are less common than positive values. They commonly occur between pairs of polymers that possess polar substituents that interact favourably. These correspond to compatible polymers that are miscible and can exist in solution together. Of course, if the interaction between the polymers is sufficiently strong, molecular association may result in coprecipitation or the formation of coacervates. Such behaviour, characterized by the formation of phases containing comparable amounts of the two polymers, should not be confused with the incompatibility noted in the preceding section for positive values for χ_{23}. In the latter case, the polymer phases formed would contain predominantly one polymer (except in the vicinity of the critical point), not comparable amounts of both polymers.

For many polymer mixtures that are known to be compatible, the extent of their compatibility decreases as the temperature increases. This diminution in miscibility with temperature is a forerunner of the occurrence of a lower critical solution temperature. At higher temperatures, the effect of the favourable interactions between the two components is reduced whereas any free volume dissimilarity difference is enhanced. The conjunction of these competing effects eventually leads to phase separation.

If the dispersion medium is a good solvent for both polymers ($\chi_{12}, \chi_{13} < \frac{1}{2}$), then the system will be stable provided that $|\chi_{23}| < 1 - \chi_{12} - \chi_{13}$. This apparently explains the heterosteric stabilization, alluded to above, that is observed on mixing aqueous polystyrene latices homosterically stabilized by poly(ethylene oxide) and polyacrylamide.

If, however, the solvency is such that this inequality is not obeyed, heteroflocculation alone will occur, provided that the dispersion medium is a better than θ-solvent for both polymers. One example of this appears to be the heteroflocculation that is observed on mixing latices homosterically stabilized by poly(acrylic acid) and poly(ethylene oxide) in 0·2 M HCl. Separately, the two dispersions are quite stable under these conditions. The packing of particles in such heteroflocculated systems should bear some similarity to the packing of ions in an ionic crystal in that particles coated by polymer 2 will be surrounded by particles coated by polymer 3, and vice versa.

If one type of particle, say 2, is present in large excess, heteroflocculation under such conditions could possibly lead to the complete removal of type 3 particles if aggregates are separated from dispersed particles. This, in a sense, is a type of selective flocculation; however, it is accomplished by heteroflocculation rather than by homoflocculation, which was the mechanism operative for the genuine case of selective flocculation discussed above for a positive χ_{23}. Of course, such heteroflocculation also implies that sterically stabilized particles can distinguish "self" from "not-self", flocculation only occurring with not-self in this instance.

If χ_{23} were only slightly negative, it may be possible to induce selective homoflocculation. This would require that the dispersion medium be made a worse than θ-solvent for, say, polymer 2 ($\chi_{12} > \frac{1}{2}$) but remain a good solvent for the second polymer ($\chi_{13} < \frac{1}{2}$). If also $|\chi_{23}| < 1 - \chi_{12} - \chi_{23}$, ΔG will be negative for 2–2 interpenetration, but will be positive for 2–3 and 3–3 interpenetration. Selective 2–2 homoflocculation would then occur.

The preceding discussion implies that compatible polymers should often induce heteroflocculation when particles homosterically stabilized by them are mixed. This should be especially evident with polymer pairs that form coprecipitates or coacervates, and in solvents that are not very good solvents for either polymer. Such behaviour contrasts markedly with the mixing of particles homosterically stabilized in the same dispersion medium by incompatible polymers, for which stability is predicted to occur. Selective flocculation seems feasible irrespective of the sign of χ_{23}, but will undoubtedly be more common in systems for which χ_{23} is positive.

The foregoing conclusions readily encompass mixtures that contain more than two types of homosterically stabilized particles since these can usually be treated in a pair-wise fashion. Of special importance is the removal of one type of coated particle in a mixture of particles by the addition of another type of coated particle that selectively heteroflocculates with it.

Zero value for χ_{23}

It is possible, at least in principle, for $\chi_{23} = 0$, even if the two polymers are not chemically identical. In that case, heterosteric stabilization would be observed if the dispersion medium were a good solvent for both polymers. If, however, the solvency were changed to worse than θ-conditions for, say, polymer 2 ($\chi_{12} > \frac{1}{2}$ but $\chi_{13} < \frac{1}{2}$), then selective flocculation of type 2 particles would occur if $1 - \chi_{13} > \chi_{12}$. If this inequality were not obeyed, both homoflocculation and heteroflocculation would be observed.

14.1.2.3. Quantitative predictions of theory

The distance dependence of the interactional free energy can be evaluated once the segment densities are specified. The requisite integrations can be performed numerically whatever the form of $\hat{\rho}_{ih}$. They can also be accomplished analytically for simple functional forms. Because these integrations become unwieldy for all but the simplest forms of the segment density distribution functions, we shall restrict our discussion to one simple type: the constant segment density. The following calculations are meant primarily to be illustrative of the interactional potentials involved in heterosteric stabilization. There exists, however, experimental evidence to suggest that a constant segment density may well be appropriate for low molecular weight stabilizers. Furthermore, we have shown that the distance dependence of the interaction is not dramatically affected by the form chosen for the segment density distribution functions unless the polymer interacts strongly with the surface.

The interpenetrational domain

In this domain, we may set $\hat{\rho}_{2\infty} = 1/L_2$ and $\hat{\rho}_{3\infty} = 1/L_3$. Evaluation of the integral in equation (14.10) using these expressions yields

$$\Delta^{FP}{}_C G_I{}^M = \omega_2\omega_3(\bar{v}_2\bar{v}_3/\bar{V}_1)N_A(1-\chi_{12}-\chi_{13}+\chi_{23})R_I{}^M kT \qquad (14.11)$$

where $R_I{}^M = (1/L_2 + 1/L_3 - h/L_2L_3)$ and the prescript C denotes a constant segment density. On performing the Derjaguin integration, we find that for spheres (denoted by the prescript S) of particle radius a

$$\Delta^S{}_C G_I{}^M = \pi a\omega_2\omega_3(\bar{v}_2\bar{v}_3/\bar{V}_1)N_A(1-\chi_{12}-\chi_{13}+\chi_{23})S_I{}^M kT \qquad (14.12)$$

In this equation, $S_I{}^M = (1 + L_3/2L_2 + L_2/2L_3 - H_0(1/L_2 + 1/L_3) + H_0{}^2/2L_2L_3)$, where H_0 = minimum distance of separation of the surfaces of the particles. Of course, the Derjaguin integration procedure is only valid if $a \gg L_2, L_3$. Note that if polymers 2 and 3 are identical, equation (14.12) reduces to the result derived previously for homosteric stabilization.

The interpenetrational-plus-compressional domain

In the interpenetrational-plus-compressional domain, we may write

$$\Delta^{FP} G_{I+C}{}^M = \sum_{2 \le i \le j = 2,3} {}^{FP}\Gamma_{i-j}{}^M R_{i-j}{}^M \qquad (14.13)$$

Here ${}^{FP}\Gamma_{i-j}{}^{M}=\omega_i\omega_j(\bar{v}_i\bar{v}_j/\bar{V}_1)N_A\,(1-\chi_{1i}-\chi_{1j}+\chi_{ij})kT/(\delta_{ij}+1)$, $R_{i-j}{}^{M}=[\int_0^h\hat{\rho}_{id}\hat{\rho}_{jd}\mathrm{d}x-\int_0^{\infty}\hat{\rho}_{i\infty}\hat{\rho}_{j\infty}\mathrm{d}x]$, and δ_{ij}=Kronecker delta. The latter adopts the value of unity if $i=j$ and zero if $i\neq j$. To this mixing term must be added the elastic free energy

$$\Delta^{FP}G^{el}_{I+C}=\sum_{i=2,3}{}^{FP}\Gamma_i^{el}R_i^{el} \tag{14.14}$$

In this equation

$$^{FP}\Gamma_i^{el}=(\omega_i/M_i)N_AkT,$$

$$R_i^{el}=\int_0^h\delta_i^{-1}\hat{\rho}_{i\infty}(x/\delta_i)\;\ln\{[\delta_i^{-1}\hat{\rho}_{i\infty}(x/\delta_i)/\hat{\rho}_{i\infty}(x)]\}\mathrm{d}x,$$

where M_i=molecular weight of polymer i and $\delta_i=h/L_i$.

Evaluation of the distance dependent terms for a constant segment density yields

$$R_{i-j}{}^{M}=1/h-\delta_{ij}/L_i,$$

$$R_i^{el}=-\ln\,\delta_i$$

These results are applicable whether or not d is greater than L_3, with the proviso that, if $d\geq L_3$, $R_{3-3}{}^{M}=R_3^{el}=0$.

The corresponding equations for spheres are

$$\Delta^{S}G_{I+C}{}^{M}=\pi a\sum_{2\leq i\leq j=2,3}{}^{FP}\Gamma_{i-j}{}^{M}S_{i-j}{}^{M} \tag{14.15}$$

$$\Delta^{S}G_{I+C}{}^{el}=\pi a\sum_{i=2,3}L_i{}^{FP}\Gamma_i^{el}S_i^{el} \tag{14.16}$$

For a constant segment density, the dimensionless distance dependent S parameters are given by

$$S_{i-i}{}^{M}=\delta_i-1-\ln\,\delta_i,$$

$$S_{i-j}{}^{M}=-\ln\,\delta_i+\delta_i/2\delta_i,$$

$$S_i^{el}=1+\delta_i\ln\,\delta_i-\delta_i.$$

Again, if $h\geq L_3$, $S_{3-3}{}^{M}=S_3^{el}=0$.

Potential energy diagrams

Displayed in Fig. 14.1 are the predicted interactional free energies for mixtures of polyacrylonitrile latex particles, of 100 nm radius, stabilized by polystyrene (polymer 2) and polyisobutylene (polymer 3), assuming a constant segment density for both polymers (Feigin and Napper, 1978). Figure 14.1a shows the

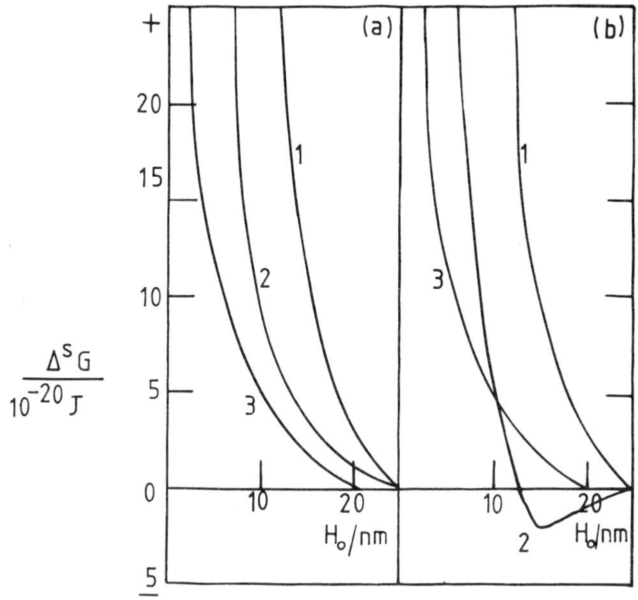

Fig. 14.1. Theoretical free energies of interaction between sterically stabilized latex particles of 100 nm radius dispersed in cyclohexane: (a) curves for 312 K, 5 K above the θ-temperature for polystyrene: (1) PS/PIB heterostabilization; (2) and (3) are for PS/PS and PIB/PIB homostabilization respectively; (b) curves for 302 K, 5 K below the θ-temperature for polystyrene. Curves numbered as in (a). Note that PS = polystyrene and PIB = polyisobutylene (after Feigin and Napper, 1979).

predicted curves for a temperature 5 K above the θ-temperature for polystyrene in cyclohexane whereas Fig. 14.1b presents the curves for 5 K below the θ-point. The data used in the calculations were: $\omega_2 = 5 \times 10^{-8}$ g cm^{-2}, $L_2 = 15$ nm, $M_2 = 6\,000$, $\theta_{12} = 307$ K, $\bar{v}_2 = 0{\cdot}93$ cm^3 g^{-1}, $\omega_3 = 3 \times 10^{-8}$ g cm^{-2}, $L_3 = 10$ nm, $M_3 = 4\,000$, $\theta_{13} = 126$ K, $\bar{v}_3 = 1{\cdot}07$ cm^3 g^{-1}, and $\bar{V}_1 = 108$ cm^3 mol^{-1}. The interactional parameters were calculated from $(\frac{1}{2} - \theta_{1i}) = \psi_{1i}(1 - \theta_{1i}/T)$, where ψ_{1i} = entropy of dilution parameter for polymer i. The values adopted in cyclohexane were $\psi_{12} = 0{\cdot}79$ and $\psi_{13} = 0{\cdot}14$ (Saeki *et al.*, 1973; Stockmayer *et al.*, 1955). In the apparent absence of experimental values for χ_{23}, we may calculate a value for the polystyrene/polyisobutylene pair from

$$\chi_{23} = \bar{V}_1(\delta_2^H - \delta_3^H)^2/RT \tag{14.17}$$

which has been shown to predict realistic values for χ_{23} in other nonpolar systems. In this equation, δ_i^H is the Hildebrand solubility parameter of polymer i and $\bar{V}_1$ appears as a scaling parameter. Use of the literature values of $\delta_2^H = 9{\cdot}1$ cal$^{\frac{1}{2}}$ cm$^{-3/2}$ and $\delta_3^H = 8{\cdot}0$ cal$^{\frac{1}{2}}$ cm$^{-3/2}$ yields a value of $\chi_{23} = 0{\cdot}22$

at 300 K. This suggests that polystyrene and polyisobutylene are incompatible, which has been confirmed experimentally (Boyer and Spencer, 1944).

Note that Croucher and Hair (1980) have claimed that according to free volume theory, χ_{23} is an order of magnitude smaller than that used by Feigin and Napper (1978). It has subsequently been shown by Roe and Zin (1980), however, that the free volume contribution to χ_{23} is relatively unimportant. These authors obtained values of $\chi_{23} \simeq 0{\cdot}14$ for polystyrene and polybutadiene by experimental measurements of the miscibility of the two polymers, whereas the free volume contribution is only ca 0·01. Roe and Zin advocated the use of the solubility parameter approach to the calculation of χ_{23}, which is what has been adopted above. This is a further instance where the free volume theory proves inadequate at the quantitative level. This gross underestimation of χ_{23} has unfortunately led Croucher and Hair (1980) to present erroneous values for the free energies of interaction for heterosterically stabilized systems.

Both of the diagrams shown in Fig. 14.1 illustrate the strong repulsive interactions that can be generated in heterosteric stabilization by incompatible polymers. Indeed it is evident in this example that the 2–3 particle interactions are stronger than either the 2–2 or 3–3 interactions. In addition, Fig. 14.1b shows the appearance of a $-5kT$ pseudo-secondary minimum in the interactional free energy of polystyrene-coated particles at 5 K below their θ-temperature. This minimum would be sufficient to ensure 2–2 homoflocculation. The 3–3 and 2–3 interactions are clearly repulsive and so the qualitative free energy calculations confirm the possibility, foreshadowed above, of the selective flocculation of one particle type in mixtures of particles sterically stabilized by different polymers.

14.2. Stabilization in polymer melts

14.2.1. Elastic steric stabilization

If the dispersion medium is a melt of a polymer of the same chemical composition as the stabilizing moieties then the mixing term is reduced to zero (we ignore here end effects, which are of necessity small for polymers of even modest molecular weight). The same conclusion is valid in θ-solvents, provided that we disregard the higher order virial coefficients. Under either of these conditions, there is no significant free energy change in the interpenetrational domain. In the interpenetrational-plus-compressional domain, the mixing term is also zero but the elastic term provides a repulsion that is able to impart colloid stability, provided that the extension of the stabilizing moieties is sufficiently large to render the van der Waals attraction negligible. Thus stabilization of colloidal particles in polymer melts or in θ-

solvents is qualitatively different from that in dispersion media that are good solvents for the stabilizing moieties. The origin of the repulsion resides solely in the loss of configurational entropy of the stabilizing moieties on compression by the surface of the opposing particle in a Brownian collision. For this reason, the generic title *elastic steric stabilization* can be usefully applied to this phenomenon. It is likely to be important in the dispersion of carbon blacks in rubbers and domain formation in certain copolymer systems.

14.2.1.1. Theories of elastic steric stabilization

Dolan and Edwards (1974) have calculated the elastic free energy of repulsion for isolated polymer chains attached terminally to parallel flat plates. It was shown in Section 11.4.1.1. that probability distribution function $G(\mathbf{r},\mathbf{r}',s,s')$ for a random flight chain obeys the diffusion equation

$$\partial G(\mathbf{r},\mathbf{r}',s,s')/\partial s-(l^2/6)\nabla^2 G(\mathbf{r},\mathbf{r}',s,s')=\delta(\mathbf{r}-\mathbf{r}')\delta(s-s') \qquad (14.18)$$

where $\mathbf{r},\mathbf{r}'$ are positions in space and s and s' measure distances along the backbone of the chain. The link length is l. At the confining surface the boundary condition $G(\mathbf{r},\mathbf{r}',s,s')=0$, so that no configurations cross the boundary. Thus $G(\mathbf{r},\mathbf{r}',s,s')$ is just the number of configurations of a chain (relative to the total number of configurations of an unconfined chain) of length $(s-s')$ that span the space from $\mathbf{r}'$ to $\mathbf{r}$ without crossing the boundary.

To calculate the entropy of an isolated chain which is confined between two parallel plane surfaces, only the x-component in the direction normal to the surfaces need be considered. The method of images yields

$$G(x, x', s, 0)=(3/2\pi l s)^{\frac{1}{2}} \sum_{m=-\infty}^{\infty} [\exp\{-3(x-x'-2mh)^2/2ls\}$$
$$-\exp\{-3(x+x'-2mh)^2/2ls\}] \qquad (14.19)$$

whereas the eigenfunction expansion method yields the entirely equivalent solution

$$G(x, x', s, 0)=(2/h)\sum_{m=1} \sin(m\pi x/h)\sin(m\pi x'/h)$$
$$\times \exp\{-\pi^2 m^2 ls/h^2\} \qquad (14.20)$$

The first of these two expressions can be used to show that for large separations ($h^2 \gtrsim 3lL$)

$$\Delta^{FP} G_{I+C}{}^{el} \simeq -kT\ln(1-2\exp\{2h^2/2lL\}) \qquad (14.21)$$

where L = contour length of the chain. The corresponding approximation for small separations ($h^2 \lesssim 3lL$) is derived from the second equation

$$\Delta^{FP} G_{I+C}{}^{el} \simeq kT(\ln\{3h/8\pi^2 lL\}+\pi^2 lL/6h^2) \qquad (14.22)$$

In elastic steric stabilization, the elastic repulsion must exceed the attractive van der Waals interaction for stability to be observed. Dolan and Edwards showed that two types of aggregation occur. The first, corresponding to close adhesion of the particles, occurs if

$$Ll < A^*/2\pi^3 \mu kT \tag{14.23}$$

where A^* = effective Hamaker constant and μ = number of chains per unit area. The second type of coagulation occurs if

$$A^*/2\pi^3 \mu kT < lL < A^* S/8\pi kT \ln(2N) \tag{14.24}$$

when a looser form of aggregation occurs, the particles being separated by distances of order of the dimensions of the chain. In equation (14.24), S is the area occupied by N tails (i.e. $\mu = N/S$).

In elastic steric stabilization, the particles would be expected to be only loosely aggregated if coagulation is induced by van der Waals attraction. Dolan and Edwards (1974) suggested that the criteria set out above can also be applied to spheres. Smitham and Napper (1976a,b) have shown that the criteria must be modified (e.g. by the Derjaguin integration) to obtain agreement between the maximum particle size that can be stabilized elastically and the size predicted theoretically. The maximum particle radius a_m that can be stabilized is given by

$$a_m \sim [\mu(2\pi^3 lL/3)^{\frac{1}{2}} \operatorname{erfc}\{(3/2lL)^{\frac{1}{2}} A^* a_m/12kT\}]^{-1} \tag{14.25}$$

where erfc is the complementary error function.

14.2.1.2. Experimental evidence for elastic steric stabilization

There is a paucity of experimental studies of elastic steric stabilization. Smitham and Napper (1976a,b) have shown that it is possible to prepare polystyrene latex particles stabilized by poly(oxyethylene) and dispersed in molten poly(oxyethylene). These experiments suggested that the maximum particle size that could be elastically stabilized was dependent upon the molecular weight of the stabilizing moieties, as would be expected intuitively. Everett and Stageman (1978a) have also reported the elastic stabilization of poly(methyl methacrylate) particles stabilized by poly(dimethylsiloxane) in liquid poly(dimethylsiloxane).

14.2.2. Stabilization in polymer melts

14.2.2.1. Theory

It is possible to generalize the preceding theories to encompass heterosteric stabilization in polymer melts. To this end, we now set out the modifications needed to allow for the polymeric nature of the dispersion medium.

As before, the subscripts 1, 2 and 3 will denote the dispersion medium (also referred to as the solvent) and the two stabilizing polymers, respectively. In the ternary polymer mixture, V_1 (previously the volume of a solvent molecule) and δn_1 (previously the number of solvent molecules in the volume element δV) are redefined as the volume per *segment* of dispersion medium and the number of segments of dispersion medium in the volume element, respectively. The segments of the stabilizing chains are taken to be of a size identical with that of the segments of the dispersion medium. All the interaction parameters χ_{ij} are now expressed on a segmental basis.

The $\delta(\Delta G_{SP,i-j})$ terms, as given by equation (14.3) are unchanged on generalizing to a polymeric dispersion medium since they depend solely upon the number of contacts and not upon the connectivity of the species involved. In contrast, the ideal mixing contribution, which is derived from a configurational entropy term, depends critically upon the actual number of solvent molecules in the volume element δV. To allow for the connectivity of the segments, we must modify equation (14.4) to read

$$\delta(\Delta G_M{}^*) = kT(\delta n_1/\bar{p}_1)\ln v_1, \tag{14.26}$$

where $\bar{p}_1$ = number of segments in a solvent molecule and v_1 = volume fraction of solvent molecules. When this revised form of the configurational entropy term is carried through the derivation, a new version of equation (14.8) results

$$\begin{aligned}\Delta^{FP}G^M = {} & \omega_2{}^2(\bar{v}_2{}^2/V_1)(1/2\bar{p}_1 - \chi_{12})kT\left[\int_0^h \hat{\rho}_{2h}{}^2 \mathrm{d}x - \int_0^\infty \hat{\rho}_{2\infty}{}^2 \mathrm{d}x\right] \\ & + \omega_3{}^2(\bar{v}_3{}^2/V_1)(\tfrac{1}{2}\bar{p}_1 - \chi_{13})kT\left[\int_0^h \hat{\rho}_{3h}{}^2 \mathrm{d}x - \int_0^\infty \hat{\rho}_{3\infty}{}^2 \mathrm{d}x\right] \\ & + \omega_2\omega_3(\bar{v}_2\bar{v}_3/V_1)(1/\bar{p}_1 - \chi_{12} - \chi_{13} + \chi_{23})kT\int_0^h \hat{\rho}_{2h}\hat{\rho}_{3h}\mathrm{d}x\end{aligned} \tag{14.27}$$

The parameters V_1 and $\bar{p}_1$ in this equation need not be specified more precisely because they can be eliminated in favour of experimentally accessible quantities. Thus another rendition of equation (14.27) is

$$\begin{aligned}\Delta^{FP}G^M = {} & \omega_2{}^2(\bar{v}_2{}^2/\bar{V}_1)(\tfrac{1}{2} - \bar{\chi}_{12})N_A kT\left[\int_0^h \hat{\rho}_{2h}{}^2 \mathrm{d}x - \int_0^\infty \hat{\rho}_{2\infty}{}^2 \mathrm{d}x\right] \\ & + \omega_3{}^2(\bar{v}_3{}^2/\bar{V}_1)(\tfrac{1}{2} - \bar{\chi}_{13})N_A kT\left[\int_0^h \hat{\rho}_{3h}{}^2 \mathrm{d}x - \int_0^\infty \hat{\rho}_{3\infty}{}^2 \mathrm{d}x\right] \\ & + \omega_2\omega_3(\bar{v}_2\bar{v}_3/\bar{V}_1)(1 - \bar{\chi}_{12} - \bar{\chi}_{13} + \bar{\chi}_{23})N_A kT\int_0^h \hat{\rho}_{2h}\hat{\rho}_{3h}\mathrm{d}x\end{aligned} \tag{14.28}$$

where $\bar{V}_1 = (V_1\bar{p}_1)$ is the molar volume of the solvent and $\bar{\chi}_{ij}(=\bar{p}_1\chi_{ij})$ is the interaction parameter for a section of polymer of size equal to that of a solvent molecule (i.e. the excess free energy associated with the transfer of a solvent molecule sized section of polymer *i* from pure polymer *i* to pure polymer *j*). Note that in deriving this equation we ignore any depletion effects (see Chapters 15–17).

As discussed above, we will assume that the interpenetrational domain controls stability. Stability will thus be determined by the sign of the term $(1/\bar{p}_1 - \chi_{12} - \chi_{13} + \chi_{23})$. Note that, if $\bar{p}_1$ is unity, we recover the expression derived above for low molecular weight dispersion media. Since $\bar{p}_1$ becomes larger as the molecular weight of the dispersion medium increases, there will be a corresponding reduction in stability on increasing the molecular weight of the dispersion medium. For high polymers, $1/\bar{p}_1$ is negligible and stability is determined by the term $(-\chi_{12} - \chi_{13} + \chi_{23})$.

Heterosteric stabilization in such melt systems thus requires that both χ_{12} and χ_{13} be nonpositive (so as to prevent homoflocculation), as well as demanding that $\chi_{23} > \chi_{12} + \chi_{13}$. Accordingly, for stability, the dispersion medium must be compatible with both polymer coatings but the coatings should not be too compatible with one another.

Homosteric stabilization in polymer melts (the 1–2–2 system)

We now discuss briefly homosteric stabilization in a melt of a different polymer, i.e. polymers 2 and 3 are identical so that $\chi_{23} = 0$. This might be termed a 1–2–2 system. For this system, stability is determined solely by the sign of $(1/\bar{p}_1 - 2\chi_{12})$. Accordingly, for high polymeric dispersion media, χ_{12} must be negative for stability. This demands that polymers 1 and 2 be compatible. Stated differently, it requires that the melt be a good solvent for the stabilizing chains. This requirement is identical with that obtained for low molecular weight dispersion medium. The crucial difference between a polymeric and a nonpolymeric dispersion medium resides in what constitutes a good solvent: for a low molecular weight dispersion medium, a good solvent will have $\chi_{12} < \frac{1}{2}$, whereas for a polymeric medium $\chi_{12} < 0$ (or, strictly $\chi_{12} < \frac{1}{2}\bar{p}_1$).

The 1–1–3 system

We next consider the case where the polymer melt and one coating are the same but the second coating is different. This means that, say, polymers 1 and 2 are the same but differ from polymer 3($\chi_{12} = 0$ and $\chi_{13} = \chi_{23}$). Interestingly, the factor determining the sign of the free energy of interpenetration is just $1/\bar{p}_1$, which for a polymer melt is effectively zero. The factor $1/\bar{p}_1$ is identical with that for the interpenetrational domain of a homosterically stabilized dispersion in a melt of its own stabilizer, which is discussed in the next section. This result is surprising at first sight. Yet it is to be expected: when interpenetration occurs, 1–3 contacts between solvent polymer a and attached polymer 3 are replaced by 1–3 contacts between the attached polymers 1 and 3. Thermodynamically, both types of 1–3 contacts are essentially identical. The ideal entropy contribution, arising from the freedom of movement of the solvent molecules, manifests itself solely through the $1/\bar{p}_1$ term, which is clearly negligible for high polymers.

If a 1–1–3 dispersion is to be formed, it requires that $(\frac{1}{2}\bar{p}_1 - \chi_{13})$ be non-negative, otherwise 3–3 homoflocculation will ensue. Essentially, what is needed for stability is for χ_{13} to be negative, i.e. polymers 1 and 3 must be compatible. There can be no heteroflocculation as all the terms in equation (14.28) are then positive.

If the conditions are altered so that $\chi_{13} > \frac{1}{2}\bar{p}_1$, 3–3 homoflocculation will result. It seems that heteroflocculation could not occur even under these conditions. Since interpenetration has a negligible effect on the free energy, heteroflocculation would require the sum of the elastic and compressional mixing free energies to be negative in the interpenetrational-plus-compressional domain. This sum, however, must always be positive because compression of a polymer coating must raise its free energy; if it were not, the coating would collapse spontaneously to the lower free energy state.

We conclude that, in the 1–1–3 system, there is either stability or selective 3–3 homoflocculation.

The 1–1–1 system

We close this discussion of stabilization in polymer melts by considering briefly the case where all three polymers are the same ($\chi_{ij} = 0$ for all i, j). The mixing free energies in all domains of close approach are proportional to $1/\bar{p}_1$. Accordingly, for high polymers, only the elastic terms, coupled with the London dispersion forces, are important in determining stability. Of course, for low molecular weight species, the $1/\bar{p}_1$ terms could conceivably become important; then particles stable in a low molecular weight melt may well flocculate when placed in a higher molecular weight homologue.

14.3. Heterosteric stabilization by mixed polymer coatings

Up to this point, we have assumed that only one type of polymer is attached to a given surface. We now take up the case where more than one type of polymer is attached to the surface of at least one of the plates. Differences between the polymers may have a chemical origin or may merely reflect differences in molecular weight. The treatment given here is highly idealized, but is nevertheless presented because the system appears not to have been treated at all previously.

We examine the approach of 2 surfaces, designated by I and II. Consider, on plate I, a small element of area, δA, which is covered solely by polymer i. As the two plates approach, this area will be aligned with an area δA on plate II. Suppose it happens to be covered solely by polymer j. In the approximation that there is no lateral overlap between regions of different polymers on the same surface, the interaction over δA is independent of the rest of the surface.

Then $\delta(\Delta G)$, the total free energy of interaction over δA for a plate separation d is,

$$\delta(\Delta G)=\delta A \Delta G_{i-j}(d) \tag{14.29}$$

where $\Delta G_{i-j}(d)$ = change in free energy per unit area associated with the approach from infinity of a plate covered entirely by i to a distance d from a plate covered entirely by j.

If f_{Ii} is the fraction of the area of plate I covered by i and f_{IIj} is the corresponding quantity for polymer j on plate II, then $f_{Ii}f_{IIj}$ is the fraction of unit area over which i on I is aligned with j on II. The total change in free energy per unit area for close approach is obtained by summing over all possible polymer interactions

$$\Delta G=\sum_{i,j} f_{Ii}f_{IIj}G_{i-j}(d) \tag{14.30}$$

where the subscripts i and j run over all polymers attached to plates I and II, respectively. $\Delta G_{i-j}(d)$ may be obtained from the equations presented previously for $\Delta G_I{}^M$, $\Delta G_{I+C}{}^M$ and $\Delta G_{I+C}{}^{el}$. Equation (14.30) is quite general in the sense that it is applicable to both high and low molecular weight dispersion media.

If only two types of polymers are attached to each plate and if the coatings on both surfaces are identical (so that the subscripts I and II become redundant), then

$$\Delta G=f_2{}^2\Delta G_{2-2}+2f_2f_3\Delta G_{2-3}+f_3{}^2\Delta G_{3-3} \tag{14.31}$$

where $f_2=1-f_3$. Equation (14.31) is only valid if the distribution of polymers in the coatings and their alignment obey random statistics. The occurrence of significant nonrandom patches of one or more polymers on the surfaces could well necessitate a different theoretical approach.

In so far as equation (14.31) is valid for flat plates, the Derjaguin integration procedure implies that it can also be applied to spheres, at least for particles whose barrier layer thicknesses are small compared with the radii of the particles. Inspection of the form of equation (14.31) reveals that, e.g. polyacrylonitrile latex particles stabilized by a mixture of polystyrene and polyisobutylene may exhibit stability in cyclohexane at temperatures well below the θ-temperature for polystyrene. Indeed, using the numerical data set forth earlier, it is readily shown, by application of equation (14.31) that the addition of small amounts (e.g. less than 1%) of polyisobutylene to the barrier would be all that is required to impart stability at 5 K below the θ-temperature for polystyrene in cyclohexane. This surprising result arises from the strong repulsions that occur between the polystyrene and polyisobutylene chains. Increasing the proportion of polyisobutylene in the barriers would increase dramatically the penetration into the worse than θ-solvent domain for polystyrene that could be tolerated before the onset of flocculation.

This penetration into the worse than θ-solvent domain before flocculation is observed is reminiscent of the enhanced steric stabilization discussed in Chapter 9. Of course, the origins of the two phenomena are quite different. In addition, the observed penetration will depend critically upon not only the composition of the barrier but also upon the relative dimensions of the different stabilizing chains. For example, in the case discussed in the preceding paragraph, the strong repulsions between the polyisobutylene and polystyrene chains were able to impart stability because the polyisobutylene was taken to be of a significantly higher molecular weight than the polystyrene. If the polyisobutylene chain dimensions were significantly smaller than those of polystyrene, extensive interpenetration of the polystyrene chains would be possible before the repulsion between the incompatible polystyrene and polyisobutylene moieties became important. Flocculation near to the θ-point for polystyrene would then be expected.

14.4. The van der Waals interaction

The van der Waals interaction between two identical colloidal particles is always attractive, irrespective of the medium in which the particles are dispersed. The same, however, cannot be said for the van der Waals force between two colloidal particles of dissimilar chemical composition immersed in a dispersion medium. In that case, the force may be attractive or repulsive depending upon the relative magnitudes of the electric polarizabilities of the different atoms or molecules that constitute the particles and the dispersion medium (Visser, 1981). This conclusion is at first sight surprising, although Hamaker (1937) was aware of the possibility of repulsion. Israelachvili (1982) has pointed out that the possibility of repulsion is not confined to van der Waals attractive force fields. The situation is closely analogous, e.g. to Archimedes' principle as applied to gravitational fields. Thus, for example, even though gravitational attraction occurs between the earth and both cork and iron in air, the net interaction force is very different in water. Cork floats on water (i.e. there is a net repulsion) whereas iron sinks (a net attraction). The reason for this difference originates in the need to displace liquid as the interacting objects approach one another. If more work is expended in displacing the liquid than is gained by reducing the potential energy of the particles as they approach it becomes energetically more favourable for the particles to move apart. This corresponds to a repulsion. In the context of the van der Waals interaction, such repulsion corresponds to a negative Hamaker constant for the two different interacting bodies in the presence of the dispersion medium. Some experimental evidence supports the existence of negative Hamaker constants (Visser, 1981). The possibility of van der Waals repulsion must be borne in mind in interpreting the results of experiments on heterosteric stabilization.

Chapter 15

The Effects of Free Polymer on Colloid Stability: Theories of Depletion Flocculation and Depletion Phase Separation

15.1. Introduction

In the foregoing chapters the effects of attached polymer chains on colloid stability have been set forth. We now turn to consider the effects of macromolecules that are not attached to the particles but rather are free in solution. Surprisingly, such free polymer is still able to affect colloid stability, being capable of generating both stability and flocculation. Stability that is imparted by free polymer is termed *depletion stabilization*. Aggregation that is induced by free polymer is called *depletion flocculation*. The latter will be discussed in this Chapter, consideration of depletion stabilization being postponed until the next Chapter.

15.2. Depletion flocculation

15.2.1. The early theories and experiments

The *creaming* of a natural rubber latex by the addition of water soluble polymers, such as gums, pectin and gelatin, has long been known, being first recognized by Traube (1924). Originally it was considered that adsorption was a prerequisite for agglomeration but Bondy (1939), using sodium alginate (0·15%) as the creaming agent, showed that adsorption was unnecessary. He also clearly demonstrated that this type of flocculation was reversed by dilution and that the latex could be taken through the aggregation-redispersion-aggregation cycle many times. Bondy claimed that this effectively disproved the hypothesis of Vester (1938a,b), who had observed an increase in the rate of creaming of natural rubber latex on addition of 0·1–1% polymers, such as pectin and sodium poly(acrylate). Higher molecular weight polymers were found to be more effective than lower molecular weight polymers. Vester had proposed that agglomeration was the consequence of a third component (perhaps proteinaceous in character) in the natural latex that produced a viscous envelope around each particle, effectively "gluing" them together.

Twiss and Carpenter (1938) had also proposed that the water soluble polymers induced creaming by a type of cross-linking mechanism. They pointed out that water soluble polymers of the type studied formed a loose network in solution, as witnessed by their exhibition of a high structural viscosity. Twiss and Carpenter argued that if the chains of the water soluble polymer were adsorbed onto the latex particles, then the particles might be interlinked by network formation between polymer chains adsorbed on different particles.

Bondy (1939) believed that his experiments had disproved this structural network hypothesis. First, he demonstrated that adsorption was not apparently necessary. Second, he showed that increasing the temperature to 90 °C had no effect on the creaming power of the water soluble polymer. Yet at this temperature, the network structure in a solution of the water soluble polymer was seemingly broken down.

Another theory that was current at that time was that aggregation was a consequence of the partial dehydration of the stabilizing layers by the addition of the hydrophilic colloid. This theory was put forward by Monaghan and White (1935) who found that gelatin agglomerated erythrocytes, even though it was not apparently adsorbed onto the surfaces of these red blood cells. Bondy (1939), and even Vester (1938a), supported this viewpoint. The dehydration mechanism was in harmony with the postulated effect of high concentrations of electrolytes on solutions of water soluble polymers: phase separation (coacervation) is still widely held to be a consequence of the dehydration of the macromolecules, although the phenomenon is undoubtedly much more complex than that (Robb, 1982; Napper, 1970b).

Bondy (1939) supported the dehydration theory by showing that oil-in-water emulsions stabilized by gelatin resisted destabilization by the addition of other water soluble polymers. This was attributed to gelatin being impervious to dehydration by other water soluble polymers. It should be pointed out, however, that a previously mentioned experiment performed by Bondy casts considerable doubt upon the dehydration explanation for flocculation. Bondy showed that heating did not reduce the creaming power of the water soluble polymer. Yet it would be expected that the interaction between the stabilizing chains and the water would be dramatically changed by such an increase in temperature, dehydration being much easier at the higher temperatures.

Subsequently, both Meyer *et al.* (1945) and Inokuchi (1951) confirmed the earlier findings of Monaghan and White (1935) with regard to the flocculation of red blood cells by hydrophilic polymers. Inokuchi (1951), for example, showed that many different polymers were effective in this regard: these included the synthetic polymer poly(vinyl alcohol), as well as the natural polymers sodium alginate, pectin, gelatin, gum arabic, dextran and sodium deoxyribonucleate. Certain proteins (e.g. ovalbumin, haemoglobin and serum albumin) were ineffectual as flocculants, however. This impotence appeared to be associated with their globular character. Inokuchi attributed aggregation to the adsorption of the hydrophilic macromolecules, coupled with what in modern parlance would correspond to a type of bridging flocculation.

Water soluble polymers have also been reported to agglomerate soils (Geoghegan and Brian, 1948). Bacterial levans and dextrans apparently lead to the aggregation of the soil particles. Pectin, however, was reported to have no effect on the soil stability.

15.2.1.1. Theories assuming complete depletion

The theory of Asakura and Oosawa

Asakura and Oosawa (1954) were the first to recognize that depletion effects could give rise to the flocculation of colloidal dispersions. They considered the specific example of two parallel flat plates immersed in solutions of rigid molecules, either spherical or rod shaped. When the distance between the plates is smaller than the diameter of the solute molecules, assumed for the purposes at hand to be spherical, none of these molecules can enter the domain between the plates. This region is then composed entirely of solvent. The solution outside the plates retains its bulk concentration of solute and so it exerts an inward force, arising from its osmotic pressure, on the plates.

This concept was subsequently extended as follows by Asakura and Oosawa (1958) to embrace spheres.

Statistical thermodynamics relates the force $\mathscr{F}$ between two particles dispersed in a polymer solution to their separation by

$$\mathscr{F} = kTN_2 \partial \ln Z / \partial d \qquad (15.1)$$

where N_2 = total number of solute macromolecules and d = centre-to-centre distance between the particles. The appropriate partition function Z is given by

$$Z = \int_V \exp\{-w(x,d)\} \mathrm{d}x \qquad (15.2)$$

where $w(x,d)$ = average free energy of a molecule at x when two particles are fixed at distance d relative to its value in the absence of the two particles and V=total volume of the solution.

Rigid spheres. If the macromolecules are rigid spheres, of diameter Δ, with no interactions between them and the suspended particles, then the configurational partition function, Z_{conf}, can be evaluated from the volume of space in which the centres-of-mass of the two particles can move freely. Stated another way, Z_{conf} can be obtained by subtracting from the total volume that volume from which the centres of mass of the particles are excluded due to the physical presence of the other particles. For two spherical particles, each of radius a, geometrical considerations lead to

$$Z_{conf} = V - V_{ex} \tag{15.3}$$

where

$$\begin{aligned} V_{ex} &= (8\pi/3)[(2a+\Delta)/2]^3, \; d \geq 2a+\Delta \\ &= (4\pi/3)\{[(2a+\Delta)/2]^3 + (3/16)(2a+\Delta)^2 d \\ &\quad - (1/16)d^3\}, \; 2a \leq d \leq 2a+\Delta \end{aligned} \tag{15.4}$$

Note that numerical errors appearing in the original paper have been corrected here.

Combining equations (15.1), (15.2) and (15.3), we obtain

$$\begin{aligned} \mathscr{F} &= kTN_2(\partial \ln Z/\partial Z)(\partial Z/\partial d) \\ &= (kTN_2/Z)(\partial Z/\partial d) \end{aligned} \tag{15.5}$$

Note that, to good approximation, $kTN_2/Z = kTN_2/V$ if the dispersion is dilute. Evaluation of the partial differential yields

$$\mathscr{F} = -\pi_0 S \tag{15.6}$$

since $\pi_0 = kTN_2/V$=limiting van't Hoff osmotic pressure of the dissolved polymer and

$$\begin{aligned} S &= 0, \; d \geq 2a+\Delta \\ &= (\pi/4)[(2a+\delta)^2 - d^2], \; 2a \leq d \leq 2a+\Delta \end{aligned} \tag{15.7}$$

The physical interpretation of equation (15.6) is straightforward. First, the negative sign indicates a force of attraction. Second, the parameter S is the area of the circular cross-section of the overlapping region in the middle of d. The following picture accordingly emerges: when the two particles approach one another such that $d < 2a+\Delta$, the two spherical volumes of diameter $(2a+\Delta)$, which exclude macromolecules, begin to overlap. The osmotic pressure of the solution can be considered to act uniformly over all the particle surface, except for that part of the surface facing the pure solvent. The total force on the outside half of one of the two particles is larger by $\pi_0 S$ than the force acting on the inside half.

The potential energy $V(d)$ of the attractive force $\mathscr{F}$ is given by

$$V(d)=0, \; d \geq 2a+\Delta$$
$$= -\pi_0(\pi/12)\{2(2a+\Delta)^3-3(2a+\Delta)^2 d+d^3\},$$
$$2a \leq d \leq 2a+\Delta \qquad (15.8)$$

Differentiation of this latter expression, as required to find $\mathscr{F}(=-\partial V(d)/\partial d)$, gives an identical result with that of equation (15.6).

Equation (15.8) is readily derived by an alternative procedure. The overlap volume, δV, of the domains of exclusion on the close approach of the particles (see Fig. 15.1) is given by

$$\delta V=(\pi/12)(2a+\Delta-d)^2(4a+2\Delta+d)$$
$$=(\pi/12)\{2(2a+\Delta)^3-3(2a+\Delta)^2 d+d^3\} \qquad (15.9)$$

The total potential energy can be calculated from

$$V(d)=-\pi_0 \delta V$$
$$=-\pi_0(\pi/12)(2a+\Delta-d)^2(4a+2\Delta+d) \qquad (15.10)$$

which is identical with equation (15.8).

If $2a \gg \Delta$ (i.e. the particle size is large compared with the size of the dissolved polymer molecules), the potential energy of interaction as given by equation (15.10) can be approximated by

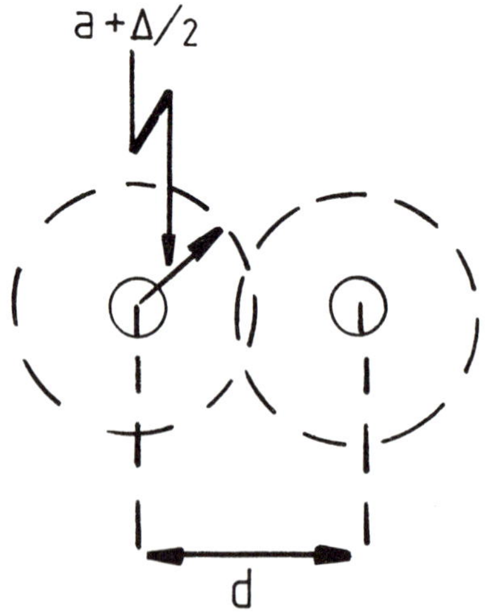

Fig. 15.1. Schematic representation of the close approach of the exclusion domains associated with two spherical particles.

$$V(d) \sim -(3/2)\phi_2\beta x^2 kT, \; 0 \leq x \leq 1 \qquad (15.11)$$

where $x=(2a+\Delta-d)/\Delta$, $\beta=2a/\Delta(\gg 1)$ and $\phi_2=(\pi/6)\Delta^3(N_2/V)$. As the latter parameter is directly proportional to the free polymer concentration, so then is the potential energy of interaction between the particles. The minimum value of $V(d)$ occurs at $x=1$ (or $d=2a$) when it adopts the value $(-3/2)\phi_2 kT$. It is also apparent from equation (15.11) that the magnitude of the attraction increases with Δ, i.e. with the molecular weight of the free polymer.

Flexible macromolecules. Calculations of the attractive potential energy according to equation (15.11) show that for spherical colloidal particles immersed in a dilute solution of rigid spheres, the attraction rarely exceeds kT. For articulated macromolecules, the configurational entropy of the chains is decreased in the neighbourhood of the interfaces and this provides a source of non-zero values for $w(x,d)$. Asakura and Oosawa (1954) approximated this entropy decrement by analysing the problem in terms of the classical theory (Carslaw, 1921) for diffusion in a vessel with walls that absorb diffusing particles. The end result for parallel flat plates of area A is

$$\mathscr{F} = -A\pi_0\{(1-f)-h\partial f/\partial h\} \qquad (15.12)$$

where h = distance between the plates and

$$f=(8/\pi^2) \sum_{v=1,3,5\,\cdots} (1/v^2)\exp\{v^2\pi^2/6)(\langle r\rangle^{\frac{1}{2}}/h)^2\}$$

Here $\langle r\rangle$ = mean end-to-end distance of the macromolecule in free solution. In equation (15.12), the first term involving $(1-f)$ is the attraction due to the osmotic effect whereas the second term is the repulsion due to the loss of configurational entropy of chain molecules confined between the plates.

Asakura and Oosawa (1958) showed that the force becomes stronger in solutions of chain molecules (or of molecules of dissymmetric shape) than in solutions of rigid spherical macromolecules (e.g. globular proteins) at the same net concentration. If the macromolecules are charged (i.e. polyelectrolytes), the force can be greatly intensified.

The aggregation process. Asakura and Oosawa (1958) also presented formulae for calculating the number $N(i)$ of small aggregates composed of i particles:

$$N(i)/N(3)=6H^{i-3}/i(i-1)(i-2), \; 4 \leq i < 10 \qquad (15.13)$$

where $H=(2^{\frac{1}{2}}6/\pi)\phi_3(K/\beta)^3$, $\beta=2a/\Delta$ and $\phi_3=(4\pi/3)(a^3N(1)/V)$, which is the particle volume fraction. The parameter K is given by

$$K=(1/t_0)\int_0^{t_0}\exp\{t^2\}\mathrm{d}t \qquad (15.14)$$

where $t_0{}^2 = -U_m/kT$, U_m being the depth of the minimum into which aggregation occurs. The expressions for calculating the numbers of flocs containing doublets and triplets are

$$N(2)/N(1) = 12\phi_3(K/\beta) \tag{15.15}$$

and

$$N(3)/N(1) = 36\phi_3{}^2(K/\beta)^3. \tag{15.16}$$

A further quantity calculated by Asakura and Oosawa (1954) is the critical volume concentration $\phi_3{}^*$ of the colloidal particles at which macroscopic aggregation takes place. They showed that for relatively small changes in the depth of the potential energy well into which the particles flocculate (e.g. an increase in well depth of ca 30%), the state of the particles changes from essentially total dispersion to complete aggregation.

15.2.2. Further developments

15.2.2.1. The polymer second virial coefficient

A trivial extension of the theory of Asakura and Oosawa (1958) has been given by Sieglaff (1959), who included the second virial coefficient (B_2) in the osmotic pressure (π^*) expression, i.e.

$$V(d) = -\pi^* \delta V \tag{15.17}$$

Here $\pi^* = \pi_0 + RTB_2c_2{}^2$, where c_2 = the polymer concentration. Note that Sieglaff employed an erroneous relationship for δV, the correct expression being provided by equation (15.9).

15.2.2.2. The particle second virial coefficient

Vrij (1976) and de Hek and Vrij (1981) have used the Asakura and Oosawa potential between colloidal particles suspended in solutions of rigid spheres to derive an expression for the second virial coefficient of the particles. They calculated the force on the particles from

$$\mathscr{F} = -\pi_0 S$$

where S = area at the centre-to-centre midplane of the circular surface of intersection of the volumes of exclusion of the two particles. Reference to Fig. 15.2, coupled with the application of Pythagoras' theorem, shows that $S = \pi\{(2a+\Delta)^2 - d^2\}$, as found by Asakura and Oosawa [equation (15.7)]. The potential of average force is given by

$$V(d) = -\int_{2a+\Delta}^{d} \mathscr{F}\,\mathrm{d}d = \int_{2a+\Delta}^{d} \pi\{(2a+\Delta)^2 - d^2\}(N_2/V)kT\,\mathrm{d}d$$
$$= -(4/3)\pi\bar{\sigma}^3\{1-(3/4)\rho + (1/16)\rho^3\}(N_2/V)kT \tag{15.18}$$

where $\bar{\sigma}=(2a+\Delta)/2$ and $\rho=d/\bar{\sigma}$. Note that this result is only applicable in the interval $2a/\bar{\sigma}\leq\rho\leq 2$ where $V(d)$ is negative. Its maximum absolute value in this interval occurs if the surfaces of the spherical particles touch ($d=2a$). When that happens

$$V(d)=-(5/12)\pi\bar{\sigma}^3(N_2/V)kT$$

The second virial coefficient B_2^c of the colloidal particles can be calculated in general from the potential of mean force via the excluded volume integral:

$$B_2^c=(\tfrac{1}{2})\int_0^\infty\{1-\exp(-V(d)/kT)\}4\pi d^2\mathrm{d}d \qquad (15.19)$$

In this instance, it is helpful to recast equation (15.18) into the form

$$V(d)=-(1+2a/\Delta)^3v_2\{1-(3/4)\rho+(1/16)\rho^3\}kT \qquad (15.20)$$

where $v_2=(1/6)\pi\Delta^3(N_2/V)$ is the volume fraction of the dissolved polymer molecules. Substitution of equation (15.20) into equation (15.19) leads to

$$\begin{aligned}B_2^c(v_2)&=(16\pi/3)a^3\{1-(3/8)(1+\Delta/2a)^3\mathrm{I}(\gamma,\Delta/2a)\}\\&=B_2^{HS}\{1-(3/8)(1+\Delta/2a)^3I(\gamma,\Delta/2a)\}\end{aligned} \qquad (15.21)$$

where B_2^{HS} = hard sphere second virial coefficient,

$$I(\gamma,\Delta/2a)=\int_{2a/\bar{\sigma}}^{2}(\exp\{\gamma[1-(3/4)\rho+(1/16)\rho 3]\}-1)\rho^2\mathrm{d}\rho$$

and

$$\gamma=(1+2a/\Delta)^3v_2$$

Equation (15.21) is quite revealing. The second term of $B_2^c(v_2)$, which expresses the deviation from hard sphere behaviour due to the addition of free polymer, is negative in sign. Accordingly, as the volume fraction of dissolved polymer v_2 increases, so the second virial coefficient of the particles is diminished. If v_2 and/or Δ are sufficiently large, $B_2^c(v_2)$ can clearly decrease to zero or even become negative. The latter corresponds to a net attraction between the colloidal particles.

Unfortunately, it does not seem possible to evaluate the integral $I(\gamma,\Delta/2a)$ analytically, although its numerical computation presents no serious difficulty. The one exception to this proscription of an analytic result occurs if v_2 is small (i.e. low polymer concentrations). In that case

$$I(\gamma,\Delta/2a)=(1/3)\gamma\{1-s^3[1-(9/16)s+s^3/32]\} \qquad (15.22)$$

with $s=2a/\bar{\sigma}$. If, e.g. $\Delta=2a$, then equation (15.21) and (15.22) result in

$$B_2^c(v_2)=B_2^{HS}[1-(17/4)v_2] \qquad (15.23)$$

which shows that B_2^c is a function that decreases monotonically with v_2.

Vrij (1976) performed numerical calculations on equation (15.21) to show that, typically, values of v_2 in the range 0·1–1 would be needed to render $B_2^c(v_2)$ negative and thus the particles attractive.

15.2.2.3. The partial depletion theories

Certain of the theories described above (e.g. that of Vrij) might be considered to be geometrical *cum* thermodynamic approaches in which the free polymer chains are considered to behave essentially as hard spheres in their interactions with the colloidal particles. Accordingly, the chains are assumed to be excluded completely from between the particles once the interparticle distance is less than the hard sphere diameter. Since polymer molecules are patently not hard spheres but relatively flexible, they are never completely excluded from the interparticle region. It is therefore hardly surprising to find that theories which take this flexibility into account have been explored in an effort to calculate the segmental concentration in the depletion zone between particles (Joanny *et al.*, 1979; Feigin and Napper, 1980b; Scheutjens and Fleer, 1982; de Gennes, 1981; 1982). These theories endeavour to estimate the potential energy of interaction of the colloidal particles immersed in a polymer solution by calculating the effects that the depletion in segmental concentration exert on the potential energy of interaction between the particles.

The theories of Feigin and Napper (1979) and Scheutjens and Fleer (1982) are qualitatively different from those of Asakura and Oosawa (1954; 1958) (and subsequent elaborations thereof) and de Gennes and coworkers (Joanny *et al.*, 1979; de Gennes, 1981; 1982) in that they predict not only depletion flocculation but also depletion stabilization. For this reason, presentation of the former two omnibus theories will be delayed until their predictions with regard to both depletion stabilization and depletion flocculation are elaborated. The de Gennes approach, which does not predict the occurrence of depletion stabilization, will be presented at this juncture.

15.2.2.4. The de Gennes theories

de Gennes and coworkers (Joanny *et al.*, 1979; de Gennes, 1981; 1982) have developed two general approaches for describing depletion flocculation: the first is a mean field theory in the spirit of Edwards (1965); the second is a more general scaling law theory. The elementary mean field theory is not correct because segmental correlations are ignored. Nevertheless, it does provide a simple physical picture of the processes involved. The scaling law approach incorporates the effects of segmental correlations but may be limited in that often it only provides information on limiting exponents.

The mean field theory

General formalism. Joanny *et al.* (1979) have used a mean field approach to calculate the concentration profile in the gap between two parallel flat plates immersed in a polymer solution. The solution is assumed to be semidilute by which is implied two almost-contradictory inequalities: first, the volume fraction of polymer (v_2) is small but, second, the chains are significantly interpenetrated.

A polymer chain at an interface is subject to a potential $U(\mathbf{r})$, which in general is composed of two components. There is an external component that arises directly from the effects of the wall. To this must be added an internal potential which reflects the mean field experienced by one segment due to the physical presence of all other segments. In the mean field approximation, this latter potential is simply proportional to the segment concentration $c(\mathbf{r})$

$$U_{\text{int}}(\mathbf{r}) = \tilde{v}c(\mathbf{r}) \tag{15.24}$$

where $\tilde{v} = l^3(1-2\chi_1)$, l^3 being the segmental volume.

To calculate the form of the segment density distribution in the depletion layers, it is first necessary to calculate $G_N(\mathbf{r}_1,\mathbf{r})$, the statistical weight of a chain of N links going from $\mathbf{r}_1$ to $\mathbf{r}$ which is subject to a purely external potential $U(\mathbf{r})$ that acts on each segment.

The conformation of a flexible polymer is closely analogous to the possible paths of a nonrelativistic particle, which is described by the time dependent Schrödinger equation.

As shown in Section 11.4.1.1, the analogous equation for the conformational statistical weight is (Edwards, 1965; de Gennes, 1969)

$$-\frac{l^2}{6}\nabla^2 G_N(\mathbf{r}_1,\mathbf{r}) + \frac{U(\mathbf{r})}{kT}G_N(\mathbf{r}_1,\mathbf{r}) = -\frac{\partial}{\partial N}G_N(\mathbf{r}_1,\mathbf{r}) \tag{15.25}$$

Here N plays the role of an imaginary time, the Laplacian operator ∇^2(del or nabla squared) acts upon the variable $\mathbf{r}$ and l^2 is the mean square length of a segment. Since a short chain must have its ends in close proximity to one another, the appropriate boundary condition is

$$\lim_{N\to 0} \quad G_N(\mathbf{r}_1,\mathbf{r}) = \delta(\mathbf{r}-\mathbf{r}_1) \tag{15.26}$$

where δ = delta function.

The statistical weight can be expanded as a series of normalized eigenfunctions (wave functions) u_n

$$G_N(\mathbf{r}_1,\mathbf{r}) = \sum_n u_n(\mathbf{r}_1)u_n(\mathbf{r})\exp\{-\varepsilon_n N\} \tag{15.27}$$

where ε_n = corresponding eigenvalue (the quantum mechanical analogue is the eigenenergy). The u_n satisfy the equation

$$-(l^2/6)\nabla^2 u_n(\mathbf{r}) + [U(\mathbf{r})/kT]u_n = Hu_n(\mathbf{r}) = \varepsilon_n u_r(\mathbf{r}) \tag{15.28}$$

where the operator $H = -(l^2/6)\nabla^2 + U(\mathbf{r})/kT$ plays the role of the Hamiltonian. Equation (15.28) has been written in this fashion to emphasise its analogy with the Schrödinger wave equation. For large $N(N \to \infty)$, it is apparent that $G_N(\mathbf{r}_1,\mathbf{r})$ will be dominated by the lowest (ground state) eigenvalue ε_0. This allows all the excited state contributions to be ignored. The corresponding ground state wave function u_0 can be written as

$$u_0 = \Xi\psi(\mathbf{r}) \tag{15.29}$$

where the normalization constant Ξ is chosen as discussed below.

In the semidilute region, with the polymer chains interpenetrated, the segment concentration has the form

$$c(\mathbf{r}) = (1/K)\sum_M \int d^3r_1 \int d^3r_2 G_M(\mathbf{r}_1,\mathbf{r})G_{N-M}(\mathbf{r},\mathbf{r}_2) \tag{15.30}$$

where K = the appropriate normalization constant. This equation merely expresses the fact that the segment concentration at a point T is made up of contributions from chains that commence at some other point $\mathbf{r}_1$, reach $\mathbf{r}$ by a series of M steps, and then terminate at some other arbitrary point $\mathbf{r}_2$ after a further $N-M$ steps. Since, to good approximation,

$$\begin{aligned} G_N(\mathbf{r}_1,\mathbf{r}) &= u_0(\mathbf{r}_1)u_0(r)\exp\{-\varepsilon_0 N\} \\ &= \Xi^2\psi(\mathbf{r}_1)\psi(\mathbf{r})\exp\{-\varepsilon_0 N\} \end{aligned} \tag{15.31}$$

the segmental concentration is given by

$$c(\mathbf{r}) = D\int\psi(\mathbf{r}_1)d^3r_1\psi(\mathbf{r})\psi(\mathbf{r})\int\psi(\mathbf{r}_2)d^3r_2 \tag{15.32}$$

where D = constant. The integrals in this equation both yield constants (c.f. the integrals of the wave functions over all space) so that $c(\mathbf{r})$ is proportional to $\psi^2(\mathbf{r})$. The normalization constant Ξ in equation (15.31) is chosen so as to satisfy the simple equality

$$c(\mathbf{r}) = \psi^2(\mathbf{r}) \tag{15.33}$$

Again the analogy with the point distribution function is obvious (crudely, the probability of finding an electron at a point is proportional to $|\psi|^2$). The wave function $\psi(\mathbf{r})$ is then normalized by

$$\int\psi^2(\mathbf{r})d^3r = NP_c \tag{15.34}$$

where P_c = number of chains in the sample.

The mean field expression for the internal potential [see equation (15.24)] when coupled with equation (15.28) yields an explicit nonlinear equation for the wave function:

$$-(l^2/6)\nabla^2\psi + \tilde{v}|\psi^2|\psi = \varepsilon_0\psi \tag{15.35}$$

This is essentially a *self-consistent* equation for ψ.

According to the variational principle, the correct (or best) value of ψ will correspond to the minimum in the (free) energy. Consider the free energy function A' (analogous to the energy integral in quantum mechanics) which has the form

$$A'/kT = \int d^3\mathbf{r}\{(l^2/6)|\nabla\psi|^2 + \tfrac{1}{2}\tilde{v}|\psi^4|\} \tag{15.36}$$

This needs to be minimized with respect to all possible variations of $\psi(\mathbf{r})$ compatible with the normalization condition, equation (15.34). Minimization can be accomplished via the introduction of the Lagrange multiplier ε_0 such that for small variations in ψ

$$\delta A'/kT = (\varepsilon_0/2)\delta(\int\psi^2 d^3r) \tag{15.37}$$

Equation (15.35) and (15.37) coincide exactly. Rewriting equation (15.37), we have

$$A = A' - (kT\varepsilon_0/2)\int\psi^2 d^3r \tag{15.38}$$

Reference to equation (15.36) shows that this may be regarded physically as a free energy with: (a) an interactional term proportional to $(\frac{1}{2})kT\tilde{v}c^2$ (cm^{-3}) and, (b) a term related to the entropy of the chain which is proportional to $|\nabla\psi|^2$.

In the bulk polymer solution far from the flat plates, the gradient term of equation (15.35) drops out and the equation reduces to

$$\tilde{v}\psi_0{}^2 = \varepsilon_0 \tag{15.39}$$

where $\psi_0{}^2 = c_0$, the bulk polymer concentration. If the factor $(6/l^2)$ is used to multiply each term in equation (15.39) then under bulk conditions we have

$$1/\xi_{MF}{}^2 = (6/l^2)\varepsilon_0 \tag{15.40}$$

where $\xi_{MF} = (l/6\tilde{v}c_0)^{\frac{1}{2}}$. The correlation length ξ_{MF} has the following interpretation: if at point 0 we impose a slight perturbation in the segment concentration ($c_0 \rightarrow c_0 + \delta c_0$), there is a shift in $c(\mathbf{r})$ at a distance $\mathbf{r}$ from 0 that decreases the further we are from 0. The shift extends, however, only over the correlation distance ξ_{MF}.

Results for flat plates. For two flat plates, we need only consider the dependence of c in the x-direction normal to the plates. Equation (15.35) then reduces to

$$-(l^2/6)\frac{d^2\psi}{dx^2} + \tilde{v}|\psi^2|\psi = \tilde{v}\psi_0{}^2\psi \tag{15.41}$$

The boundary conditions are $\psi(0) = \psi(\mathrm{h}) = 0$, where h = distance between the plates. A further boundary condition applies at the midplane ($x = h/2$)

between the plates: there $\partial\psi/\partial x=0$ by symmetry. Integrating equation (15.41) once and applying these boundary conditions yields

$$(l^2/6)\left(\frac{\partial\psi}{\partial x}\right)^2=(\tilde{v}/2)(\psi^4-\psi_1{}^4)-\tilde{v}\psi_0{}^2(\psi_0{}^2-\psi_1{}^2) \qquad (15.42)$$

where $\psi_1=\psi(h/2)$, i.e. the value of ψ at the midplane ($x=h/2$).

The complete solution of equation (15.42) depends critically on the ratio of the plate separation to the correlation length ξ_{MF}. If $h/\xi_{MF}<\pi$, the solution is $\psi=0$, i.e. the chains are completely excluded from the gap between the plates. This nonphysical solution points to the caution needed in interpreting the quantitative results based on the assumption of infinite molecular weight but the general qualitative picture that emerges appears to be correct. Actually, for $h<\xi_{MF}$, the solution between the plates is dilute rather than semidilute, so that the foregoing theory does not strictly apply in any event.

For $h\gg\pi\xi_{MF}$, the solution for ψ is nonzero between the plates. Its value can be calculated from

$$x/2\xi_{MF}=(1+k^2)^{\frac{1}{2}}\int_0^{\alpha}\frac{\mathrm{d}u}{(1-k^2\sin^2 u)^{\frac{1}{2}}}=(1+k^2)F(k,\alpha) \qquad (15.43)$$

where $\alpha=\sin^{-1}(\psi/\psi_1)$, $F(k,\alpha)$ is the first order elliptical integral (Gradshtein and Ryznik, 1965) and the modulus $k=\psi_1{}^2/(2\psi_0{}^2-\psi_1{}^2)$. The concentration profile between the plates calculated in this fashion is shown in Fig. 15.2.

Note that under these conditions, the maximum value (c_1) of c, which occurs at the midplane, approaches the bulk value exponentially

$$c_1\sim c_0[1-8\exp\{-2^{\frac{1}{2}}h/\xi_{MF}\}] \qquad (15.44)$$

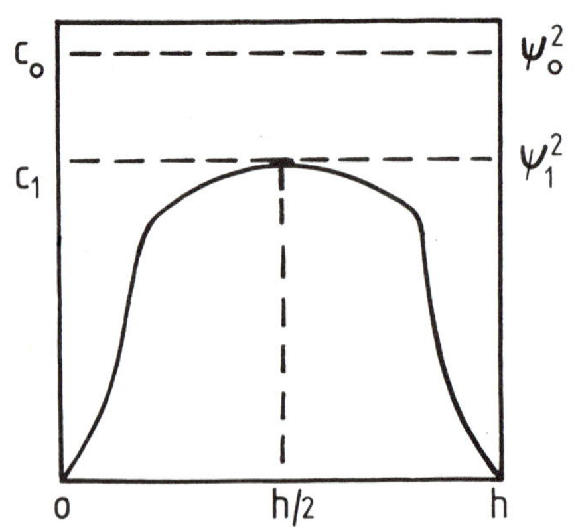

Fig. 15.2. The concentration profile between two flat plates according to the self-consistent field approach (after Joanny *et al.*, 1979).

The attraction between the plates can be evaluated from the surface free energy, γ_S, given by

$$\gamma_S = \int dx[A(\psi) - A(\psi_0)] \tag{15.45}$$

The free energy can be found from equation (15.36) using the relationships inherent in equation (15.42). These yield, after some algebra,

$$\begin{aligned} A' &= 2kT\int[(l^2/3)(\partial\psi/\partial x)^2 + (\tilde{v}/2)(\psi_0{}^2 - \psi_1{}^2)^2]dx \\ &= (kTc_0l^2/6\xi_{MF})g(h/2\xi_{MF}) \end{aligned} \tag{15.46}$$

where $g(h/2\xi_{MF})$ is a universal function whose functional form is shown in Fig. 15.3.

At distances of separation less than the correlation length ($h < \xi_{MF}$), $g(h/2\xi_{MF})$ increases linearly with h and so the force of attraction between the plates ($-\partial\gamma_S/\partial h$) is constant. Its actual magnitude is $-l^2c_0kT/12\xi_{MF}{}^2$ per unit area of plate. At distances of separation greater than the correlation length, the universal function reaches a limiting value. The surface energy then adopts a value equal to twice that of a single plate with its own depletion layer (i.e. no attraction is evident).

Subsequently, de Gennes (1982) further elaborated the effects of free polymer on colloid stability using the mean field approximation. He assumed full reversibility of the exchange equilibrium for the polymer chains between the weakly adsorbing surfaces and the bulk solution. The force between the plates was found to be always attractive, whether or not the dispersion medium was a good or bad solvent for the polymer. This, however, is unlikely to relate to depletion effects because the polymer segment concentration was predicted to be greatest at the surface and least at the midplane (i.e. the reverse of that for depletion layers). As shown by Scheutjens and Fleer (1982), this is likely to be a consequence of polymer bridging.

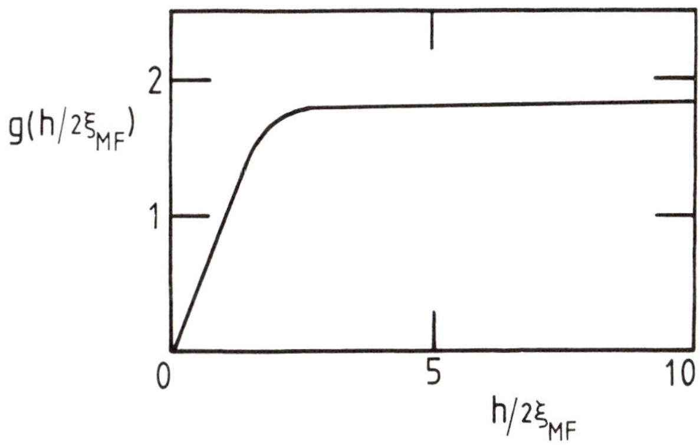

Fig. 15.3. The universal distance dependent function for the free energy of interaction according to the self-consistent field approach (after Joanny *et al.*, 1979).

The scaling theory

The qualitative features predicted by the mean field theory that were set forth in the preceding section seem likely to be correct. Quantitatively, however, the theory is suspect for the reasons already outlined. The mean field approximation is based upon considerations of the average conformation. All fluctuations from the mean positions are disregarded so that correlations between the positions of the segments are simply ignored.

An alternative approach utilizes scaling theory methods. Scaling theory might be described as a "back of an envelope" type of semi-intuitive approach with quite extraordinary powers for predicting certain "universal" properties in polymer solutions and melts. Of special interest are the characteristic exponents of the scaling laws (i.e. the laws that govern the dependence of a parameter upon the number of segments N composing the polymer). This central interest in characteristic exponents leads to the unusual propensity, perhaps justifiable in the early stages of the development of a theory, for completely disregarding the numerical coefficients in most formulae, especially if they would obscure the main line of thought (de Gennes, 1979).

In the semidilute regime, the polymer chains undergo interpenetration. Because of this overlap, the interactions between segments separated by distances greater than the correlation length ξ are screened. In the domains in the bulk solution where interactions occur, each domain is of volume ξ^3. The thermal energy is kT and so the energy per unit volume is simply kT/ξ^3. The total energy is obtained by summing these individual contributions over the entire volume of solution.

The energy per unit volume for the polymer solution between the plates would be expected to be modified by some function $f(\xi/h)$ (dependent only upon ξ/h), as suggested by the mean field theory. Summing over the entire slab between the plates, the total surface energy becomes $\gamma_S \sim (hkT/\xi^3)f(\xi/h)$. It is only necessary now to evaluate the functional form of $f(\xi/h)$.

For large h, γ_S must be independent of h. This can only be achieved if $f(\xi/h)$ is proportional to ξ/h, i.e. $\gamma_S \sim kT/\xi^2$. On the other hand, for sufficiently small h, the polymer chains are completely excluded from the space between the plates. The free energy per unit volume (the bulk polymer solution is now the slab under study) is kT/ξ^3. The surface free energy is then proportional to the width of the slab so that $\gamma_S \sim hkT/\xi^3$. This, too, implies that the attractive force per unit area between the plates is constant and proportional to T/ξ^3 (which is just the osmotic pressure of the polymer solution, as shown in Section 4.6.3.4).

Joanny *et al.* (1979) then used a simple Derjaguin-type integration to examine the repulsion between two spherical particles, each of radius $a(a \gg \xi)$. It was assumed that the interaction between two rings on the surfaces of the spheres was the same per unit area as that for two infinite plates separated by a distance z. Following the mean field result, they wrote for spheres that

$\gamma_S \propto (kT/\xi^3)(\pi\xi - x)$, which implies that γ_S vanishes when $x \gtrsim \pi\xi$. The total free energy A is thus found from

$$A = \int_0^{\pi\xi} \gamma_S \pi a \mathrm{d}x$$

$$\sim -(akT/\xi^3)(\pi\xi - H_0)^2,\ H_0 < \pi\xi \tag{15.47}$$

where H_0 = minimum distance between the surfaces of the spheres. Note that $A = 0$ for $H_0 \geq \pi\xi$.

de Gennes (1982) has presented additional scaling law arguments to describe the interaction between two flat plates immersed in a polymer solution. The dispersion medium was assumed to be a good solvent for the polymer. For large separations, the polymer concentration at the mid-plane (c_m) between the plates is comparable (to within a factor of order 2) to the concentration due to a single plate. The latter had previously been shown by de Gennes (1981) to be given by $\sim (l/h)^{4/3}$. The interaction free energy is known from a scaling law analysis of the semidilute region to be given by $\sim (kT/l^3)c_m{}^{9/4}$. It follows that the force per unit area (i.e. the osmotic pressure) is given by $-kT/h^3$. An identical result was also found to hold in θ-solvents.

A simple physical picture of the scaling law result was presented by de Gennes as follows: in all cases we are considering a fluctuating medium where the size of the concentration fluctuations is truncated by the extent of the gap h. The only energy available is thermal energy kT and the only relevant distance is h. Thus the interactional energy per unit area must be of the form kT/h^2 (corresponding to a force per unit area of $-kT/h^3$). The attraction was claimed to arise from the following cause: when one flat plate polarizes the surrounding medium, the second plate, being identical with the first, lowers its energy by being placed in a polarized region. Note that these results are only applicable in the limit of infinite dilution of the polymer in the bulk phase.

de Gennes recognized that his conclusions were seemingly at odds with the results of experiment. The latter show clearly that the interactional force may be either attractive or repulsive, depending upon the polymer concentration. de Gennes claimed that this discrepancy between theory and experiment was due to departures from equilibrium. This explanation may well be correct: it will be shown in Chapter 17 that depletion stabilization appears to correspond to thermodynamic metastability.

Segment density profile in the depletion layers. Scaling laws also enable the segment density profile close to the surface to be established as follows. Let the segment density at the surface be ρ_s. Assume that the wall can be represented by a short range repulsive step potential $U(x)$ acting on all monomers within the range x_{eff}. Then ρ_s is the concentration just beyond this range, i.e. $\rho_s = \rho(x_{eff} + \delta)$ where $\delta \to 0$.

To find ρ_s as a function of the bulk concentration c_2, we recall the result of Section 4.6.3.3 where it was shown that the osmotic pressure is given by

$$\pi \simeq kT/\xi^3 \tag{15.48}$$

Now π can be related to the force near the wall surface by

$$\pi = (-\partial U/\partial x)\rho(x)\mathrm{d}x \tag{15.49}$$

The scaling postulate relates $\rho(x)$ to ρ_s by

$$\rho(x) = \rho_s f(x) \tag{15.50}$$

where $f(x)$ depends upon $U(x)$ but is independent of ρ_s. It is therefore possible to write

$$\pi = \rho_s \int f(x)(-\partial U/\partial x)\mathrm{d}x = B\rho_s \tag{15.51}$$

where B=constant. Thus ρ_s scales like the osmotic pressure. As shown in Table 4.4, π scales in a good solvent as $c_2^{9/4}$. Thus, in a good solvent,

$$\rho_s \sim c_2^{9/4} \tag{15.52}$$

In a θ-solvent, ρ_s scales as c_2^3.

To find the segment density profile for free chains close to an inert impenetrable surface, we follow the procedure adopted in Section 4.6.4.1 for the concentration dependence:

$$\rho(x) = c_2 f(x/\xi) \tag{15.53}$$

Here $f(z)$ is unity at large z whereas for small z, $\lim_{z\to 0} f(z) = z^m$. Since $\rho_s = \rho(x_{eff})$,

$$\rho_s = c_2(x_{eff}/\xi)^m \tag{15.54}$$

As $\rho_s \sim c_2^{9/4}$, $\xi \sim c_2^{-3/4}$ and x_{eff} is determined solely by the wall, the exponent equation becomes

$$9/4 = 1 + (3/4)m \tag{15.55}$$

so that $m = 5/3$ in a good solvent. In a θ-solvent, m is found to be 2, which is identical with the mean field theory result for a good solvent. Thus $\rho \sim x^{5/3}$ close to the surface for mobile chains.

15.2.2.5. Other theories for flocculation by free polymer

Solvation layer theories

Sato in a series of papers (Sato, 1971; 1979; Sato and Sieglaff, 1980), has propounded a very different mechanistic basis for the exclusion of polymer chains from between colloidal particles than that outlined above. His studies were sparked by the experimental observation that the stability of iron oxide particles dispersed in isopropyl alcohol was adversely affected by the presence of nonadsorbed polyamides dissolved in the continuous phase (Sato, 1971). Each iron oxide particle, according to Sato, is surrounded by a thick (ca 10–20 nm) multilayer of adsorbed isopropyl alcohol. All layers in this multilayer, except perhaps the first, are physically adsorbed. The solvation layers are assumed to be impenetrable to polymer. No attempt was made to justify this unorthodox assumption of very thick solvation layers.

In the first version of the theory expounded by Sato (1979), these solvation layers were assumed to be irreversibly adsorbed. This was subsequently modified (Sato and Sieglaff, 1980) to permit reversible adsorption. The latter allows the solvation layers to undergo mutual interpenetration on close approach of the particles. Interpenetration was postulated to produce desorbed solvent that dilutes the polymer in the bulk phase. This dilution is reminiscent of that first postulated by Asakura and Oosawa (1954) for quite different reasons. Dilution leads to an attraction between the particles.

Sato and Sieglaff (1980) argue that three free energies determine whether or not desorption of the solvation layers can take place. These are the free energy of dilution of the bulk polymer solution (ΔG_{dil}), the thermal energy of the solvent (ΔG_{the}) and the free energy of solvent desorption (ΔG_{des}). These were evaluated as follows:

(i)
$$-\Delta G_{\text{dil}} = -(\mu_1 - \mu_1^{\ominus})\Delta n_1 = \pi_{OP}\bar{V}_1\Delta n_1$$
$$= RT\{c_2/M + B_2c_2^{\,2} + B_3c_2^{\,3} + \ldots\}(\delta V) \tag{15.56}$$

where μ_1 = chemical potential of the solvent, Δn_1 = number of moles of solvent in the overlap volume (δV) and $\bar{V}_1$ = partial molar volume of the solvent;

(ii)
$$\Delta G_{\text{the}} = (\delta V/\bar{V}_1)\ RT \tag{15.57}$$

since Sato believed that each solvent molecule possesses thermal energy kT;

(iii)
$$\Delta G_{\text{des}} = (\delta V/\bar{V}_1)\ q_1 \tag{15.58}$$

where q_1 = isosteric heat of adsorption of the solvent.

The criterion that was proposed for the desorption of the solvation layers was

$$\Delta G_{\text{dil}} + \Delta G_{\text{the}} \geq \Delta G_{\text{des}} \tag{15.59}$$

For polyacrylamide of molecular weight 900 in isopropyl alcohol, this criterion showed that an energy of adsorption of ca 2·4 kJ mol^{-1} was necessary to prevent desorption. It was argued that the desorption energies would be less than this and so desorption could occur on close approach of the particles.

It would be expected on the basis of the preceding model that the solvation layers around the particles, which were only 50 nm in size, would be independent of the free polymer molecular weight. Yet the thickness of the solvation layers calculated from the model varied strongly with molecular weight. The values of the thickness of the solvation layers so calculated were very much greater than the chain dimensions. That such thick solvation layers as postulated by Sato could exist is extremely doubtful and no direct experimental evidence for their existence exists as yet. Furthermore, an energy of adsorption of less than 2·4 kJ mol^{-1}, claimed to be the upper limit for desorption, corresponds to less than thermal energy ($3/2\ RT = 3{\cdot}7$ kJ mol^{-1}). Accordingly, the multilayers postulated by Sato would not be thermodynami-cally stable. There seems little evidence to support the solvation layer notions expounded by Sato, especially as the phenomenon is totally explicable by depletion effects.

The steric layer theory

A radically different theory for the origin of the flocculation of colloidal dispersions by dissolved polymers has been proposed by Vincent, Luckham and Waite (1980). This involves the presence of a steric layer at the particle/dispersion medium interface, and its interactions with the free polymer.

The essence of this theory can be grasped from Fig. 15.4. In the semidilute region, the free polymer molecules can be assumed to overlap the polymer in the steric layers. However, these unattached interpenetrated chains are displaced into the bulk phase on the close approach of two sterically stabilized particles. The free energy change that accompanies this displacement is simply

$$\Delta G = (n/2)G^{LL} + (n/2)G^{PP} - nG^{PL} \tag{15.60}$$

where n = number of free polymer-steric layer contacts displaced and the superscripts LL, PP and PL denote steric layer-steric layer, free polymer-free polymer and free polymer-steric layer contacts respectively. Equation (15.60) merely expresses the stoichiometry of the displacement process: n free polymer-steric layer contacts are replaced by $n/2$ steric layer-steric layer contacts and $n/2$ free polymer-free polymer contacts, the total number of contacts being conserved in this model.

Vincent, Luckham and Waite assumed that $G^{PP} = G^{PL}$, as the free polymer and the stabilizing polymer were of the same chemical composition in their experiments. They further assumed that $(n/2)G^{LL} = G^{LL,0}$, where the latter is the steric interaction free energy in the absence of free polymer. Equation (15.60) can be simplified under these assumptions to

$$\Delta G = G^{LL,0} - (n/2)G^{PP} \tag{15.61}$$

According to this approach, ΔG can decrease to zero and become negative,

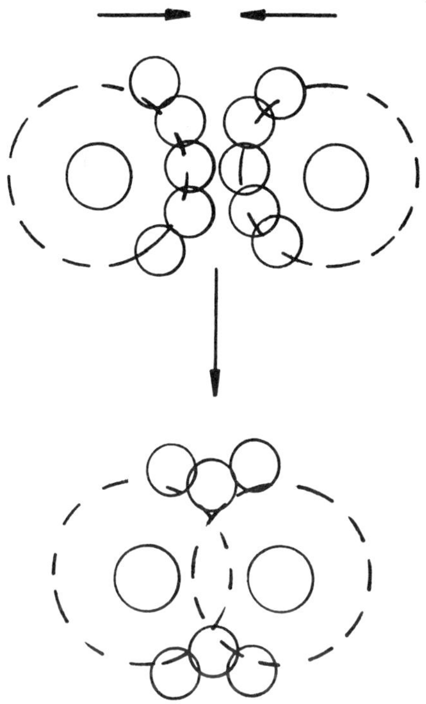

Fig. 15.4. Schematic representation of the steric layer theory (after Vincent *et al.*, 1980).

even if $G^{LL,0}$ is positive. In other words, even though colloidal particles may be sterically stabilized in a good solvent, addition of free polymer can induce flocculation, provided, of course, that G^{PP} is positive. This proviso will be satisfied when the steric layer polymer and the free polymer are identical in their chemical composition and the dispersion medium is a better solvent for the stabilizing moieties than a θ-solvent.

Note that reference to equation (15.60) for the case where the polymer composing the steric layer is different from that in free solution shows that, under some circumstances, ΔG cannot become negative. Specifically, this requires that G^{PL} be negative in a dispersion medium that is a better than θ-solvent for both of the polymers. A negative value for G^{PL} corresponds to the free polymer and the stabilizing moieties being compatible (Flory, 1953). If equation (15.60) is correct, it shows that free polymer that differs in chemical composition from the polymer in the steric layers cannot induce flocculation if the two polymers are compatible. No experimental evidence is as yet available to ascertain the validity of this prediction.

There are several criticisms that can be levelled at the theory of Vincent, Luckham and Waite (1980). The first, and most obvious in light of the discussion presented above, is the complete disregard by the theory of any depletion effects associated with the free polymer. These will occur whether or not the particles are coated with polymer. This neglect could be justified to some extent if the free polymer were compatible with that in the steric layers. In that instance, however, the theory of Vincent, Luckham and Waite, as we have seen, is unable to provide an explanation for flocculation. A second criticism arises if the free polymer is of significantly higher molecular weight than that of the steric layers. In that case, many more free polymer chains will be displaced from the region between the particles than is calculated from the zone of interaction of the steric layers shown in Fig. 15.4. Thirdly, no allowance is made for the presence of free polymer in the steric layers. This effect will be especially important if the molecular weight of the free polymer is much smaller than that of the polymer composing the steric layers. Finally, we note that the Vincent, Luckham and Waite explanation for the flocculating effects of free polymer could not possibly be operative in the absence of steric barriers. Yet, as we shall see, flocculation that is induced by free polymer has been observed even when the steric layers are essentially nonexistent (Sperry *et al.*, 1981).

Chapter 16

The Effects of Free Polymer on Colloid Stability: Experimental Studies of Depletion Flocculation and Depletion Phase Separation

16.1. Model experiments on depletion association

The experiments that were described in the introduction to the preceding chapter demonstrated that the addition of polymer can induce the flocculation of colloidal dispersions. The interpretation of these studies, however, was not clearcut. The experiments reviewed therein utilized systems, such as natural rubber latices, erythrocytes and soil, which, whilst being of considerable practical importance, could scarcely be described as being model colloidal dispersions. It seems likely that for this reason the results of these experiments have been ignored to some degree. Flocculation in such systems, for example, could unwittingly be attributed to bridging, even though the polymer concentrations necessary are massive compared with those (a few parts per million) usually required for bridging flocculation. It has therefore been crucial to the establishment of the effects of free polymer on colloid stability that experiments have been performed on model or near-model systems.

16.2. The physical nature of the association induced by free polymer

The question of the physical nature of the association phenomenon caused by the presence of free polymer will recur constantly in the following discussion. The reason for this is that there appear to be at least two phenomena that can be observed: one is a genuine phase separation that resembles the coacervation which eventuates when the solvent for a noncrystallizing polymer is made suitably poor; the other is an aggregation of particles, similar to that manifest in an electrostatically stabilized system at high ionic strengths. The detailed reasons why one mode of association is favoured over the other are simply not understood as yet. It would be expected, however, that the difference in free energy between the two final states would be relatively small.

16.2.1. Phase rule considerations

The introduction of free polymer into the dispersion medium appears to make no significant difference to the number of phases accessible to the system. This can be discerned from the following simplistic considerations using the Phase Rule, which will be assumed to be applicable. We consider a dispersion of colloidal particles sterically stabilized by terminally grafted polymer chains. Free polymer is dissolved into the dispersion medium.

The number of components (C) will be taken to be 3: latex particles, free polymer and dispersion medium. The number of degrees of freedom (F) is 2, the latex particles concentration being taken as an extensive variable. The two intensive variables are the temperature and the concentration of free polymer, both of which can be varied independently. In accordance with the Phase Rule, the number of phases is given by $P = C - F + 2 = 3$, one of which is the vapour phase. This means that two liquid phases are possible. These will be separated by a macroscopic liquid/liquid interface. Note that no information is obtained from this relationship as to the dispersed or flocculated nature of the latex particles in either liquid phase, nor to the respective concentrations of each component in the two phases.

The foregoing interpretation of the nature of the two phases is valid at concentrations of free polymer above the critical value necessary to induce the separation into two phases. Below this critical value, the interpretation of the two phases is quite different. One phase is a liquid, the dispersion medium; the other is the latex particles. Again, no information is available from the phase rule *vis-à-vis* the state of aggregation of the dispersed particles in the dispersion medium. This latter analysis is presumably also applicable to dispersions subject to flocculation by the advent of free polymer.

16.2.2. Microgel particles

The first attempts to investigate the consequences of the addition of free polymer to model colloidal systems were undertaken by Sieglaff (1959). He studied microgel particles composed of polystyrene cross-linked by 0·3% divinylbenzene dispersed in toluene. Free polystyrene of weight average molecular weight 5×10^5 or 2×10^5 was dissolved into the dispersion medium.

The cross-linking of the polystyrene latex particles prevented their complete dissolution when dispersed in toluene, although significant swelling still ensued (hence the designation "microgel").

Sieglaff observed what he termed "phase separation" on adding free polystyrene to the toluene. Presumably, this meant the formation of a distinct interface on standing, although nowhere is that explicitly reported. What is stated is that on standing, all of these ternary systems, with the exception of those with low free polystyrene concentrations, formed two phases "*the top phase of the samples was clear, while the lower phase had a grainy and turbid appearance*". The latter observation in fact suggests that the microgel particles may have undergone flocculation. If this were so, then it seems unlikely that two genuine phases, separated by a distinct interface, were actually formed. Sieglaff himself believed that the microgel particles were stable after phase separation and formed a close-packed array that contributed to the grainy appearance of the microgel phase.

It should be noted that Sieglaff used toluene as the dispersion medium because he considered that toluene was an athermal solvent for polystyrene. It was argued that if enthalpy effects were eliminated, entropy effects alone must be responsible for the observed phenomenon. Unfortunately, it is now well established that toluene is not an athermal solvent for polystyrene (Lewis and Johnson, 1969; Gaeckle *et al.*, 1973). The heat of dilution of this system is unquestionably negative, its absolute magnitude increasing with increasing polymer concentration. This is, of course, contrary to the predictions of the simple equation of state theory. This discrepancy may perhaps be associated with preferential orientations of the anisotropic toluene molecules in the neighbourhood of the polystyrene chains. Whatever its origin, the negative heat of dilution invalidates Sieglaff's original entropy arguments as to the purely entropic origin of the phenomenon.

Note that it is implicit in Sieglaff's results that a critical concentration of free polymer was essential for phase separation to occur. Sieglaff actually determined the phase diagrams for these three component systems. Examples of his results are displayed in Figs 16.1a and b.

In more recent times, Clarke and Vincent (1981b) have extended the studies of Sieglaff on microgel particles. They used ethyl benzene and *n*-butyl formate, rather than toluene, as the dispersion medium. Their various samples of relatively narrow molecular weight free polystyrene spanned the molecular weight range $1{\cdot}8 \times 10^3$ to 2×10^6. The microgel particles when swollen had a diameter of between 500 and 1150 nm, as determined by photon correlation spectroscopy.

Addition of polymer to the dispersion medium raised the LCFT in all cases, i.e. the microgel was rendered inherently less stable. For example, the addition of polystyrene of molecular weight 11 300 (at a volume fraction of 0·05) to a microgel of swollen diameter 750 nm in *n*-butyl formate (at a microgel volume fraction of $\phi_3 = 7 \times 10^{-4}$) raised the LCFT from -9 ± 1 °C, in the absence of polymer, to -3 °C.

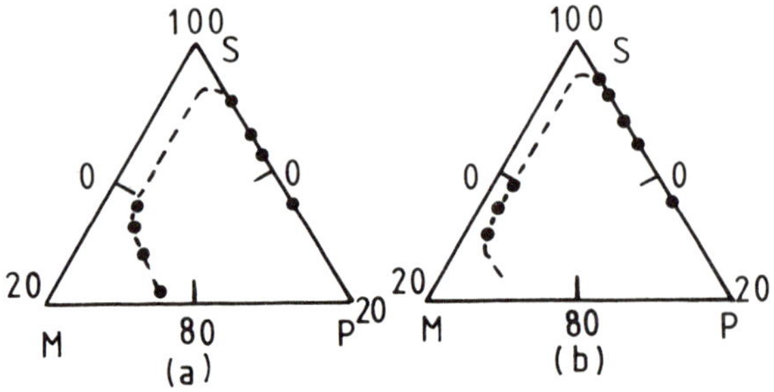

Fig. 16.1. The three component diagram for polystyrene microgel (M) particles immersed in toluene (S) in the presence of free polystyrene (P) of different molecular weights: (a) $5{\cdot}6 \times 10^5$; (b) $2{\cdot}3 \times 10^5$ (after Sieglaff, 1959).

Clarke and Vincent in these studies established the phase diagrams for the free polystyrene-microgel-ethyl benzene systems, two of which are displayed in Fig. 16.2a and b. In all cases, only the stability/instability boundary is shown. These lines correspond to the locus of the critical volume fraction of free polymer required to induce flocculation ($v_2{}^*$) as a function of the volume fraction of the microgel particles (ϕ_3).

Examination of Fig. 16.2a shows that the critical volume fraction of free polymer decreased with increasing molecular weight (at constant ϕ_3). Higher molecular weight polymers were therefore more effective in promoting flocculation than were those of lower molecular weight. Further, as shown in Fig. 16.2b, values of $v_2{}^*$ for larger particles were less than those for smaller particles. This implies that larger particles are more readily flocculated by free polymer than are smaller ones.

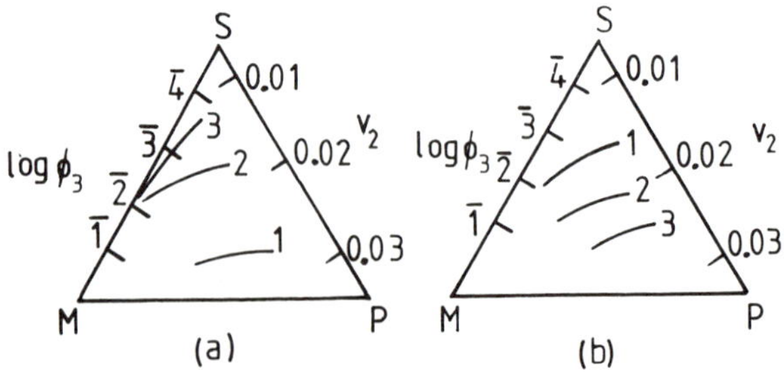

Fig. 16.2. Three component diagrams at 25 °C for polystyrene microgel particles (M) in ethylbenzene (S) in the presence of free polystyrene (P); (a) free polystyrene of various molecular weights: curves 1, 1 800; 2, 18 000; 3, 2×10^6; (b) different particle sizes: curves 1, 520 nm; 2, 460 nm; 3, 403 nm (after Clarke and Vincent, 1981b).

In all cases where flocculation was induced by free polymer, it was weak and reversible: dilution of the continuous phase below v_2^* led to rapid redispersion (although in some cases mild agitation was required to accomplish this). In these experiments, Clarke and Vincent (1981b) clearly observed flocs that could be detected visually. The separation of these flocs of microgel particles, as reported by Sieglaff in his experiments, only occurred with the highest molecular weight (2×10^6) polystyrene sample. Interestingly, the volume of flocculum generated in this way rose from zero at v_2^* up to a maximum and then diminished. This may be interpreted as evidence indicative of the onset of depletion stabilization effects at higher free polymer concentrations. The turbidity of the supernatant layer, moreover, followed an inverse pattern, passing through a minimum. These experiments demonstrated the equilibrium nature of the weak flocculation produced by free polymer.

16.2.3. Nonaqueous dispersions

Two groups of workers, de Hek and Vrij (1979; 1981) and Clarke and Vincent (1981a) have independently investigated the effects of free polystyrene on the stability of silica dispersions in nonaqueous media. Although their results show many parallels, there is one over-riding difference between them: de Hek and Vrij on exceeding the critical concentration of free polymer apparently observed the formation of two translucent liquid phases that were separated by a sharp (although sometimes barely discernible) interface; Clarke and Vincent, on the other hand, apparently observed bona fide flocculation of the silica particles. This apparent difference is neither merely academic nor semantic. The silica particles in the separated phases in the experiments of de Hek and Vrij were presumably quite stable; those in Clarke and Vincent's experiment underwent agglomeration. It is as if two quite distinct phenomena are observable in these systems: one is a flocculation process, the other is phase separation. It is at least conceivable that the two processes are interlinked in the sense that they represent different options that the systems may follow.

16.2.3.1. Phase separation studies

de Hek and Vrij (1979; 1981) studied the effects of free polystyrene on two different types of silica particles. One type was sterically stabilized by polyisobutene of relatively low molecular weight ($1{\cdot}3 \times 10^3$ or $6{\cdot}5 \times 10^3$) and the other was coated with stearyl alcohol. The silica particle concentrations were relatively small (0·01 or 0·05 w/v). Cyclohexane was used as the dispersion medium at 34 °C because under these conditions it is a θ-solvent for polystyrene. This was a requirement of the theoretical treatment developed by these authors, which demanded that the free chains be perfectly permeable to one another.

A minimum concentration of free polymer was found necessary in order for phase separation to occur. When polymer was added to the silica dispersions above this critical concentration, the mixture became turbid. Within a short period of time (usually less than one hour), the turbidity disappeared and two translucent phases could be distinguished. These were separated by a sharp interface. The lower phase was rich in silica whereas the upper phase was rich in polystyrene.

The effect of molecular weight of free polymer

The effect of varying the molecular weight of the free polystyrene on the critical concentration of free polymer needed for phase separation (c_2*) is summarized in Table 16.1. These results refer to silica particles stabilized by stearyl alcohol. It is apparent that the critical concentration decreased with increasing molecular weight, the dependence being approximately an inverse one-fourth power. It should be noted, however, that the molecular weight dependence was significantly stronger with the polyisobutene stabilized silicas.

TABLE 16.1
Effect of molecular weight on the critical concentration of polystyrene required for phase separation of silica particles

Polystyrene molecular weight	c_2^*/g dm^{-3}	$10^{-2}\,(c_2^*)\,(MW)^{\frac{1}{4}}$
8 000	82 ± 1	7·8
36 000	48 ± 1	6·6
130 000	35 ± 1	6·6
194 000	31 ± 1	6·5
2 400 000	17 ± 1	6·7

The effects of other variables

The value of the critical polystyrene concentration was found by de Hek and Vrij (1981) to decrease when the particle concentration was increased from 1 to 5%. The effect was relatively small (e.g. c_2* decreased by ca 10% for a five-fold increase in the particle concentration). The results are consistent with the hypothesis that the particles are more stable in the more dilute systems because of their greater configurational entropy.

The fact that de Hek and Vrij observed phase separation rather than flocculation in their experiments suggests that it would be unlikely that depletion stabilization could be observed in these systems, if only because the dispersed particles are stable after phase separation. There is, however, another possible explanation: if phase separation occurs because the free polymer chains cannot fit into the space between particles without significant loss of configurational entropy, this severe constraint will not be removed by increasing the polymer concentration. Indeed, that increase would, in contrast, provide a somewhat greater driving force for phase separation.

16.2.3.2. Flocculation

The systems studied by Clarke and Vincent (1981a) were closely analogous to those of de Hek and Vrij mentioned previously. Polystyrene was grafted on to silica particles by an anionic polymerization technique. Its molecular weight lay in the range 5 000 to 24 000. The diameter of the monodisperse silica particles was varied from ca 100 to 300 nm. Ethyl benzene was chosen as the dispersion medium.

Clarke and Vincent presented their results in the form of what they termed triangular, three component phase diagrams, one of which is reproduced as Fig. 16.3.

Their results can be summarized as follows:

(i) high molecular weight free polymers were more effective in inducing flocculation than were low molecular weight chains;
(ii) larger particles were more readily flocculated (i.e. required a lower v_2^*) than were smaller ones;
(iii) the stability of the particles increased with increasing surface coverage by grafted polystyrene.

Conclusions (i) and (ii) clearly parallel those reported by de Hek and Vrij (1979; 1981) for phase separation in nonaqueous media. Conclusion (iii) shows that the presence of steric layers influences depletion flocculation significantly.

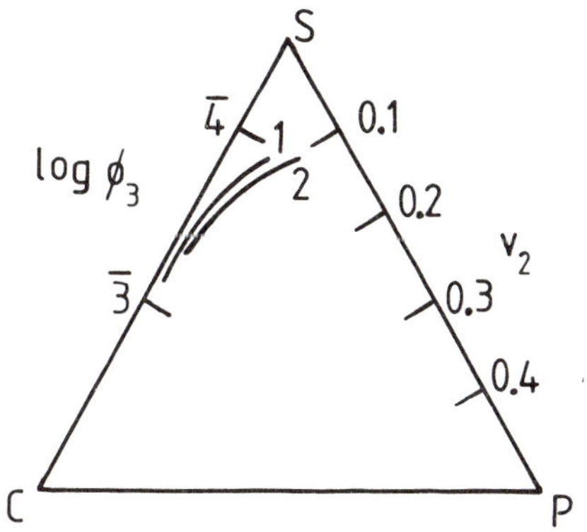

Fig. 16.3. Three component diagrams for silica particles (C) in ethylbenzene (S) in the presence of free polystyrene (P); molecular weight of free polystyrene, curves 1, 28 000; 2, 6 000 (after Clarke and Vincent, 1981a).

16.2.4. Aqueous dispersions

16.2.4.1. Phase separation studies

Reports of the occurrence of phase separation induced by free polymer in aqueous dispersion media seem curiously rare. Garvey (1982) has found that when highly concentrated dispersions of silica particles are mixed with a solution of ca 0·1% xanthan gum (an extracellular polysaccharide from *Xanthomonas campestris*), phase separation is observed on standing. It is tempting to associate this phenomenon with the high molecular weight (ca 2×10^6) and high extension [perhaps rods of length 500 nm, according to Wellington (1981)] of this gum in solution but de Hek and Vrij (1979; 1981) have shown in nonaqueous media that such requirements are apparently unnecessary for phase separation. Carboxymethyl cellulose also exhibits the same phase separation behaviour with highly concentrated aqueous silica dispersions.

The experiment described by Garvey (1982) vividly illustrates the difference between phase separation and flocculation. In the former instance, the coacervated phase of xanthan gum is readily creamed by gentle centrifugation, leaving a stable silica dispersion as the lower phase. Had the silica particles been flocculated, centrifugation would have transported the aggregated silica particles to the bottom of the centrifuge tube, leaving a stable xanthan gum solution as the supernatant serum. This provides an unequivocal criterion for distinguishing between phase separation and flocculation in this system.

One possible explanation for the phase separation in both aqueous and nonaqueous systems is the very high occupancy of the space by the sterically stabilized particles. This would mean that the free polymer cannot diffuse into the dispersion media without a significant loss of configurational entropy. The exigencies created by such severe volume restrictions at high dispersed phase concentrations could be responsible for phase separation. The fact that the polymer chains cannot physically diffuse into the dispersion would prevent the chains from inducing either depletion flocculation or depletion stabilization.

The foregoing concepts seem reasonable because Flory (1978a,b) has shown theoretically that a ternary system consisting of solvent, a rigid rod solute and randomly coiled polymer chains may undergo phase separation into an anisotropic phase and an isotropic phase. The anisotropic phase consists almost exclusively of dissolved highly-oriented rods whose mutual alignment alleviates obstructions between neighbouring species. The isotropic phase exhibits a somewhat greater tolerance for the rodlike component but is composed predominantly of dissolved randomly coiled polymer chains.

16.2.4.2. Flocculation studies

Sterically stabilized dispersions

The first clear-cut demonstration that unattached water soluble polymers can flocculate model sterically stabilized dispersions was provided by Li-in-on *et al.* (1975). Subsequently, Vincent, with these and other coworkers, has further elaborated the phenomenology of flocculation by free polymer in aqueous dispersions (Cowell *et al.*, 1978; Vincent *et al.*, 1980; Cowell and Vincent, 1982a).

All the dispersions studied by Vincent, with but one exception (Cowell and Vincent, 1982a), were sterically stabilized. Typically they consisted of relatively large polystyrene particles that were stabilized by relatively low molecular weight (e.g. 750 or 2000) poly(oxyethylene). Consequently, none of the dispersions displayed the correlation between the UCFT and the corresponding θ_L-temperature. Instead, the dispersions, in the absence of free poly(oxyethylene), underwent weak coagulation in dispersion media that were significantly better solvents than θ-solvents for the poly(oxyethylene) (i.e. UCFT $<\theta_L$).

The apparent stability ratio

The relative stability of colloidal dispersions near to incipient coagulation can be characterized by their stability ratio W where (Fuchs, 1934)

$$W = 2a \int_{2a}^{\infty} \exp\{V(r)/kT\}/r^2 \mathrm{d}r \tag{16.1}$$

where r = centre-to-centre distance between the colloidal particles, $V(r)$ = potential energy of interaction between the particles and a = particle radius. For stability, $V(r)$ should be large and positive which will render W large and positive. An alternative way of viewing W is through

$$W = k_0/k \tag{16.2}$$

where k = coagulation rate constant and k_0 = Smoluchowski hypothetical rate constant for irreversible fast flocculation in the complete absence of interparticle interactions ($k_0 = 4k_BT/3\eta$, where η – viscosity of the dispersion medium and k_B = Boltzmann constant; at 25 °C in water, k_0 adopts the value $6{\cdot}2 \times 10^{-15}$ dm^3 s^{-1}). Clearly, if k, the coagulation rate constant, is small, W is large. Note that just as W is a measure of kinetic stability, its reciprocal, $1/W$, is a measure of instability. In polymer solutions the measured values of $1/W$ must be interpreted with caution because of their possible subversion by hydrodynamic and entanglement phenomena associated with the polymer chains.

Cowell *et al.* (1978) measured the apparent $1/W$ (actually $s = 2k_BT\phi_3{}^0/\pi a^3 W$, where $\phi_3{}^0$ = the initial particle volume fraction) for sterically stabilized polystyrene latices in the presence of free poly(oxyethylene). Their results, for a wide range (200–300 000) of molecular weights of free polymer, are displayed in Fig. 16.4.

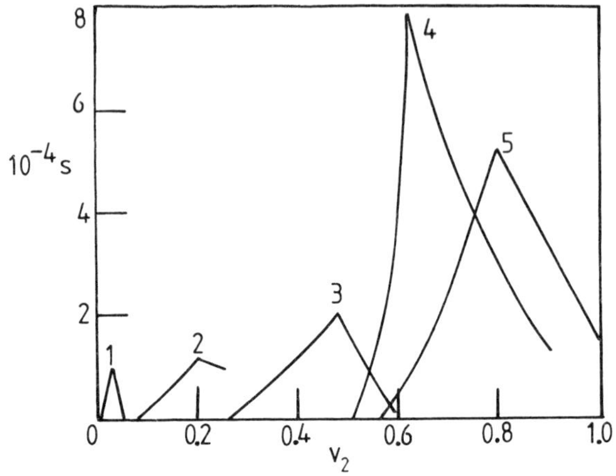

Fig. 16.4. The reciprocal of the stability ratio of poly(oxyethylene) stabilized latex particles in the presence of free poly(oxyethylene) of different molecular weights: curves 1, 300 000; 2, 10 000; 3, 4 000; 4, 600; 5, 200 (after Cowell *et al.*, 1978).

It is apparent that up to a critical volume fraction $v_2{}^*$ of free polymer, the particles displayed long-term stability ($1/W = 0$). Any further increase in the free polymer concentration resulted in the onset of instability, which was manifested by an increase in $1/W$. In all cases, however, $1/W$ reached a maximum value and then declined. Flocculation was thus not evident at high concentrations of free polymer. It usually occurred only over a finite range of free polymer concentrations. This lack of flocculation at higher concentrations of free polymer is the phenomenon of depletion stabilization mentioned in the prefatory remarks to this chapter. It will be considered in detail in the next chapter.

The effect of molecular weight of the free polymer. Figure 16.5 presents the characteristic dependence of the critical volume fraction of free polymer required to induce flocculation as a function of the molecular weight of the added polymer. For the particles stabilized by poly(oxyethylene) 750, two different temperatures (25 °C and 45 °C) are shown.

It is obvious that, as with nonaqueous dispersions, the critical polymer concentration required for flocculation decreases with increasing molecular weight. Increasing the temperature rendered the latices more susceptible to flocculation (Fig. 16.6).

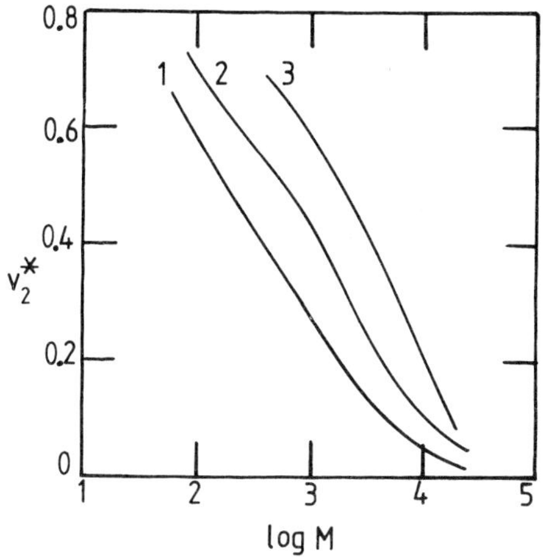

Fig. 16.5. The critical volume fraction of free poly(oxyethylene) required for depletion flocculation as a function of its molecular weight: curves 1, stabilizer POE 750, 45 °C; 2, stabilizer POE 750, 25 °C; 3, stabilizer POE 2 000, 25 °C (after Cowell *et al.*, 1978).

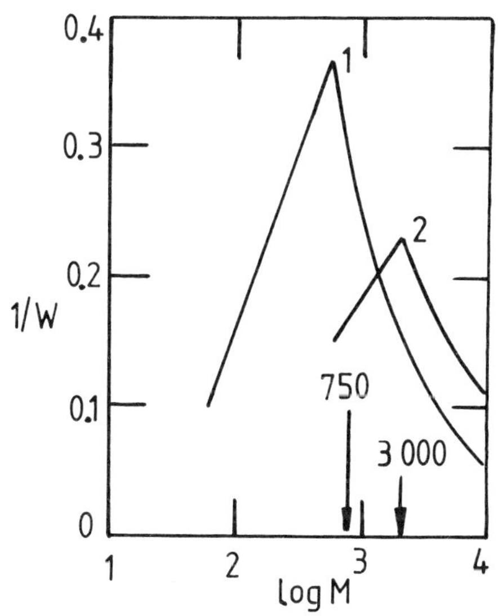

Fig. 16.6. The reciprocal of the stability ratio of poly(oxyethylene) stabilized latex particles in the presence of free poly(oxyethylene) of different molecular weights; curves 1, stabilizer POE 750; 2, stabilizer POE, 2 000 (after Cowell *et al.*, 1978).

Cowell *et al.* also found that the maximum in the instability (i.e. in $1/W$) corresponded to the free polymer being of a molecular weight comparable to that of the anchor polymer. Even under these conditions, however, the dispersions only flocculated at a rate of about one-third of the hypothetical Smoluchowski diffusion controlled rate. This rate is significantly less (ca one-half) than that expected when allowance is made for hydrodynamic interactions between the approaching particles. This rate reduction, if real, may indicate the presence of a small repulsive energy barrier that must be surmounted during the flocculation process, as predicted by the theories of depletion stabilization.

Three component triangular diagrams. Both Cowell *et al.* (1978) and Vincent *et al.* (1980) presented what they termed three component phase diagrams for aqueous systems composed of water (or electrolyte solution), free poly(oxyethylene) and polystyrene latex particles sterically stabilized by poly(oxyethylene). Such a diagram is reproduced in Fig. 16.7.

It should be noted that the axes of this diagram are quite different from those of a conventional Roozeboom triangular phase diagram. The lengths of the sides of the latter are usually scaled from 0–100% of a particular component. This is not practicable for dispersions because the volume

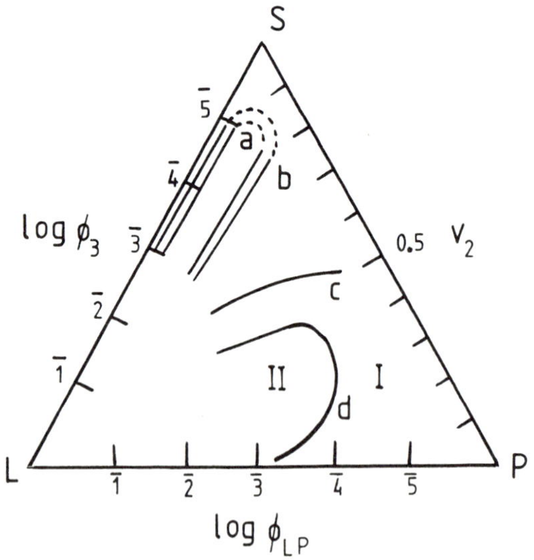

Fig. 16.7. The three component diagram at 25 °C for poly(oxyethylene) stabilized latex particles (L) in aqueous (S) poly(oxyethylene) (P) solutions; (a) stabilizer POE 750 with 10 000 free POE in water; (b) same as (a) but in 0·065 M $MgSO_4$; same as (a) but with 1500 free POE; (d) same as (a) but with 400 free POE. The numerals I and II denote dispersed particles only and dispersed particles plus flocs respectively (after Cowell *et al.*, 1978).

fraction of the particles accessible by, e.g. light scattering, is only in the range 10^{-5}–10^{-3}. Such an interval would be minuscule on the conventional diagram. This difficulty is circumvented by the subterfuge of plotting the logarithm of the volume fraction of particles. Any scaled representation of zero volume fraction of particles is then impossible, although it is readily achieved in an unscaled fashion.

The apices of the triangular diagrams presented by Vincent and coworkers represent 100% polymer, 100% latex and 100% dispersion medium (which itself may contain more than one component). That side of the triangle representing the complete absence of latex (i.e. from 0–1 volume fraction of free polymer) can be scaled as in a conventional triangular diagram, if that proves advantageous. The remaining sides are scaled logarithmically using an identical scale, which represents the logarithm of the volume fraction of latex particles. The side of the triangle joining the apices depicting pure polymer and 100% latex corresponds to the absence of dispersion medium. The remaining side, connecting the points designating 100% dispersion medium and 100% latex, connotes the complete absence of polymer.

The contour lines shown in Fig. 16.7 represent the boundaries between completely stable dispersions and those that are flocculated. Note, however, that the flocculated particles were in equilibrium with those in the dispersed state, i.e. a domain exists (designated by II for case d in Fig. 16.7) in which both flocs and singlet dispersed particles coexist. Outside this domain, only singlet particles are observed (domain I for case d).

The UCFT of the latex stabilized by poly(oxyethylene) 750 in 0·065 M magnesium sulphate solution was measured as 48·5 °C at a particle volume fraction $\phi_3 = 5 \times 10^{-4}$ in the absence of free polymer. The corresponding three component diagram lines at this temperature were constructed for the separate addition of two molecular weight free polymers, one of molecular weight 1000 and the other molecular weight 10 000 (see Fig. 16.8). Both lines when extrapolated to the pure latex-pure solvent axis (i.e. no free polymer present) intersected that axis at a common particle volume fraction (ca 5×10^{-4}).

Electrostatically stabilized dispersions. Perhaps the most conclusive experiment that has been performed to-date to demonstrate the existence of depletion flocculation has been disclosed by Sperry *et al.* (1981). Other workers [e.g. de Hek and Vrij (1979) and Vincent and coworkers] studied dispersions whose particles were enveloped in a steric layer. Accordingly, the presence of the steric layer could have been mandatory for the occurrence of flocculation. Indeed, both Vincent *et al.* (1978) and Vrij (1976) have proposed that the steric layers are either wholly or partially responsible for the observed phenomena. What Sperry *et al.* (1981) have succeeded in doing is to generate flocculation in a latex system from which steric layers were seemingly absent.

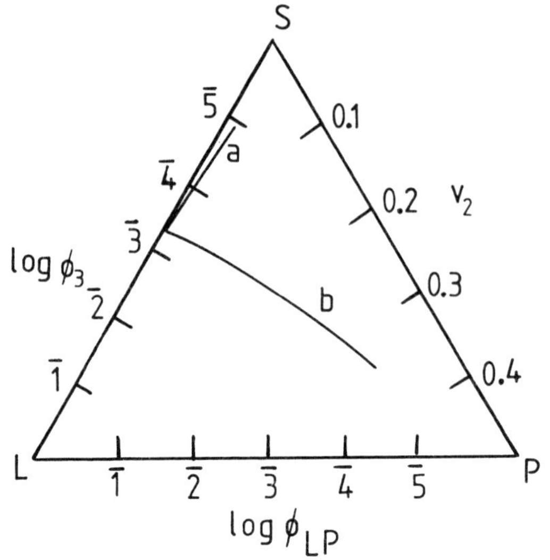

Fig. 16.8. Three component diagram at 48·5 °C for poly(oxyethylene) stabilized latices (L) in aqueous (S) solutions of free poly(oxyethylene) (P): (a) stabilizer POE 750 with free POE 10 000 in 0·065 M $MgSO_4$; (b) stabilizer POE 750 with free POE 1 000 in 0·065 M $MgSO_4$ (after Vincent *et al.*, 1980).

The latices studied were composed primarily of either polystyrene or a copolymer of *n*-butyl acrylate and methyl methacrylate. Their diameters were of order 0·4 μm. Hydroxyethyl cellulose was the water soluble polymer subjected to detailed investigations, but poly(oxyethylene) and polyacrylamide exhibited a qualitatively similar pattern of flocculation behaviour. Six different samples of hydroxyethyl cellulose were studied, their molecular weights varying from 68 500 to 887 000.

Flocculation was observed to be reversible, either by shear or by dilution of the dispersion medium to below the critical concentration of free polymer. In the former instance, redispersion was only temporary in that the flocs reformed on cessation of the shearing action. In the latter case, redispersion was permanent.

Sperry *et al.* found that the addition of either the anionic surfactant sodium dodecyl sulphate or the nonionic surfactant Triton X-405 completely desorbed any hydroxyethyl cellulose from the surface of the latex particles. This meant that, even in the presence of free hydroxyethyl cellulose in the continuous phase, none of the flocculating polymer was attached to the surface. The latex particles in the presence of the sodium dodecyl sulphate (0·5%) were thus electrostatically stabilized whereas the nonionic Triton surfactant conferred steric stabilization.

Hydroxyethyl cellulose did not cause flocculation of the latex particles unless present in excess of a critical concentration (c_2*). For the electrostatically stabilized system $c_2^* \sim 0{\cdot}12\%$ (w/w) for hydroxyethyl cellulose of molecular weight 225 000. The corresponding value for the sterically stabilized particles was somewhat higher ($c_2^* \sim 0{\cdot}16\%$).

The critical concentration of free polymer that must be added to cause flocculation decreased with increasing molecular weight (see Fig. 16.9). The value of c_2^* depended upon the molecular weight with an exponent of $-0{\cdot}7$. This value is quite different from the exponent ($-0{\cdot}25$) originally reported by de Hek and Vrij (1979) for nonaqueous silica dispersions stabilized by stearyl alcohol and flocculated by polystyrene. The value measured by Sperry *et al.* is

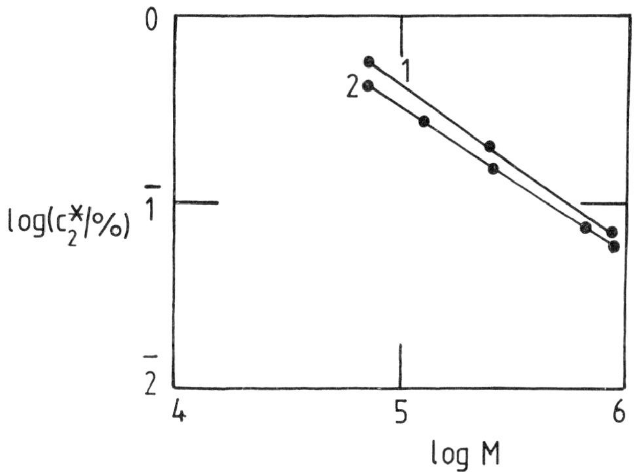

Fig. 16.9. The dependence of the critical concentration of free hydroxyethyl cellulose upon its molecular weight for the depletion flocculation of polystyrene latex particles; curves 1, in the absence of anionic surfactant; 2, in the presence of nonionic surfactant (after Sperry *et al.*, 1981).

closer to that ($-0{\cdot}5$) observed by Vincent *et al.* (1978) for sterically stabilized aqueous systems (Feigin and Napper, 1980a,b) and the exponent (ca $-0{\cdot}5$) measured by de Hek and Vrij (1981) for nonaqueous dispersions of silica particles stabilized by poly(isobutylene). Note, however, that Sperry *et al.* observed a significantly different exponent ($-0{\cdot}3$) for latices with strongly adsorbed hydroxyethyl cellulose on the surface. This is closer to the original exponent of de Hek and Vrij (1979). These wide variations in the measured exponents reflect the complexity of the phenomena involved.

Terminology. Both de Hek and Vrij (1979; 1981) and Sperry *et al.* (1981) attribute their observations to the effects of volume restriction, referring to the phenomenon as *volume restriction flocculation* or *phase separation*. Strictly speaking, this terminology is misleading. As will be seen, free polymer chains, in contrast to attached polymer chains, tend to escape from the zone of interaction so that these chains are not actually compressed, as would be implied by the "volume restriction" terminology. The reality is that it is the depletion of the polymer chains from the regions near the particle that gives rise to the observed phenomena. Hence the strong preference in this book for this nomenclature.

16.3. General discussion of the association phenomena

The results of the experiments detailed above suggest that whether the dispersion medium is aqueous or nonaqueous, two quite different phenomena can eventuate on adding free polymer to colloidal dispersions: phase separation and particle flocculation. These two phenomena are radically different in the sense that the colloidal particles apparently remain stable if they undergo phase separation yet appear to be aggregated after undergoing flocculation. Despite these obvious differences, the two phenomena seem to follow the dictates of the same general guiding principles. Both phenomena are readily reversible. Each is more readily induced by higher molecular weight free polymer and by the presence of larger particles. Neither is especially sensitive to the particle number concentration; what dependence is observed implies that dilute dispersions are marginally more stable than the corresponding concentrated systems.

In the light of these parallels, it seems reasonable to assert that fundamentally the same phenomenon is the root cause of both the phase separation and the flocculation that follows from the addition of free polymer to the dispersion medium. Put simply, both phenomena appear to originate in the fact that the polymer molecules are forced to compete with the colloidal particles for the available space. On the one hand, any compression of the polymer chains by the impenetrable surfaces of the colloidal particles results in their losing configuration entropy. On the other, the colloidal particles, either on aggregation or phase separation, also lose configurational entropy. The delicate balance *inter alia* between these two effects presumably determines which phenomenon, if any, is observed. It is tempting to suggest that the promotion of flocculation is the first option that is open to the system to counteract the effects of free polymer. If flocculation proves impossible, e.g. if the repulsion between the particles is too great, then the alternative possibility, phase separation, which would be little different in free energy, ensues. There is, however, no firm experimental evidence to support these tentative conjectures. Indeed, there is even some preliminary evidence which can be interpreted as refuting them.

Whatever the mode of association of the particles, both phenomena appear to require the existence of a net attraction between the particles when free polymer is present. The consensus of opinion, which is certainly not unanimous, is that this attraction arises as a consequence of the depletion of the polymer chains from the region between the colloidal particles. It should be pointed out in this context that if concentrated polymer solution is added to the dispersion, complete mixing may never occur. Two phases would always be evident. Particle flocculation would not be possible because the concentration of polymer in the continuous phase would never reach a sufficiently high value to induce flocculation.

16.3.1. Phase separation

de Hek and Vrij (1981) have attempted to establish a criterion for phase separation in colloidal dispersions containing free polymer. They followed the example of Zeman and Patterson (1972) who used the spinodal condition to describe phase separation in ternary polymer-polymer-solvent systems. The spinodal represents the absolute limit to the stability of polymer solutions.

The spinodal condition for the colloidal particles is given by

$$\left(\frac{\partial \mu_1}{\partial \phi_3}\right) = 0 \tag{16.3}$$

where μ_1 = chemical potential of the dispersion medium and ϕ_3 = volume fraction of the colloidal particles. The osmotic pressure arising from the particles (π^c) is given by

$$\pi^c = -(\mu_1 - \mu_1^{\ominus})/\bar{V}_1 \tag{16.4}$$

where $\bar{V}_1$ = effective molar volume of the dispersion medium. Accordingly, the spinodal condition becomes

$$\left(\frac{\partial \mu_1}{\partial \phi_3}\right) = \left(\frac{\partial \pi^c}{\partial \phi_3}\right) = 0 \tag{16.5}$$

Physically this means that the limit to thermodynamic stability is reached when the system no longer resists long range fluctuations in concentration.

Since the osmotic pressure of the colloidal particles can be expanded in virial form

$$\pi^c = kT(\phi_3 + B_2^c \phi_3^2 + \ldots) \tag{16.6}$$

where B_2^c = second virial coefficient of the particles. Thus equation (16.5) leads to the equality

$$1 + 2B_2^c \phi_3 + \ldots = 0 \tag{16.7}$$

Provided that the colloidal dispersion is dilute ($\phi_3 \ll 1$), only the first two terms in equation (16.7) need be considered. The spinodal point is given in that instance by

$$B_2{}^c = -1/2\phi_3 \tag{16.8}$$

It was shown in Section 15.2.2.2 that the second virial coefficient of the colloidal particles was a function of the volume fraction of free polymer

$$B_2{}^c/B_2{}^{HS} = 1-(3/8)(1+\Delta/2a)^3 I(\gamma,\Delta/2a) \tag{16.9}$$

where $B_2{}^{HS}$ = hard sphere second virial coefficient, a = particle radius, Δ = diameter of the free polymer chains (assumed in their interactions with the particles to be hard spheres). The integral $I(\gamma,\Delta/2a)$ is given by

$$I(\gamma,\Delta/2a) = \int_{4a/12a+\Delta)}^{2} (\exp\{\gamma[1-(3/4)\rho$$

$$+(1/16)\rho^3]\}-1)\rho^2 d\rho \tag{16.10}$$

in which $\gamma = (1+2a/\Delta)^3 v_2$, where v_2 = volume fraction of free polymer. It is an inexorable consequence of equation (16.9) that the addition of free polymer decreases the second virial coefficient of the particles. If enough free polymer is added to the dispersion medium, the second viral coefficient of the particles can become sufficiently negative to satisfy equation (16.8). Phase separation should then take place.

Combination of equations (16.8) and (16.9) leads to an expression for the critical volume fraction of colloidal particles (ϕ_3*) at the spinodal point

$$\phi_3{}^* = -(3/32\pi a^3)[1-(3/8)(1+\Delta/2a)^3 I(\gamma,\Delta/2a)]^{-1} \tag{16.11}$$

Unfortunately, the absence of an analytic expression for $I(\gamma,\Delta/2a)$, which features the volume fraction of free polymer in the continuous phase (v_2) as one of its several variables, precludes the inversion of equation (16.10) to provide an analytic expression for v_2 as a function of ϕ_3 (which is the relevant value of ϕ_3*), Δ and a. Notwithstanding this, equation (16.10) allows ϕ_3* to be calculated, given values for v_2, a and Δ. In this way, a comparison can be made between the results of experiment and the predictions of theory, the only additional arbitrary assumption required relating to the depth of potential energy well necessary for phase separation.

The theory, which is really only qualitative in character, was found to predict correctly that v_2* should decrease with increasing molecular weight of the free polymer, increasing particle size and increasing particle concentration. Indeed, by appropriate adjustment of the arbitrary parameters, semi-quantitative agreement could be achieved between the predictions of theory and the results of experiments. Of course, any theory that treats polymer chains as hard spheres is foredoomed to quantitative inaccuracy.

16.3.2. Flocculation

16.3.2.1. General discussion

The first point that must be stressed with regard to flocculation is that it is relatively weak. Indeed, it is in many ways analogous to the weak coagulation described in Section 8.4.5. In that case, coagulation resulted from the long range London attraction operative between the core particles. In these systems, any London attraction is supplemented by the attraction arising from effects of the free polymer. Just as weak coagulation is dependent upon the particle concentration and particle size, so too weak flocculation depends *inter alia* upon the same parameters. The dependence upon the particle concentration arises from precisely the same cause: the configurational entropy of the particles ensures that close to the critical concentration of free polymer required for flocculation, dilute dispersions are somewhat more stable than more concentrated systems. The dependence upon particle size is, however, of quite a different origin from that for weak coagulation (*vide infra*).

The precise nature of the curve depicting the potential energy of interaction between two colloidal particles in the presence of free polymer should depend critically upon whether the particles are sterically or electrostatically stabilized. Qualitatively, the simplest potential energy curves would be those shown in Fig. 16.10a and b. Curve a would be relevant to certain electrostatically stabilized dispersions. The curve displays a secondary minimum whose depth is at least a few k_BT. Close in, it exhibits the usual features associated with electrostatic stabilization: a maximum and a primary minimum. Curve b in Fig. 16.10 would apply to some sterically stabilized systems (e.g. with the free polymer of much higher molecular weight than that of the stabilizing moieties). It displays neither a maximum nor a primary minimum, only a pseudo-secondary minimum. A potential energy curve of type c could in principle apply to sterically stabilized systems in which the free

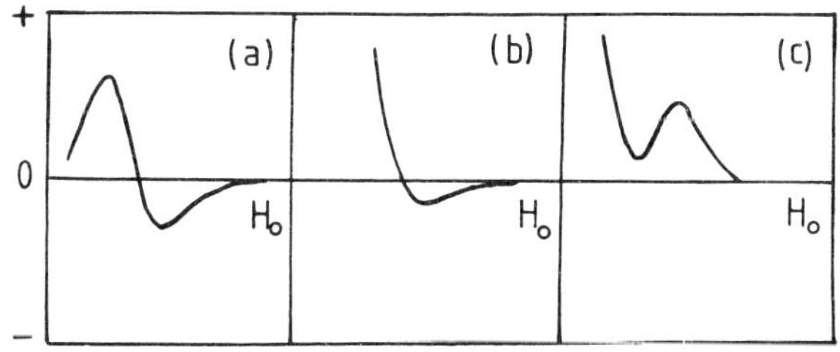

Fig. 16.10. Schematic representation of possible potential energy diagrams for depletion stabilization.

polymer is of somewhat lower molecular weight than that of the stabilizing chains. It is characterized by a pseudo-secondary minimum that does not fall below the total free energy at infinite separation. The pseudo-secondary minimum in this instance represents a metastable energy state from which escape would be kinetically relatively rapid compared with entry (i.e. the equilibrium state would favour single particles rather than the flocs). A similar type of curve could also in principle be obtained with electrostatic systems.

16.3.2.2. Comparison of experiments with the predictions of the depletion theories

Depletion flocculation theories all predict qualitatively the observed dependence of the critical concentration of free polymer required for flocculation on both the molecular weight of the free polymer and the particle radius. All of the observations published to-date point to (i) higher molecular weight free polymer being a more potent flocculant than lower molecular weight species, and (ii) larger particles being more readily flocculated than smaller ones.

The qualitative explanation for these results is illustrated schematically in Fig. 16.11. The block labelled a compares two pairs of particles, all of the same radius, at the same distance of separation in the presence of free polymer of differing molecular weight. That labelled b compares pairs of particles, each pair being of different radius, in identical polymer solutions (i.e. the same molecular weight). Varying the molecular weight changes the extent of overlap of the depletion layers adjacent to the surfaces of the particles, from which the centres of mass of the free polymer molecules are excluded. The

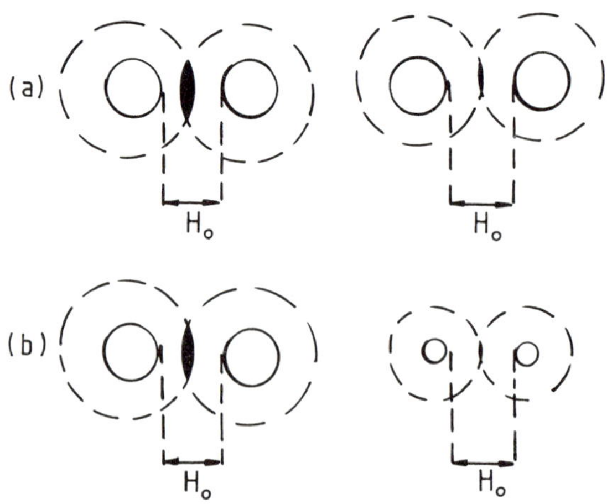

Fig. 16.11 Diagrammatic representation of the dependence of the overlap volume upon (a) the depletion layer thickness at constant particle radius and (b) particle radius at constant depletion layer thickness. Note that the separation between the particles (H_0) is held constant.

magnitude of the attraction between the particles is directly related to this volume, at least according to the approach of Asakura and Oosawa (1954; 1958). Clearly, higher molecular weight chains result in a greater overlap volume, at constant surface separation of the spheres, than do lower molecular weight species. In a like fashion, larger particles, at constant surface separation, obviously result in a larger overlap volume than do smaller particles (see Fig. 16.11b). The greater the overlap volume in each case, the greater is the attractive interaction produced. Hence large particles and high molecular weight free polymer should both promote association of the particles.

The foregoing comparisons were all made at the same minimum distance of surface-to-surface separation (H_0). This basis for comparison is quite arbitrary. A more realistic basis might perhaps be constant reduced distance H_0/Δ, where Δ = the effective diameter of the free polymer. None of the qualitative conclusions reached in the preceding paragraph would be altered by such a change, however.

In concluding this section, we note parenthetically that the particle size and molecular weight dependencies elaborated above follow immediately from the over-simplified viewpoint that the polymer chains and the particles compete for space. The higher is the polymer molecular weight and the larger are the particles, the more intense is that competition for space.

Dependence upon the molecular weight of free polymer

It is possible to provide a crude estimate of the exponent involved in the dependence of the critical volume of free polymer (v_2*) required for flocculation on the molecular weight of the free polymer. This is in the spirit of scaling law theory.

The volume of overlap (δV) of the excluded zones is given by

$$\delta V = (\pi/12)(2a + \Delta - d)^2(4a + 2\Delta + d) \tag{16.12}$$

where a = particle radius and $H_0 = d - 2a$. On substitution, this becomes

$$\delta V = (\pi/12)(\Delta - H_0)^2(6a + 2\Delta + H_0) \tag{16.13}$$

If $a \gg \Delta, H_0$, this can be approximated by

$$\delta V \sim (\pi/2)a(\Delta - H_0)^2 \tag{16.14}$$

The term in $(\Delta - H_0)$ introduces the molecular weight dependence of the attraction. If, for example, the extreme case is taken that $H_0 = \Delta/2$, then δV is proportional to Δ^2. An identical result follows if $H_0 \ll \Delta$. Accordingly, it appears that the overlap volume is proportional to $\langle r^2 \rangle$, since the effective diameter of a linear polymer is proportional to its rms end-to-end distance. The latter is proportional to $M^{0.5+\varepsilon}$, where M = molecular weight and ε, which allows for the polymer intramolecular excluded volume, lies in the range

$0 \leq \varepsilon \leq 0{\cdot}1$. Note that $\varepsilon = 0$ in a θ-solvent. Therefore the overlap volume can be taken proportional to $M^{1{\cdot}0+2\varepsilon}$, as can the free energy of interaction.

An identical result was obtained by Feigin and Napper (1980b) using an alternative line of reasoning. They considered first a pair of parallel flat plates at the same reduced distance of separation ($\bar{h} = h/\langle r^2 \rangle^{\frac{1}{2}}$, where h = plate separational distance). Geometrical considerations lead to an overlap volume that is proportional to $\langle r^2 \rangle^{\frac{1}{2}}$. The Derjaguin method for converting flat plate potentials into sphere potentials introduces an additional distance dependence, again leading to a multiplying term also proportional to $\langle r^2 \rangle^{\frac{1}{2}}$. The overall dependence of the potential energy of interaction thus varies as $\langle r^2 \rangle$. This corroborates the dependence of the potential upon molecular weight derived above.

Suppose a fixed, minimum value of the depth of the potential energy well (ΔG^*) is required for flocculation. This will be proportional to both the overlap volume and the osmotic pressure of the polymer solution. For very high molecular weight free polymer, the latter is determined by the second virial coefficient term and so is proportional to $v_2{}^2 M^{-0{\cdot}2}$, where v_2 = volume fraction of the polymer. Here we have taken the second virial coefficient as being proportional to $M^{-0{\cdot}2}$ (Morawetz, 1965). Hence we can write for critical behaviour.

$$\Delta G^* \propto (v_2{}^*)^2 M^{0{\cdot}8+2\varepsilon} \qquad (16.15)$$

For the assumed constant value of ΔG^*, clearly

$$v_2{}^* \propto M^{-(0{\cdot}4+\varepsilon)} \qquad (16.16)$$

where $0 \leq \varepsilon \leq 0{\cdot}1$.

Equation (16.16) suggests that $v_2{}^*$ should depend approximately upon the molecular weight of the free polymer with an inverse power of ca 0·4–0·5. This is significantly less than the exponent of $-0{\cdot}7$ measured experimentally (Sperry *et al.*, 1981) for the flocculation of aqueous latices, sterically stabilized by poly(oxyethylene), using hydroxyethyl cellulose. The latter, however, are extremely stiff chains.

Shown in Table 16.2 are some of the experimental results reported by Li-in-on *et al.* (1975) and Cowell *et al.* (1978) for the flocculation of polystyrene latex particles sterically stabilized by poly(oxyethylene) of two different molecular weights (750 and 2000). It is apparent that with high molecular weight free poly(oxyethylene) as the stimulus to flocculation, $v_2{}^*$ was approximately proportional to $M^{-\frac{1}{2}}$. This is in reasonable agreement with the theoretical predictions set forth above. A similar dependence of $v_2{}^*$ on $M^{-\frac{1}{2}}$ was observed by Clarke and Vincent (1981b) for polystyrene microgel particles flocculated by free polystyrene in ethyl benzene (Table 16.3). In both sets of experiments, however, the molecular weight of the free polymer was scarcely high enough to justify the exclusive use of just the second virial

TABLE 16.2
Molecular weight dependence of v_2^* for flocculation by polyoxyethylene

Free polymer molecular weight M	Latex stabilizer molecular weight 750 v_2^*	$v_2^* M^{\frac{1}{2}}$	2 000 v_2^*	$v_2^* M^{\frac{1}{2}}$
800	0·47	13	—	—
1 000	0·38	12	0·57	18
1 500	0·28	11	0·42	16
4 000	0·22	13	0·32	20
6 000	0·16	12	0·27	21
10 000	0·055	6	0·18	18
20 000	0·03	4	0·11	16
300 000	0·01	5	0·03	16

TABLE 16.3
The molecular weight dependence of v_2^* for the flocculation of microgel particles by polystyrene in ethyl benzene

Polystyrene molecular weight M	Particle diameter (nm) 490 v_2^*	$v_2^* M^{\frac{1}{2}}$	750 v_2^*	$v_2^* M^{\frac{1}{2}}$
1 800	0·22	9	0·14	6
6 400	0·17	14	0·11	9
11 300	0·11	12	0·07	7
18 000	0·10	13	0·03	4
51 000	0·04	9	0·02	5
2 000 000	0·005	7	0·004	6

coefficient term. An analysis, which is analogous to that outlined above, that merely incorporates the first virial coefficient term for the osmotic pressure yields that v_2^* should be proportional to $M^{-(1\cdot 0+2\varepsilon)}$. As $\varepsilon \sim 0$ for low molecular weight chains (i.e. the chains are too short for the occurrence of significant long-range repulsion between the segments, even in a good solvent), v_2^* might be expected to be proportional to $M^{-1\cdot 0}$. Clearly, any intrusion of the ideal contribution to the osmotic pressure must increase the absolute magnitude of the negative exponent.

It should further be noted that any agreement between theory and experiment presented in this section may be somewhat fortuitous in that the dispersions studied experimentally were sterically stabilized. Steric layers, unless very thin, can influence profoundly the quantitative effects of free polymer. This is vividly illustrated by the exponents reported by de Hek and

Vrij (1979; 1981) for nonaqueous silica dispersions: these were $-0{\cdot}25$ for particles stabilized by stearyl alcohol but ca $-0{\cdot}5$ for particles stabilized by polyisobutylene. The latter is in fair agreement with the theoretically predicted exponent but the former is not.

Finally, we note that Sperry (1982) has recently used the theory of Asakura and Oosawa (1954; 1958) to explain the results obtained by Sperry *et al.* (1981) for the flocculation of polystyrene latex particles by hydroxyethyl cellulose. Although semi-quantitative agreement was claimed between theory and experiment, it should be noted that the theory contains too many adjustable parameters to allow a satisfactory comparison to be made with experiment.

Dependence upon the particle size

Equation (16.14) of the preceding section shows that for large particles, the attractive potential energy generated by free polymer is proportional to the particle radius a. Therefore, at constant molecular weight of free polymer, we have by analogy with equation (16.15)

$$\Delta G^* \propto (v_2^*)^2 a \tag{16.17}$$

For constant ΔG^*, this yields

$$v_2^* \propto a^{-\frac{1}{2}} \tag{16.18}$$

This predicts that v_2^* should be inversely related to the square root of the particle radius. This relationship holds with the caveat that the particle size must be large in comparison with the size of the polymer chains. We further note that the presence of steric layers may dramatically influence any predicted dependence upon particle size.

There exists only fragmentary data in the literature for the experimental dependence of v_2^* on the particle radius. For example, Table 16.4 shows the data of Clarke and Vincent (1981a) for the flocculation of silica particles, sterically stabilized by polystyrene in ethyl benzene, by free polystyrene of molecular weight ca 7 000. An approximate dependence of v_2^* on the inverse square root of the particle size is indicated.

This dependence is not borne out, however, by the results published by de Hek and Vrij (1981), also for silica particles in the presence of polystyrene (see

TABLE 16.4

Dependence of v_2^* for flocculation on the particle radius for sterically stabilized silica particles

Particle radius a (nm)	v_2^*	$v_2^* a^{\frac{1}{2}}/\text{nm}^{\frac{1}{2}}$
47	0·15	10
144	0·064	8

Table 16.5). In this instance, however, the particles were dispersed in cyclohexane rather than ethyl benzene. Moreover, the dispersions were said to undergo phase separation rather than particle flocculation. The data reported by de Hek and Vrij for two particle sizes suggest that v_2^* is inversely related to the particle radius, rather than the square root of the particle radius. The radius dependence observed for this system is that which would be predicted if the ideal van't Hoff term was predominantly responsible for the osmotic pressure of the polymer solution, since ΔG^* would then be proportional to $v_2^* a$, i.e. $v_2^* \propto 1/a$. This result is scarcely surprising since cyclohexane is a poor solvent for polystyrene ($\theta = 34$ °C). In those circumstances, only the ideal contribution to the osmotic pressure need be considered. The different radius dependences observed in such experiments is accordingly explained.

TABLE 16.5
Dependence of v_2^* for phase separation on the particle radius for silica particles

Molecular weight	Particle radius a (nm) 42 v_2^*	42 $v_2^* a$	92 v_2^*	92 $v_2^* a$
8 000	0·082	3·4	0·042	3·9
36 000	0·048	2·0	0·025	2·3
156 000	0·031	1·3	0·017	1·6
1 900 000	0·017	0·7	0·012	1·1

Chapter 17

The Effects of Free Polymer on Colloid Stability: Theories of Depletion Stabilization and Depletion Flocculation

17.1. Introduction

It was pointed out in the penultimate chapter that the flocculating effects of free polymer have been known for almost fifty years. The correct mechanistic explanation for this phenomenon was first proposed by Asakura and Oosawa (1954) some thirty years ago. This occurred almost contemporaneously with the publication by Heller and Pugh (1954) of their pioneering studies on steric stabilization. Yet, whereas the phenomenology and theory of steric stabilization have been the subject of intense investigations since that time, the role of free polymer in influencing colloid stability has not been actively pursued until recently. Indeed, the first experimental demonstration that free polymer can prevent the flocculation of colloidal particles was reported by Vincent and coworkers less than a decade ago (Li-in-on *et al.*, 1975). An explanation for this type of stabilization in terms of depletion effects was propounded by Feigin and Napper even more recently (Feigin and Napper, 1980a,b).

17.2. Experimental evidence for depletion stabilization

17.2.1. Aqueous systems

Li-in-on *et al.* (1975) were the first to publish compelling experimental evidence which showed that relatively high concentrations of free polymer could prevent particles from undergoing flocculation. These observations have subsequently been corroborated by Cowell *et al.* (1978) and by Dodd (1980). The expression "relatively high concentrations of free polymer" in this context refers to concentrations greater than those needed to cause depletion flocculation. The absolute values of the free polymer concentration need not, however, be very large: e.g. with polymers of molecular weight of over, say, 100 000, volume fractions of free polymer of only a few per cent may be sufficient to prevent the onset of flocculation. A diagram illustrating the dependence of the reciprocal stability ratio ($1/W$) on the volume fraction of free polymer, as reported by Li-in-on *et al.* (1975) and Cowell *et al.* (1978), was presented in Fig. 16.4 and will not be discussed further.

17.2.1.1. Dependence upon molecular weight and particle radius

The critical volume fraction of free polymer (v_2^{**}) required to prevent flocculation at high polymer concentrations was found to depend *inter alia* upon the molecular weight of the free polymer and the particle radius. The crude arguments developed in Section 16.3.2.2 for the dependence of the depth of the minimum potential energy curve on these two parameters also hold without change for the height of the maximum required for stability. It follows that v_2^{**} might be expected to decrease with molecular weight with an exponent of order $-0{\cdot}4$ to $-0{\cdot}5$ and vary with the particle radius with an exponent of order $-0{\cdot}5$.

Table 17.1 shows a test of this predicted molecular weight dependence for polystyrene latex particles, sterically stabilized by thin layers of poly(oxyethylene) of molecular weight 750, using poly(oxyethylene) of different molecular weight as the free polymer. These experimental results were reported by Cowell *et al.* (1978). It is apparent that v_2^{**} decreased with molecular weight with an exponent of ca $-0{\cdot}5$, as expected from the crude theory.

TABLE 17.1
The molecular weight dependence of the critical polymer concentration for depletion stabilization

Free polymer molecular weight	v_2^{**}	$v_2^{**}M^{\frac{1}{2}}$
4 000	0·55	35
6 000	0·30	23
10 000	0·21	21
20 000	0·14	20
300 000	0·04	22

17.2.2. Nonaqueous dispersions

Little work has been published on depletion stabilization in nonaqueous dispersion media. Clarke and Vincent (1981a) have noted that it is possible to prevent silica particles stabilized by polystyrene in ethylbenzene from undergoing depletion flocculation by adding a high concentration of polystyrene ($v_2{}^b = 0{\cdot}015$ for a free polystyrene molecular weight of ca 2×10^6). This was the first reported observation of depletion stabilization in nonaqueous dispersion media.

17.3. Qualitative features of depletion stabilization

17.3.1. Résumé of depletion flocculation

The origins of depletion flocculation, which were first enunciated by Asakura and Oosawa (1954; 1958) were outlined in Section 15.2.1.1. Briefly, it was pointed out that when the surfaces of colloidal particles dispersed in a polymer solution approach to distances closer than the "diameter" of the polymer molecules, the chains are excluded from the space between the particles. The reason for this exclusion is quite straightforward: for the chains to enter the interparticle space, their conformations would need to be drastically distorted. Since the polymer chains minimize the free energy of the bulk solution by adopting the lowest free energy conformation, any distortion of this conformation must inevitably increase the total free energy of the system. Such distortions are thermodynamically disfavoured. Between the particles, therefore, there exists a microreservoir of almost pure solvent. Even closer approach of the particles pushes some of this pure solvent into the bulk polymer solution. This is thermodynamically favoured in a good solvent and so the total free energy of the system is lowered by closer approach. Depletion flocculation can ensue if the depth of the minimum in the potential energy *vs* distance curve is sufficiently great (i.e. at least several k_BT).

17.3.2. Origins of depletion stabilization

A subtle, logical difficulty presents itself on closer scrutiny of the preceding argument. This analysis takes as its starting point particles that are in such close proximity that microreservoirs of solvent exist between them. The plates can crudely be considered as acting as semi-permeable membranes so that they are forced together by the osmotic pressure of the bulk polymer solution, resulting in depletion flocculation. No consideration is given in the theory of Asakura and Oosawa to the free energy changes involved in bringing the particles from infinite separation to the separation where microreservoirs are formed. This is not surprising because once the space between the plates becomes accessible to the polymer chains, the plates can no longer be considered to act as a semi-permeable membranes. The simple osmotic pressure approach is thus no longer applicable. Yet to calculate the profile of the free energy of close approach of two colloidal particles as a function of their distance of separation, it is mandatory to consider the energy changes involved in their approach from infinite separation.

Recognition of this deficiency in the approach of Asakura and Oosawa provides a startling corollary: the formation of these microreservoirs of pure solvent must increase the total free energy of the system. Thus free polymer can generate not only attraction but also repulsion between colloidal particles. If sufficiently strong, such repulsion may even prevent the flocculation that might normally be expected with this concentration of free polymer. This is the origin of depletion stabilization, as first pointed out by Feigin and Napper (1980a,b).

The process of bringing two parallel flat plates immersed in polymer solution from infinite separation to a distance of close approach somewhat less than the rms end-to-end distance of the polymer chains is shown schematically in Fig. 17.1. The polymer molecules are assumed to be monodisperse in molecular weight. At infinite separation, all of the space, and thus all of the solvent, between the two plates is accessible to the macromolecules. The only caveat on this statement relates to the domains adjacent to the flat plates: the polymer segments have only partial accessibility to these regions because of the presence of the impenetrable surfaces. Nevertheless, accessibility to these domains is considerably greater than that when the plates are close together. In the latter case, there is almost total exclusion of the chains from between the particles. In a manner of speaking, it is as if the polymer chains can cope with the presence of one impenetrable interface by making the necessary conformational adjustments but are overwhelmed by the presence of two impenetrable surfaces and simply retreat from the interparticle domain.

Whenever a good solvent is rendered inaccessible to free polymer, the total free energy of the system must be raised. This concept can most readily be grasped by considering the reverse process: if pure solvent is layered on to the surface of a polymer solution, mixing occurs spontaneously because the total free energy is lowered. The reverse demixing process is not spontaneous and is accompanied by an increase in the total free energy of the system. Yet the formation of microreservoirs must involve the demixing of polymer segments from the solvent molecules that ultimately appear as microreservoirs of pure solvent between the particles. This demixing must increase the free energy of the system and work must be expended if it is to be accomplished. This work requirement is a manifestation of the repulsion that is generated on bringing the particles from infinite separation to a distance of separation where almost pure solvent exists between the particles.

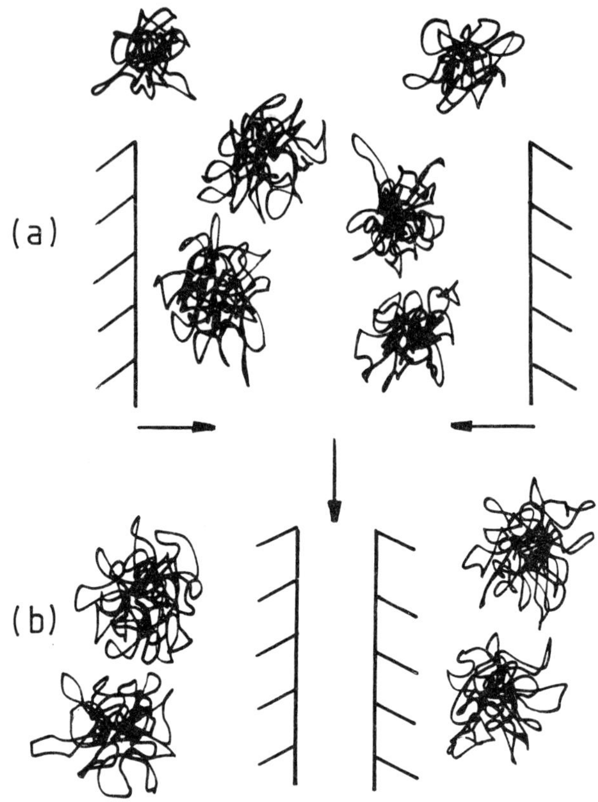

Fig. 17.1. Diagrammatic representation of the close approach of two flat plates in a polymer solution; (a) corresponds to a distance of separation greater than the size of the polymer chains whereas (b) corresponds to a plate separation less than the polymer chain size (Feigin and Napper, 1980a).

17.3.3. The potential energy diagram for depletion stabilization alone

The preceding discussion suggests that for two naked uncharged colloidal particles immersed in a polymer solution, their potential energy diagram should display the qualitative features shown in Fig. 17.2. The most striking feature of this curve is its general shape: it is radically different from that expected for steric stabilization by attached polymers. Paradoxically, the curve resembles that for two electrostatically stabilized particles. It displays a maximum far out and a primary minimum closer in. It is stressed that this diagram is for depletion effects alone and that the presence of steric and/or electrostatic barriers may complicate the diagram enormously.

The existence of a maximum in the potential energy curve is crucial for the

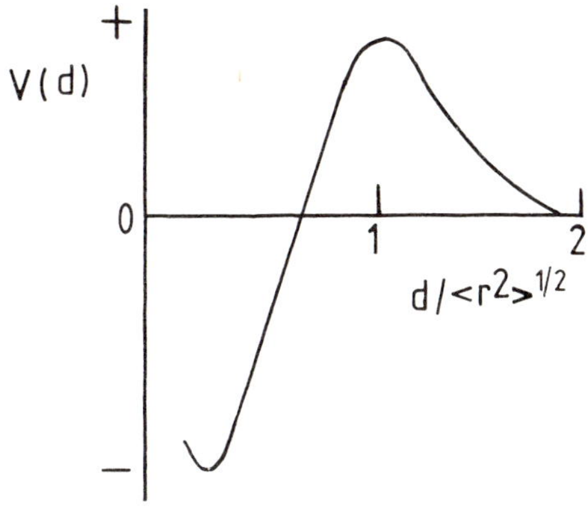

Fig. 17.2. Schematic representation of the potential energy diagram for two colloidal particles immersed in a polymer solution (after Feigin and Napper, 1980a).

occurrence of depletion stabilization. Already it is possible to see why depletion flocculation invariably precedes depletion stabilization as the concentration of free polymer in the continuous phase is progressively raised. In order for weak flocculation to occur, a well depth of only a few $k_B T$ is required; but to impart stability, the maximum in the potential energy curve must be an order of magnitude larger, say 20–25 $k_B T$. Clearly, much higher concentrations of polymer are needed to generate the latter than the former. Depletion flocculation should therefore be evident at significantly lower concentrations of free polymer than depletion stabilization.

If the stability of particles at relatively high concentrations of free polymer arises from the presence of a repulsive maximum in the potential energy curve, then it follows that any stability that is imparted by free polymer must be of a kinetic type. Depletion stabilization thus corresponds to thermodynamic *metastability*. In this regard, it obviously resembles electrostatic stabilization rather than steric stabilization, which usually corresponds to thermodynamic stability.

17.3.4. An alternative approach to depletion stabilization

As mentioned previously, the presence of the maximum in the potential energy curve is essential for the occurrence of depletion stabilization. The origin of this maximum can be envisaged in an alternative, but entirely equivalent, fashion from that set forth in the penultimate section.

Consider two particles, assumed for the purposes at hand to be flat plates, that have approached to the point where the polymer chains have just been excluded from the interparticle region (see Fig. 17.3). A microreservoir of inaccessible solvent thus exists between the plates.

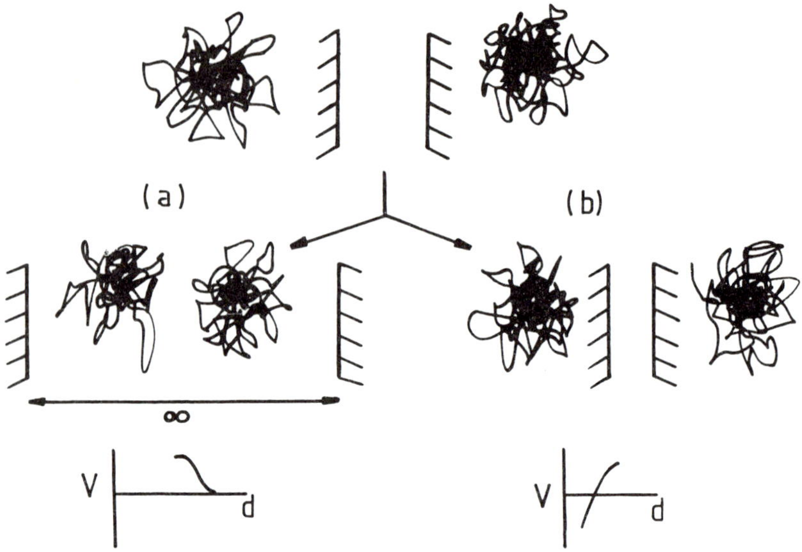

Fig. 17.3. Diagrammatic representation of the origin of the maximum in the potential energy diagram for colloidal particles immersed in a polymer solution.

Now the separation between the plates is changed. This may be accomplished in two ways; the plates can either be moved further apart [option (a) in Fig. 17.3] or closer together [option (b)]. As the plates are moved further apart, the solvent that was previously inaccessible to the polymer molecules becomes, to a greater or less degree, accessible. In the limit of infinite separation, it becomes completely accessible, subject only to the restrictions imposed by the impenetrable nature of the particle surface. If the plates are moved closer together, previously inaccessible solvent mixes with the bulk solution, again lowering the free energy of the system. In either case, movement of the plates, whether inwards or outwards, causes inaccessible solvent to become more accessible. As this necessarily lowers the free energy of the system, it is clear that a maximum must exist in the potential energy *vs* distance curve.

17.4. Quantitative theories of depletion stabilization and depletion flocculation

It was mentioned in Section 15.2.2.2 that both Feigin and Napper (1980a,b) and Scheutjens and Fleer (1982) had developed portmanteau theories that comprehend both depletion stabilization and depletion flocculation. These theories will now be reviewed and their predictions compared with experiment.

17.4.1. The theory of Feigin and Napper

The first attempt to develop a theory of depletion stabilization and depletion flocculation was published by Feigin and Napper (1980b). These authors coupled the rotational isomeric state (RIS) scheme with a Monte Carlo (MC) procedure to predict the nature of the depletion layer at single flat interfaces and between two parallel flat plates. The free energy required to form these layers was calculated using the Flory-Huggins theory. Flat plate potentials were then converted into sphere potentials by means of the Derjaguin integration procedure (Derjaguin, 1934).

17.4.1.1. The segmental density depletion zones

The model adopted

Feigin and Napper (1980b) considered two parallel flat plates that were uncoated by polymer and uncharged. These plates were assumed to be immersed in a polymer solution and to be inert to the free polymer. Feigin and Napper took the free polymer to be poly(oxyethylene) so that the results of their calculations could be compared with the experimental data of Vincent and coworkers (Li-in-on *et al.*, 1975; Cowell *et al.*, 1978). The method for calculating the conformation of polymers in θ-solvents was detailed in Section 4.3.1. The RIS-MC scheme allows the positions of all segments in a particular simulated molecule to be specified. The solution adjacent to the wall was divided into a series of layers, all of equal thickness and all parallel to the surface (see Fig. 17.4). The layer thickness typically was chosen to be, say, one-twentieth of the rms end-to-end distance of the free polymer. Halving or doubling the thickness of the layer did not significantly alter the results obtained. Each simulated polymer chain was also divided into layers of the same thickness. The number of segments in each layer of each simulated chain was found using the RIS-MC scheme.

It would be expected for polymers in free solution or in the melt, that, in the absence of an interface, there would be on average an equal number of centres of mass of the polymer molecules in each layer. This is in conformity with the principle of entropy maximization. The presence of the interface was introduced by excluding those conformations that cross the interface. The total segment density in each layer was obtained by summing the segment density contributions from all allowed conformations. This was normalized by recalling that far from the interface, the calculated segment density is simply the bulk volume fraction of polymer ($v_2{}^b$).

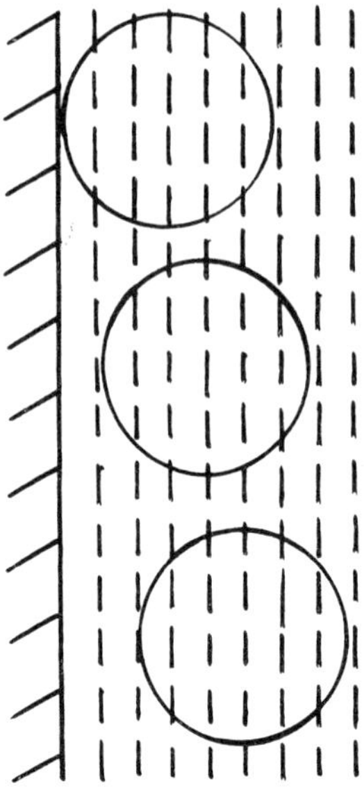

Fig. 17.4. Diagrammatic representation of the segmental layers for a polymer at an impenetrable interface.

A single flat plate

Plots of the normalized segment volume fraction (v_2/v_2^b) as a function of the reduced distance $d/\langle r^2\rangle^{\frac{1}{2}}$ (where d=distance normal to the surface and $\langle r^2\rangle^{\frac{1}{2}}$=free solution rms end-to-end distance of the chains) are presented in Fig. 17.5. Results for both a short (27 bonds) and a longer (272 bonds) poly(oxyethylene) chain are given, together with data for a random flight chain of 4 000 bonds.

It is apparent that the segment density of the free polymer is small near to the interface but that it rises smoothly to its bulk value at a distance from the interface of order equal to the rms end-to-end distance of the chain, which is, roughly speaking, the "diameter" of the chain.

The depletion of polymer segments from the layers adjacent to the wall can be attributed to at least three causes: first, fewer molecules contribute to the polymer concentration near the wall; second, those molecules that do contribute do so from their low segment density peripheral regions; third, layers near the wall, which are populated primarily by the peripheral regions of the chains, are occupied on average by a smaller volume of the chains than are layers more distance from the wall.

The reason why the segment density realizes its bulk value at a distance $d \sim \langle r^2\rangle^{\frac{1}{2}}$ is readily grasped from Fig. 17.6. Crudely speaking, the segment density in any layer is made up of contributions from chains whose centres of mass lie both to the right and to the left of the layer. For layers closer in to the

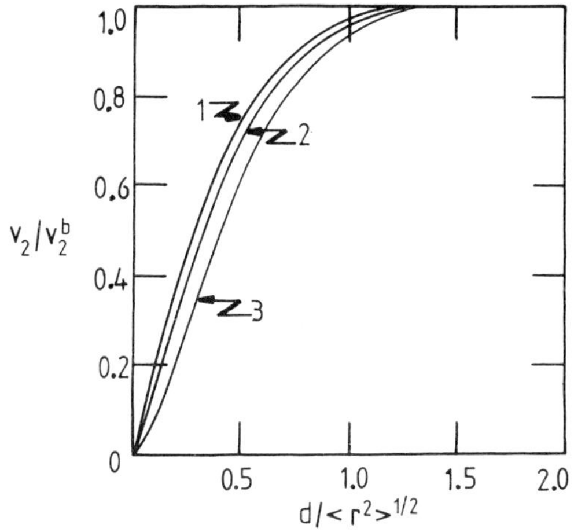

Fig. 17.5. The distance dependence of the relative segment density for tails: (1) poly(oxyethylene) of molecular weight 400; poly(oxyethylene) of molecular weight 4 000; 3, random flight chain of 4 000 bonds (after Feigin and Napper, 1980b).

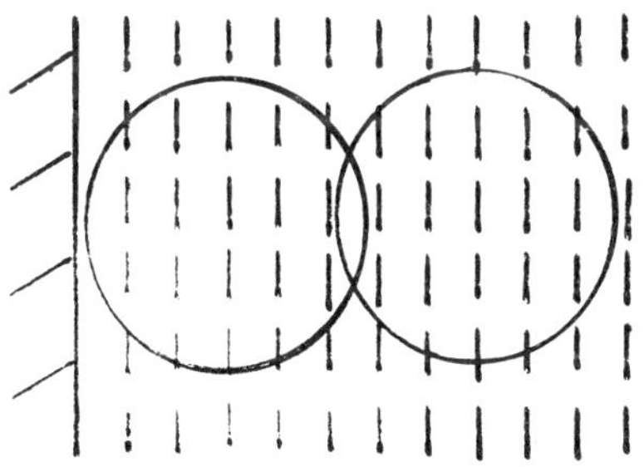

Fig. 17.6. The segmental layers of free polymers at an impenetrable interface.

surface than $d \sim \langle r^2 \rangle^{\frac{1}{2}}$, the contributions to their overall segment density are reduced because of the elimination in the presence of the wall of conformations whose centres of mass lie to the left of the layer. Only when $d \gtrsim \langle r^2 \rangle^{\frac{1}{2}}$ do the full complement of chains, both to the right and to the left of the layer, contribute to the segmental concentration in the layer.

Also apparent from Fig. 17.5 is the close agreement between the 272 bond poly(oxyethylene) and the 4 000 bond random flight chains when plotted on the reduced distance scale. As the characteristic ratios C_n of these two chains are quite different (4 *vs* 1, respectively), the results suggest that the molecular structure of the chain does not drastically influence the segment density *vs* distance curves plotted in this reduced fashion. This conclusion requires further confirmation.

Two flat plates

Maximum entropy approach for θ-solvents. The volume fraction of polymer in the region between two parallel plates separated by a distance h was determined in a like fashion by introducing a second surface, parallel to the first, and rejecting all conformations that cross it. The relative volume fractions of polymer so calculated are shown in Fig. 17.7 for poly(oxyethylene) of molecular weight 4 000 (272 bonds).

It is apparent that if the distance between the plates exceeds twice the rms end-to-end distance of the polymers ($h > 2\langle r^2\rangle^{\frac{1}{2}}$) then the maximum value for the segmental volume fraction is not dramatically reduced below that in the bulk solution. The distance between the plates is clearly sufficient to permit the polymer segment density of each of the depletion layers to rise to near its bulk value, the rise length being ca $\langle r^2\rangle^{\frac{1}{2}}$.

Approach of the plates to closer than twice the rms end-to-end distance of the polymers ($h < 2\langle r^2\rangle^{\frac{1}{2}}$) results in a significant reduction in the polymer segment density compared with that in the bulk phase. At a plate separation, for example, of one rms end-to-end distance of the polymer, the maximum in the volume fraction of the polymer is less than half the bulk value. Even closer approach ($h < 0{\cdot}6\langle r^2\rangle^{\frac{1}{2}}$) results in the polymer segment density between the plates being reduced effectively to zero.

Similar results were obtained by Feigin and Napper for random flight chains. For chains over a thousand bonds, the distance dependence of the

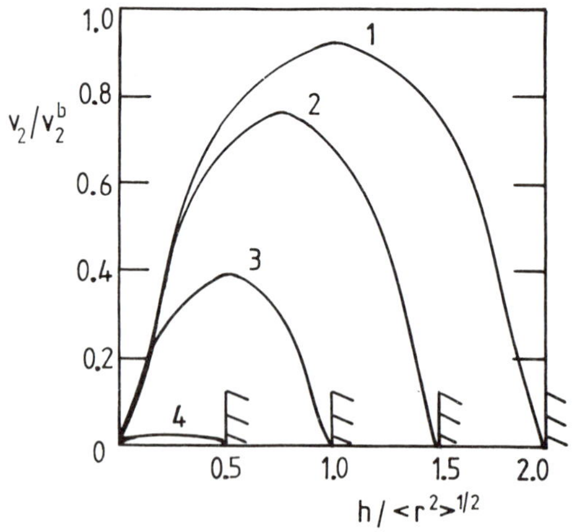

Fig. 17.7. The distance dependence of the relative segment density for poly(oxyethylene) between two flat plates: the reduced distance of separation: 1, 2·0; 2, 1·5; 3, 1·0; 4, 0·5 (after Feigin and Napper, 1980b).

segment density between two flat plates becomes almost a "universal" curve if plotted on a reduced scale $h/\langle r^2 \rangle^{\frac{1}{2}}$.

Free energy minimization procedure. The distribution of free polymer near the interface was calculated on the assumption that the polymer chains had an equal probability of occurrence wherever they would fit. This corresponds to maximization of the configurational entropy of the centres of mass of the polymer chains, which is the appropriate criterion for polymer chains in a θ-solvent. The correct criterion in general is, of course, the minimization of the free energy of the system. The presence of the depletion layers near an interface, or between two plates, implies that in a good solvent, polymer chains will tend to move into the depletion zone in an endeavour to counteract the uneven distribution of polymer segments set up in the system. This partially nullifies the depletion effects calculated according to the entropy maximization principle. It does, however, introduce a dependence of the segment density distribution function upon the bulk concentration of free polymer, which seems physically plausible.

The rigorous calculation of the segmental distribution near an interface in a good solvent seems out of the question at present if the criterion chosen is free energy minimization. Feigin and Napper adopted a simplified procedure to provide an approximate solution to the problem. The method involved was essentially a variational procedure whereby the concentration of free polymer in the interplate zone that minimized the free energy of the system was determined by trial and error. The free energy that was minimized included not only the free energy of mixing of the segments with the solvent but also the loss of configurational entropy for the chains placed between the plates. The results obtained in this way are displayed in Fig. 17.8.

Comparison of the curves in Fig. 17.8 shows, as expected, that the entropy maximization procedure over estimates the magnitude of the depletion of chains between two plates compared with the predictions of the free energy minimization procedure. Furthermore, the extent of the depletion is a sensitive function of the volume fraction of free polymer in the bulk solution. There is an alternative, but equivalent, way of envisaging what is happening in these systems: the polymer molecules between the plates undergo conformational distortions in order to render uniform, as far as possible, the segmental concentration over the entire system. Note that the two maxima and the intervening minimum calculated at high $v_2{}^b$ for the volume fraction of polymer between the plates are likely to be artefacts of the calculation procedure.

The three domains of close approach. Three domains of close approach of two flat plates immersed in polymer solution can be usefully distinguished. These are represented schematically in Fig. 17.9, assuming θ-solvency conditions. The three domains will be discussed in turn.

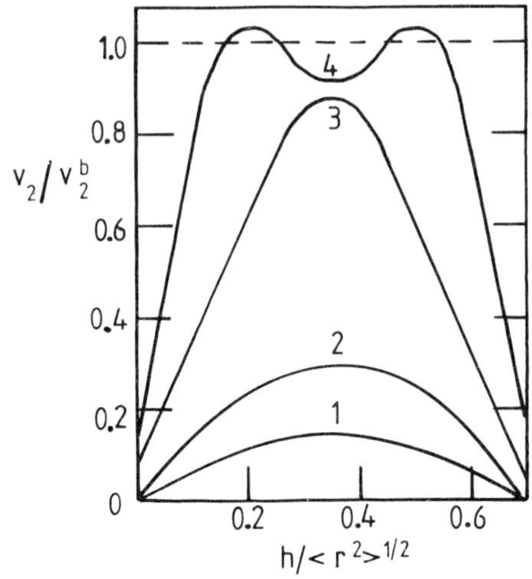

Fig. 17.8. The distance dependence of the relative segment density for poly(oxyethylene) between two flat plates at a reduced distance of separation of 0·7 using the free energy minimization procedure: the volume fraction of free poly(oxyethylene) was curve 2, 0·1; 3, 0·4; 4, 0·9. Curve 1 is for no free energy minimization (after Feigin and Napper, 1980b).

A. $h > 2\langle r^2\rangle^{\frac{1}{2}}$. In this domain, the segment density in the depletion layer associated with each plate has sufficient distance to rise to the bulk value. Closer approach of the plates simply squeezes out solution at the bulk concentration into the bulk solution. Clearly, no free energy change is involved in this process and the free energy of the system is constant in this regime.

B. $\langle r^2\rangle^{\frac{1}{2}} \leq h \leq 2\langle r^2\rangle^{\frac{1}{2}}$. The segmental volume fraction in each of the depletion layers in this regime has insufficient distance to rise to the bulk value because of the presence of the opposite plate (see Fig. 17.9b). The segmental concentration between the plates is thus everywhere less than that in the bulk solution.

Closer approach of the plates now pushes out polymer segments from the lower concentration between the plates to a higher concentration in the bulk solution. Work must be done to accomplish this since it corresponds to forcing the segments to be less evenly distributed over the system. This corresponds to a repulsion between the plates in this domain.

An alternative way of viewing this domain, which reaches the same conclusion, was given by Feigin and Napper (1980a). They noted that the distribution of the centres of mass of the macromolecules was simply narrowed by close approach of the plates, the height of the distribution remaining constant. This can only be achieved if the polymer excluded from between the plates has approximately the bulk concentration. This demands

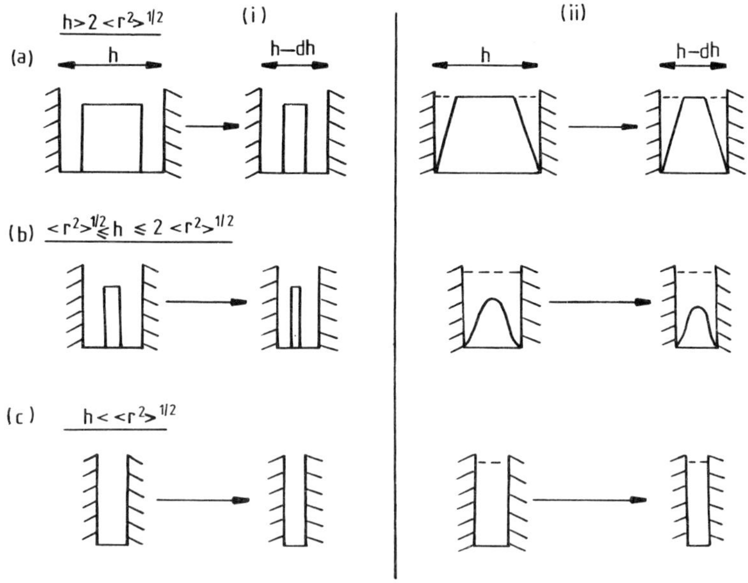

Fig. 17.9. The three domains of close approach for flat plates immersed in a polymer solution: (i) the distribution of the centres of mass of the free polymer molecules; (ii) the average segmental concentration (after Feigin and Napper, 1980a).

that on closer approach the polymer solution between the plates be split into a more dilute and a more concentrated solution. Such a process is not spontaneous. It corresponds to an increase in the free energy of the system, implying repulsion between the plates.

C. $h < \langle r^2 \rangle^{\frac{1}{2}}$. In this domain, it will be assumed that there is essentially no free polymer between the plates merely reservoirs of solvent (see Fig. 17.9c). Closer approach mixes pure solvent with bulk polymer solution which decreases the free energy of the system. This is equivalent to an attraction between the particles in this domain. Another way of viewing this exclusion of the solvent from between the plates is that it results in a more uniform distribution of the segments over the entire system. It is therefore thermodynamically favoured.

These three domains of close approach corroborate the qualitative picture presented in Section 17.3.3 of the shape of the potential energy curve for the interaction between colloidal particles. It will exhibit a repulsive interaction in the domain $\langle r^2 \rangle^{\frac{1}{2}} \leq h \leq 2\langle r^2 \rangle^{\frac{1}{2}}$, followed by attraction in the domain $h < \langle r^2 \rangle^{\frac{1}{2}}$.

Quantitative prediction of the depletion free energy of interaction. Feigin and Napper (1980b) were unable to predict analytically the free energy of interaction between colloidal particles immersed in a polymer solution. They were forced to resort to numerical computation of the potential energy curves.

Once the volume fraction of polymer in each layer was determined by one or other of the preceding methods, the free energy of mixing in each layer could be calculated via the Flory-Huggins expression

$$(\Delta G_M) = (kT/V_1)\delta V[v_1 \ln v_1 + (v_2/x)\ln v_2 + \chi^* v_2 v_1] \qquad (17.1)$$

where δV = volume of the layer, v_2 = volume fraction of segments, $v_1 = 1 - v_2$, χ^* = interaction parameter (which may be a function of v_2), x = ratio of the molar volume of the polymer to that of the solvent and V_1 = volume of a solvent molecule. This was summed for every layer associated with each plate, both for infinite separation and for closer approach, and their difference taken.

The other contribution to the free energy of close approach that must be incorporated into the theory derives from the change in free energy of the bulk solution on addition of the polymer solution that is excluded from between the plates on close approach. Assuming that there is a large reservoir of bulk solution, the free energy of a small volume $(\delta V)^P$ of polymer to the bulk may be approximated by

$$\delta\,(\Delta G_M)^P \sim \frac{\partial(\Delta G_M)(\delta V)^P}{\partial V^P}$$

$$\sim \frac{\partial(\Delta G_M)}{\partial n_2}\,\frac{\mathrm{d}n_2\,(\delta V)^P}{\mathrm{d}V^P} \qquad (17.2)$$

where the superscript P refers to polymer.

Evaluation of the requisite differentials from the appropriate Flory-Huggins expressions yields

$$\delta(\Delta G_M)^P \sim (kT/V_1)\delta V^P\left\{-v_1 + v_1{}^2\left(\chi^* + v_2\frac{\mathrm{d}\chi^*}{\mathrm{d}v_2}\right)\right.$$

$$\left. + (\ln v_2 + v_1)/x\right\} \qquad (17.3)$$

The analogous expression for solvent, denoted by superscript SO, is given by

$$\delta(\Delta G_M)^{SO} = (kT/V_1)(\delta V)^{SO}\left\{\ln v_1 + v_2 + v_2{}^2\left(\chi^* - v_1\frac{\mathrm{d}\chi^*}{\mathrm{d}v_2}\right) - v_2/x\right\} \qquad (17.4)$$

The total free energy that accompanies the expulsion of a volume element δV of polymer solution of volume fraction Q into the bulk phase is obtained by summing $\delta(\Delta G_M)^P + \delta(\Delta G_M)^{SO}$

$$\delta(\Delta G_M)^{P+SO} = (kT\delta V/V_1)\left\{\ln\ v_1 + v_2 - Q(1+\ln\ v_2) + (Q + Q\ln v_2 - v_2)/x + \chi^*(Qv_1{}^2 - Qv_2{}^2 + v_2{}^2) + v_1 v_2(Qv_1 + Qv_2 - v_2)\frac{d\chi^*}{dv_2}\right\} \quad (17.5)$$

As noted above, Q must be near bulk concentration (i.e. $Q \sim v_2{}^b$).

Results for two flat plates. The potential energy of interaction per unit area of two parallel flat plates immersed in an aqueous solution as a function of the reduced distance ($h/\langle r^2\rangle^{\frac{1}{2}}$) is displayed in Fig. 17.10. Three concentrations of poly(oxyethylene), of molecular weight 20 000, were studied: these were volume fractions 0·02, 0·04 and 0·07. The value of the Flory-Huggins parameter used in these calculations was

$$\chi^* = m(-[\ln(1-v_2)]/v_2) + b \quad (17.6)$$

with $m=0{\cdot}66$ and $b=0{\cdot}10$.

It is immediately apparent from Fig. 17.10 that the potential energy curves

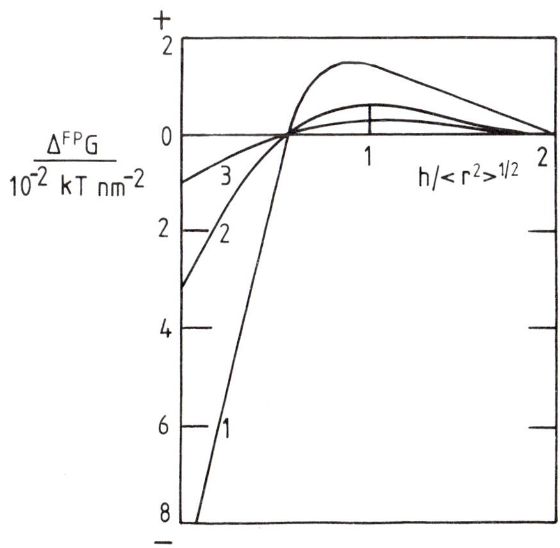

Fig. 17.10. The distance dependence of the depletion free energy for two flat plates immersed in a poly(oxyethylene) solution of molecular weight 20 000 for difference bulk polymer volume fractions ($v_2{}^b$): curves 1, $v_2{}^b=0{\cdot}07$; 2, $v_2{}^b=0{\cdot}04$; 3, $v_2{}^b=0{\cdot}02$ (after Feigin and Napper, 1980b).

display the qualitative features anticipated in Section 17.3.3. They display a maximum at a distance of separation comparable to the rms end-to-end distance of the polymer and a minimum near the interface. Both the height of the maximum and the depth of the minimum increase in absolute magnitude with increasing volume fraction of free polymer. For flat plates, it so happens that the depth of the minimum exceeds significantly the height of the maximum.

Sphere-sphere interactions. The Derjaguin integration of the flat plate potential can be used to predict sphere-sphere interactions provided that the range of interaction is small compared with the particle radius. Feigin and Napper (1980b) evaluated this integration numerically in the absence of an analytic theory. Some curves obtained for the free energy of interaction for two spheres (ΔG^S), each of radius 77·5 nm, when immersed in an aqueous poly(oxyethylene) (molecular weight 20 000) solution as a function of the reduced distance $H_0/\langle r^2 \rangle^{\frac{1}{2}}$ are presented in Fig. 17.11. Here H_0 is the minimum distance of surface separation of the particles.

The potential energy curves for spheres show the same qualitative features as were displayed by the flat plates. For spheres, however, the height of the maximum is comparable to the depth of the minimum. The primary purpose of presenting these potential energy curves here is to demonstrate that under suitable conditions both the height of the maximum and the depth of the

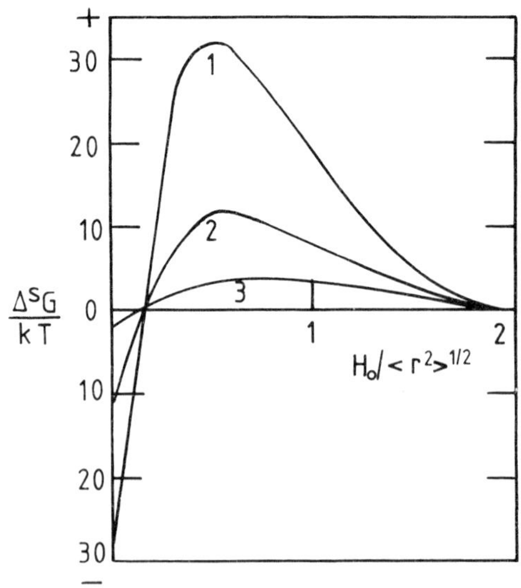

Fig. 17.11. Distance dependence of the depletion free energy of two spheres of radius 77·5 nm immersed in a poly(oxyethylene) solution of molecular weight 20 000 for different bulk polymer volume fractions (v_2^b): curve 1, $v_2^b = 0{\cdot}07$; 2, $v_2^b = 0{\cdot}04$; 3, $v_2^b = 0{\cdot}02$.

minimum can achieve values sufficient in magnitude to impart stability or induce flocculation. As flocculation requires a well depth of only a few $k_B T$ whereas stability requires, say, 10 times this value, depletion flocculation will be induced at a lower concentration of polymer than is depletion stabilization.

Flocculation rates. Stability ratios for spherical particles dispersed in polymer solution were calculated numerically by Feigin and Napper (1980b) from

$$W = 2a \int_{2a}^{\infty} [\exp\{\Delta G^S/kT\}/d^2]\mathrm{d}d \qquad (17.7)$$

where d = centre-to-centre particle distance. Strictly this result only applies for a deep primary minimum and for a potential barrier that extends a large distance compared with the distance required for the particles to be decelerated by the medium. This treatment is thus only qualitative in character.

Figure 17.12 displays some illustrative plots of $1/W$ as a function of the volume fraction of free polymer for particles of radius 77·5 nm immersed in aqueous poly(oxyethylene) solutions of three different molecular weights (4 000, 10 000 and 20 000). Steric barriers were assumed to be absent. Comparison of Figs 16.4 and 17.12 shows that the theoretical curves bear a qualitative resemblance to those measured experimentally. However, the

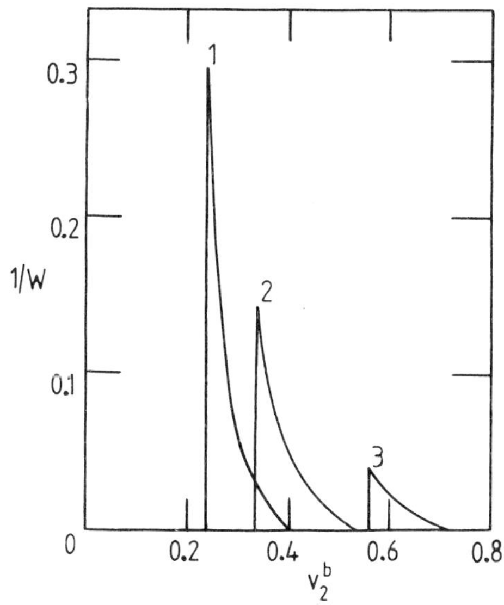

Fig. 17.12. Plots of the reciprocal of the stability ratio for spheres of radius 77·5 nm as a function of the volume fraction of free poly(oxyethylene) of different molecular weights: curves 1, 20 000; 2, 10 000; 3, 4 000 (after Feigin and Napper, 1980b).

sharp rise to the maximum displayed by the theoretical results would be expected to be blunted somewhat by the presence of steric barriers.

Comparison of theoretical values for v_2 and v_2** with experiment*

The effects of steric barriers. The methodology whereby the depth of the minimum and the height of the maximum in the potential energy curves for spheres can be computed has been elaborated in the preceding sections. By assuming critical values for the minimum ($-3k_BT$) and maximum ($15k_BT$), it was possible to predict the onset of depletion flocculation (at v_2^*) and depletion stabilization (at v_2^{**}).

The results listed in Table 17.2 compare the theoretical values of v_2^* and v_2^{**} with the experimentally observed values for a range of molecular weights of free polymer. The experiments were performed on latices stabilized by poly(oxyethylene) of 750 molecular weight. It is apparent that the experimental values of v_2^* and v_2^{**} are significantly greater than the theoretical values, which were calculated assuming that the steric layers were absent.

TABLE 17.2
Comparison of the experimental values of v_2^* and v_2^{**} with the theoretical values computed in the absence of steric barriers

Molecular weight	v_2^* Experiment	v_2^* Theory	v_2^{**} Experiment	v_2^{**} Theory
200	0·52	0·35	>0·95	0·50
400	0·51	0·23	>0·95	0·35
600	0·50	0·18	>0·95	0·28
800	0·47	0·15	>0·8	0·24
1 000	0·38	0·13	>0·6	0·22
1 500	0·28	0·10	>0·55	0·18
4 000	0·22	0·06	0·55	0·11
6 000	0·16	0·04	0·30	0·07
10 000	0·055	0·03	0·21	0·05
20 000	0·03	0·02	0·14	0·04
40 000	—	—	—	0·03
80 000	—	—	—	0·02
300 000	0·01	—	0·04	0·012

The data presented in Table 17.2 were obtained by the entropy maximization procedure. Recalculation adopting the free energy minimization approach yielded the results shown in Table 17.3. This leads to somewhat better agreement between theory and experiment but there is still a strong tendency for the theoretical values of both v_2^* and v_2^{**} to lie below the experimentally observed values. This suggests that the steric layers, which were present in the experiment but ignored in the calculations to date, may have an effect on the observed values of v_2^* and v_2^{**}. Such an effect could be

TABLE 17.3
Comparison of the experimental values for ν_2^* and ν_2^{**} with the theoretical values calculated using free energy minimization

Molecular weight	ν_2^* Experiment	ν_2^* Theory	ν_2^{**} Experiment	ν_2^{**} Theory
200	0·52	0·35	>0·95	>0·90
400	0·51	0·20	>0·95	0·87
600	0·50	0·15	>0·95	0·82
800	0·47	0·12	>0·8	0·70
1 000	0·38	0·10	>0·6	0·65
1 500	0·28	0·08	>0·55	0·56
4 000	0·22	0·04	0·55	0·38
6 000	0·16	0·04	0·30	0·32
10 000	0·055	0·025	0·21	0·25
20 000	0·03	0·02	0·14	0·17
40 000	—	0·013	—	—
80 000	—	0·010	—	—
300 000	0·01	0·006	0·04	—

understood as follows. The segments of the steric barrier tend to even out the distribution of segments over the entire system and thus to counteract the effects of the depletion layers. Alternatively, the steric layers can be considered to repel the free polymer chains at infinite separation, which must dampen the effects of free polymer. Note that Vrij (1976) has argued that the presence of the steric layers enhances depletion effects. This seems, however, to be unlikely if the steric stabilizing moieties are of the same chemical composition as the free polymer.

Some results obtained using the free energy minimization procedure but including the effects of the poly(oxyethylene) steric barrier are shown in Table 17.4. Only free polymer of somewhat higher molecular weight than the stabilizing moieties were studied because of the problems associated with

TABLE 17.4
Comparison of experimental values of ν_2^* and ν_2^{**} with the theoretical values in the presence of steric layers using free energy minimization

Free polymer molecular weight	ν_2^* Experiment	ν_2^* Theory	ν_2^{**} Experiment	ν_2^{**} Theory
1 500	0·28	0·30	>0·55	0·54
4 000	0·22	0·17	0·38	0·38
6 000	0·16	0·11	0·30	0·32
10 000	0·055	0·07	0·22	0·25
20 000	0·03	0·04	0·16	0·17
40 000	—	0·02	—	—
80 000	—	0·01	—	—
300 000	0·01	0·006	—	—

entry of polymer chains into the steric layers. It is apparent that the agreement between theory and experiment is satisfactory for both v_2* and v_2**. Thus, for example, with poly(oxyethylene) of molecular weight 20 000, the theoretical region of depletion flocculation is 0·04–0·16, compared with the experimentally determined range of 0·03–0·14. Similarly, for 300 000 molecular weight, the respective ranges are 0·006–0·034 and 0·01–0·04.

The nonadditivity of steric and depletion effects

Potential energy diagrams. Figure 17.13 displays the free energies of close approach for steric and depletion stabilization by low molecular weight polymers when acting separately. Also shown is the free energy of close approach when the steric and depletion effects occur together. It is immediately apparent that the free energies need not simply be additive but the two phenomena may interact strongly with one another. The major effect in this case appears to be associated with the interaction of the steric layer on one surface with the depletion layer adjacent to the opposing surface. This produces a significant reduction in free energy for the particles when well separated, an effect that is absent when the phenomena are manifest separately.

Of course, the interaction between the steric layers and the free polymer depends critically upon their relative molecular weights. For low molecular

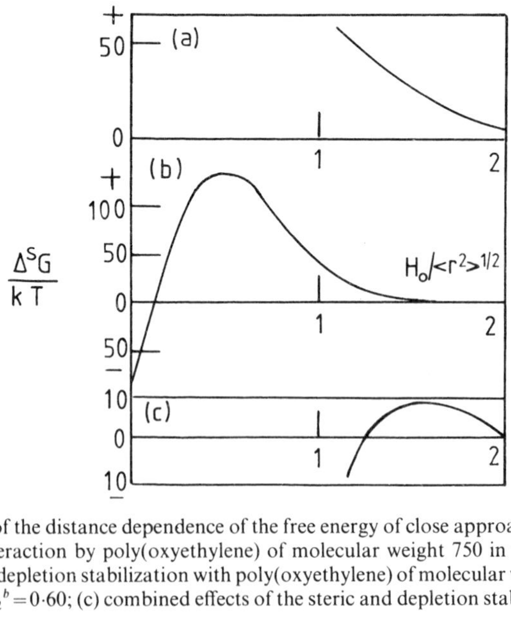

Fig. 17.13. Plots of the distance dependence of the free energy of close approach for two spheres: (a) pure steric interaction by poly(oxyethylene) of molecular weight 750 in the absence of free polymer; (b) pure depletion stabilization with poly(oxyethylene) of molecular weight 600 at a bulk volume fraction $v_2^b = 0{\cdot}60$; (c) combined effects of the steric and depletion stabilization presented in (a) and (b).

weight stabilizing moieties and high molecular weight free polymer, the effects will be minimized because the steric layer thickness is small in comparison with the depletion layer thickness. Thick steric layers and thin depletion layers will display primarily steric stabilization perturbed by a change in solvency.

Typical potential energy diagrams that might be expected for simple combinations of depletion stabilization with steric or electrostatic stabilization are displayed in Fig. 17.14. The minimum in the potential energy curve need not be below the zero energy level in order to induce weak flocculation, although it would be expected that this type of behaviour would be rare and depend upon the specific details of the system in question.

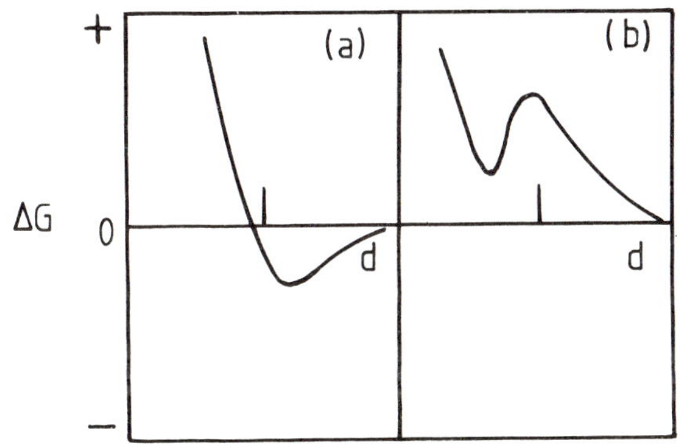

Fig. 17.14. Schematic representation of the potential energy diagrams for combinations of depletion stabilization with steric and/or electrostatic stabilization.

17.4.2. The theory of Scheutjens and Fleer

Scheutjens and Fleer (1982) have developed a theory for depletion stabilization and depletion flocculation based upon their statistical thermodynamic approach to polymer adsorption and steric stabilization. This theory is cast in terms of the most primitive model for a polymer molecule, the random flight chain. This weakens the theory in so far as providing quantitative predictions at the fundamental level for real systems is concerned. The theory does, however, offer qualitative results over a wide range of conditions, being especially powerful in establishing the various trends involved.

The qualitative predictions reported by Scheutjens and Fleer in general parallel those obtained by Feigin and Napper (1980b) outlined above. The explanation advanced by Scheutjens and Fleer was essentially that first propounded by Hall (1972) and Ash *et al.* (1973).

There appear, however, to be some important differences between the predictions of the theory of Scheutjens and Fleer and those of Feigin and

Napper (1980b). For example, Scheutjens and Fleer claim that at low concentrations of polymer the interaction between particles dispersed in a polymer solution is attractive at all separations. Only at high volume fractions does very strong repulsion appear at distances comparable to the rms end-to-end distance of the macromolecule. The physical basis for such a qualitative change in the sign of the free energy of interaction at larger separations seems difficult to understand. Perhaps it is an artefact of the procedures adopted.

The Scheutjens and Fleer theory predicts an interesting, if at first sight unexpected, effect of the quality of the solvent on the free energy of interaction. The barrier opposing flocculation is predicted* by them to be enhanced by decreasing the solvency of the dispersion medium. Moreover, it is shifted to larger separations. Just before the critical point, the effects are dramatically amplified. Strong stabilization is thus predicted in poor solvents at high free polymer concentrations unless, of course, phase separation of the polymer intervenes.

Feigin and Napper (1980b) predicted quite different behaviour with respect to the solvency of the system: the better is the solvent for the free polymer, the larger are the free energy changes involved and thus the smaller are the values of $v_2{}^*$ and $v_2{}^{**}$. These predictions, however, only apply to naked surfaces. The situation is much more complicated when steric barriers are present, as they would be in the theory of Scheutjens and Fleer and in many practical situations. In that case, the steric repulsion is greater in the absence of free polymer in a better solvent. It is the subtle interplay of the effects of solvency change on both steric stabilization and depletion stabilization that determines the overall trend observed. At present there does not appear to be any data in the literature that permit a clear cut statement to be made of the effects of solvency on depletion stabilization in the presence of steric layers.

One indication of the effects of solvency comes from the phase separation studies of silica particles reported by Pathmamanoharan *et al.* (1981). They found that in a good solvent $v_2{}^*$ for polystyrene was decreased by a factor of 3 when the solvent was changed from cyclohexane at the θ-point to the good solvent toluene. This is suggestive that depletion effects may be significantly greater in good solvents than in θ-solvents for some systems.

In contrast, Dodd and Napper (1980) found that the values of $v_2{}^*$ for aqueous poly(oxyethylene) stabilized polystyrene latices were significantly greater in a better solvent. Free poly(oxyethylene) was used to induce depletion flocculation in these experiments. Some data are presented in Table 17.5 for particles stabilized by poly(oxyethylene) of molecular weight 1500. The values of $v_2{}^*$ in the poorer solvent (0·39 M $MgSO_4$) were, in the main, somewhat less than those in pure water.

The simplest explanation for the foregoing results invokes the presence of a significant repulsive contribution from the steric layers. This repulsion is

* See footnote on p. 239.

TABLE 17.5
The effects of solvency on depletion stabilization in the presence of steric barriers

Molecular weight of free polymer	Particle radius (nm)	v_2^* water	v_2^* 0·39 M $MgSO_4$
20 000	98	0·03 ± 0·01	0·03 ± 0·01
6 000	98	0·15 ± 0·02	0·03 ± 0·01
6 000	64	0·08 ± 0·01	0·03 ± 0·01

reduced when the solvency of the dispersion medium is reduced. As a result, the competition that occurs between the steric repulsion on the one hand and the depletion attraction on the other is moved in the direction to favour the depletion attraction. A lower concentration of polymer is therefore required to induce flocculation.

It would appear that the effects of solvency on depletion flocculation and depletion phase separation are quite complex. The presence of steric layers means that depletion effects alone do not determine stability and proper allowance must be made for the effects of solvency on the steric interactions.

17.4.3. The theory of Clark and Lal

Richmond and Lal (1974) considered the pressure exerted on two parallel flat plates by a random flight chain confined between them. Adsorption on the plates was not allowed. Not surprisingly, the loss of configurational entropy of the chains on close approach resulted in the generation of a repulsion. This situation is quite unphysical for colloidal systems since the polymer chains in free solution can escape from the compressional zone. Accordingly, the predictions of Richmond and Lal do not relate to depletion stabilization.

Clark and Lal (1981), however, have simulated a polymer molecule on a tetrahedral latice using a Monte Carlo procedure. Three rotational conformational states were assumed to be accessible to each bond. The excluded volume condition was incorporated into the model by assuming that two or more segments cannot occupy the same position in space. These calculations permitted the evaluation of the change in configurational free energy of the polymer chains as a function of the interplane separation.

The results obtained by Clark and Lal for the variation in the net pressure due to free polymer, both with and without excluded volume effects, are presented in Fig. 17.15. According to this model, only the excluded volume polymer chain provides repulsion at larger distances of separation, the non-excluded volume chain generating only attraction at all distances of separation.

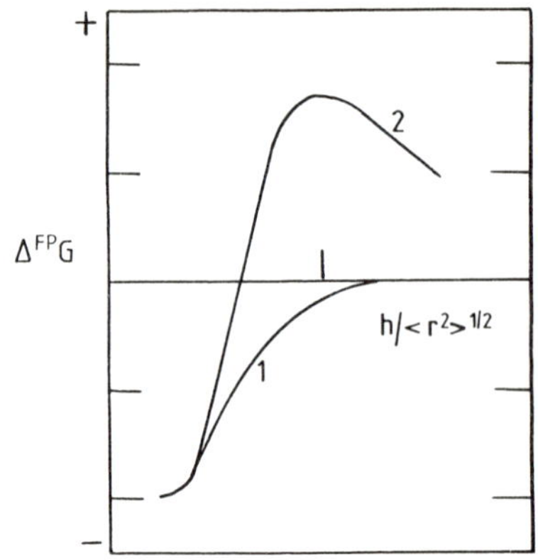

Fig. 17.15. Schematic representation of the depletion free energy for two flat plates immersed in free polymer solutions according to Clark and Lal (1981): curve 1, no excluded volume; curve 2, with excluded volume.

17.4.4. The non-prediction of depletion stabilization

In the light of the foregoing discussion of the portmanteau theories of depletion stabilization and depletion flocculation, it seems reasonable to raise the question as to why the theories of Vrij (1976) and Joanny *et al.* (1979) fail to predict stability at high polymer concentrations. Indeed, Vrij and coworkers (Pathmamanoharan *et al.*, 1981) also failed to observe stability at higher polymer concentrations, although their evidence is inconclusive because of the difficulties associated with preparing polymer solutions at high polymer concentrations.

First, it should be recognized that for systems that undergo phase separation, the colloidal particles are evidently quite stable after phase separation. This is in sharp contrast to the aggregation of the particles in systems that undergo depletion flocculation. Obviously, if the particles do not undergo flocculation at lower polymer concentrations, the question of their being stabilized at higher polymer concentrations scarcely arises. The observation of phase separation at higher polymer concentrations is entirely expected.

There is, however, a second point of some theoretical importance that bears upon this question. Vrij (1976) modelled polymer chains as hard spheres, which they decidedly are not. The potentials assumed are those appropriate to hard spheres. The second virial coefficient was then calculated from

$$B_2{}^c = (\tfrac{1}{2})\int_0^\infty [1 - \exp\{-V(r)/kT\}]4\pi r^2 \mathrm{d}r \tag{17.8}$$

In any calculation of the second virial coefficient in this manner, the integral will be dominated by those terms involving negative potentials, positive potentials being entirely overwhelmed. This is illustrated in Fig. 17.16 where the potential energy of interaction of the type expected for free polymer systems is portrayed. For the purposes of illustration, both the repulsive barrier height and well depth have been arbitrarily set at 40 kT. Examination, term by term, of the integrals involved shows that the value of B_2^c calculated will be determined almost entirely by the term involving e^{+40}, which will have a negative sign. Hence B_2^c will be predicted to be negative. Yet, because of the strong repulsive potential energy at large separations, the stability ratio,

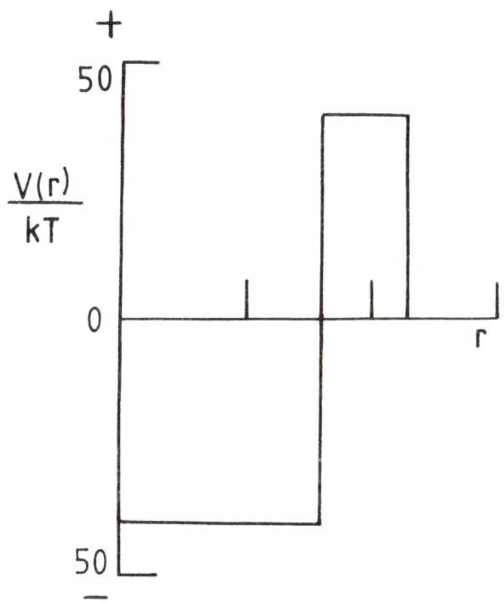

Fig. 17.16. Schematic representation of an arbitrary potential energy diagram appropriate to depletion stabilization.

determined by calculating an appropriate integral of $e^{V(r)/kT}$, will be very large. It follows that a dispersion of particles subject to such a potential would display kinetic stability, despite the occurrence of a negative second virial coefficient. Stated differently, equilibrium thermodynamic calculations of B_2^c appear to be unable to comprehend this type of metastability. The equilibrium thermodynamic method of Joanny *et al.* (1979) may well contain a comparable debilitating integral over all space that prevents it from comprehending thermodynamic metastability.

17.5. Thermodynamic formalism for depletion stabilization and depletion flocculation

17.5.1. Conspectus

Considerable progress has been made in treating the effects of the adsorption of molecules on colloid stability by the general formalism of equilibrium thermodynamics (Mackor and van der Waals, 1952; Hall, 1972; Ash *et al.*, 1973; Everett, 1976). Given the recognition of two important provisos, these theories can be considered to provide a general thermodynamic basis for depletion stabilization and depletion flocculation. Of course, as shown previously (Section 10.3), these theories also comprehend in a general sense, steric stabilization. They thus provide a unifying approach that in principle encompasses all aspects of the effects of polymers on colloid stability.

The two provisos mentioned above relate to the meaning of the term "adsorption" in this context. First, in what follows, adsorption does not refer to the attachment of polymer chains to the particle/dispersion medium interface. This narrow interpretation must be replaced by the much broader concept of polymer chains, free in solution, existing in a surface phase bounded by the particle surface and an arbitrarily chosen Gibbs dividing surface. Second, it must also be recognized that polymer segments, and perforce polymer molecules, undergo *negative* adsorption near an impenetrable interface. This, as explained above, arises from the constraints that such a surface places on the conformations of the polymer molecule. These two points are emphasized because depletion stabilization and depletion flocculation arise from polymer chains that are free in solution. Attachment is quite unnecessary and merely serves to complicate the problem acutely. Depletion stabilization and flocculation can in principle be observed even if the particle surfaces are repulsive to the polymer which, in one sense, they always are.

17.5.2. The effects of gas adsorption on colloid stability

It may seem strange to introduce the thermodynamic discussion of depletion effects by reference to the influence of gas adsorption on colloid stability. It so happens that the generalizations that evolve from an equilibrium thermodynamic treatment of the effects of the adsorption of gases on the force between two flat plates can be readily expanded to encompass solutions (Ash, 1974; Everett, 1976). Since the treatment of gas adsorption is of utmost simplicity, it will be considered first in some detail to illustrate the principles involved. The subsequent elaboration of this theory to comprehend binary solutions does no significant conceptual violence to the integrity of the original approach, which is therefore presented in some detail. The theory must be applied to polymers with some caution, however, because it is assumed that equilibrium is maintained at all times. This seems likely to be so with totally unattached polymer, although it is not necessarily correct if the polymer undergoes adsorption/desorption phenomena.

The following discussion follows closely the seminal contribution of Ash *et al.* (1973). They considered the system represented schematically in Fig. 17.17.

This system, designated I, contains an amount, n, of a single component vapour. Adsorption is supposed to occur only on the inner faces of the plate, each of which is of area A. The plates are held apart at an equilibrium distance h by an externally applied force $\mathscr{F}$. This is given a positive sign when the plates repel one another and a negative sign when they attract. It is related to the force per unit area f by $\mathscr{F}=Af$. The system is assumed to be able to participate in processes that involve both $P\mathrm{d}V$ and $\mathscr{F}\mathrm{d}h$ work terms. It is compared with a reference system (II) of the same volume and at the same pressure but containing no adsorbing surfaces. As a result, system II contains a different amount, n^o, of vapour. The surface excess amount of vapour in system I is thus $n^\sigma = n - n^o$.

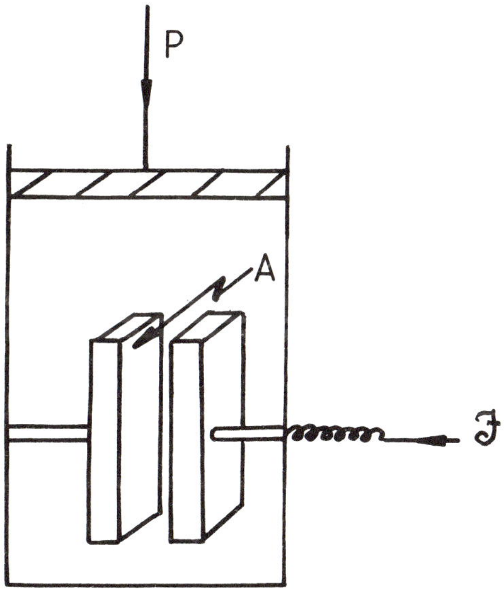

Fig. 17.17. Diagrammatic representation of the effect of the adsorption of gases on the interaction between flat plates.

The total differential of the internal energy of system I is given by

$$\mathrm{d}U = T\mathrm{d}S - P\mathrm{d}V + \mu\mathrm{d}n + 2\sigma\mathrm{d}A - Af\mathrm{d}h \qquad (17.9a)$$

where σ = surface tension of the gas/solid interface and μ = chemical potential of the vapour. That for system II is

$$\mathrm{d}U^o = T\mathrm{d}S^o - P\mathrm{d}V + \mu\mathrm{d}n^o \qquad (17.9b)$$

Subtraction yields

$$\mathrm{d}(U - U^o) = T\mathrm{d}(S - S^o) + \mu\mathrm{d}n^\sigma + 2\sigma\mathrm{d}A - Af\mathrm{d}h \qquad (17.10)$$

The Gibbs free energy is now defined such that $G = U + PV - TS$. It follows that

$$d(G-G^o) = d(U-U^o) - Td(S-S^o) + (S-S^o)dT \qquad (17.11)$$

Combination of equation (17.10) and (17.11) yields

$$d(G-G^o) = -(S-S^o)dT + \mu dn^\sigma + 2\sigma dA - Afdh \qquad (17.12)$$

which on integration at constant T, μ, σ and h leads to

$$G - G^o = \mu n^\sigma + 2\sigma A \qquad (17.13)$$

Differentiation of this equation and subtraction from equation (17.12) gives a Gibbs-Duhem type of equation

$$-(S-S^o)dT - Afdh = n^\sigma d\mu + 2Ad\sigma \qquad (17.14)$$

which at constant temperature can be re-expressed as

$$-2d\sigma = fdh + 2\Gamma d\mu \qquad (17.15)$$

where $\Gamma(=n^\sigma/2A)$ is the surface excess concentration of adsorbate. Cross-differentiation of this equation gives

$$\left(\frac{\partial f}{\partial \mu}\right)_{T,h} = 2\left(\frac{\partial \Gamma}{\partial h}\right)_{T,\mu} \qquad (17.16)$$

which on integration yields

$$f - f^o = 2\int_{-\infty}^{\mu} \left(\frac{\partial \Gamma}{\partial h}\right)_{T,\mu} d\mu \qquad (17.17)$$

The potential energy of interaction between unit areas of plate (V_p) relative to zero at infinite separation can be found from

$$V_p - V_p{}^o = -\int_{\infty}^{h} (f - f^o)dh \qquad (17.18)$$

Since equation (17.15) above yields

$$f = -2\left(\frac{\partial \sigma}{\partial h}\right)_{T,\mu} \qquad (17.19)$$

the potential energy can be evaluated from

$$V_p - V_p{}^o = 2\int_{\infty}^{h} d(\sigma - \sigma^o)$$

$$= 2[(\sigma-\sigma^o)_h - (\sigma-\sigma^o)_\infty] \qquad (17.20)$$

Physically, two opposing effects arise when h is decreased. First, the adsorption fields of the two plates overlap increasingly, promoting adsorption. At the same time, the volume between the plates decreases, tending to reduce adsorption. Thus $(\partial\Gamma/\partial h)_{T,P}$ may be positive (and contribute a repulsion) or negative (and contribute an attraction).

The foregoing theory leads to the following qualitative generalization. If moving the plates together causes a larger number of molecules to be accommodated at the lower potential energy of the adsorbed state, then there will be an attraction between the plates. Contrariwise, if moving the plates together reduces the space available for adsorption, leading to the squeezing out of molecules, then repulsion will ensue (we ignore here the van der Waals attraction between the plates). The former may be considered to arise from an energy effect whereas the latter originates in entropy effects.

17.5.3. Extension to liquid mixtures

Ash *et al.* (1973) extended the preceding theory to the adsorption from liquid mixtures. It follows very similar lines to the gas/solid situation and so, in the interests of concision, it will not be reproduced here. For a binary mixture of solute 2 in solvent 1, the analogue of the gas/solid equation (17.17) is

$$f_2-f_1=2\int_{-\infty}^{\mu_2^*}\left(\frac{\partial\Gamma_2^{(1)}}{\partial h}\right)_{\mu_2}\mathrm{d}\mu_2 \qquad (17.21)$$

where f_2 = force in the binary solution, f_1 = force in pure solvent, μ_2^* = chemical potential of pure component 2 at the temperature and pressure concerned. The relative surface excess of the solute 2 (with respect to the solvent 1), $\Gamma_2^{(1)}$, is defined by

$$\Gamma_2^{(1)}=\Gamma_2-(c_2/c_1)\Gamma_1 \qquad (17.22)$$

where $c_i (i=1,2)$ is the bulk concentration of component i. One advantage of $\Gamma_2^{(1)}$ defined in this manner is that its value is independent of the arbitrary choice of the position of the Gibbs dividing surface.

The potential energy per unit area of the plates in a binary mixture relative to its value at infinite separation, $V_p^{(2)}$, is given by

$$V_p^{(2)}-V_p^{(1)}=2\int_{-\infty}^{\mu_1}[\Gamma_2^{(1)}(\infty)-\Gamma_2^{(1)}(h)]\mathrm{d}\mu_2 \qquad (17.23)$$

where $V_p^{(1)}$ = the corresponding potential energy difference for the solvent and $[\Gamma_2^{(1)}(\infty)-\Gamma_2^{(1)}(h)]$ is the increment in the relative adsorption of the solute caused by bringing the plates together from infinite separation to h at constant μ_2.

Ash (1974) has subsequently elaborated the corresponding theory for regular binary mixtures when the adsorption forces extend beyond the first lattice layers adjacent to the plates.

17.5.3.1. Thermodynamic interpretation of depletion flocculation and depletion stabilization

The most relevant formula to carry forward from the preceding sections is equation (17.21) for it provides some thermodynamic insight into depletion stabilization and depletion flocculation. If the van der Waals attraction between the plates is ignored and solvent adsorption is negligible, it may be written as

$$f_2 = 2\int_{-\infty}^{\mu_2^*} \left(\frac{\partial \Gamma_2^{(1)}}{\partial h}\right)_{\mu_2} d\mu_2 \tag{17.24}$$

where the factor of 2 merely allows for the presence of two plates. The integration limits for the chemical potential of the solute are taken between infinite dilution (i.e. zero solute concentration) and pure solute (corresponding to μ_2^*). It is obvious that given this variation in μ_2, the sign of f_2, as represented by the area under the curve of the plot of $(\partial \Gamma_2^{(1)}/\partial h)_{\mu_2}$ versus μ_2, must necessarily be determined by the sign of $(\partial \Gamma_2^{(1)}/\partial h)_{\mu_2}$ and any variation thereof.

Depletion stabilization

If the relative excess $\Gamma_2^{(1)}$ of the free dissolved polymer decreases as the colloidal particles approach (i.e. h decreases), then the sign of $(\partial \Gamma_2^{(1)}/\partial h)_{\mu_2}$ must be positive. This implies that f_2 is positive and a repulsion is operative as the polymer molecules are squeezed out of the space between the particles, just as in the gas/solid case. This is the situation that pertains in the early stages of the close approach of colloidal particles immersed in free polymer solutions. Thus a repulsive force must be operative under those conditions. This is the thermodynamic basis of depletion stabilization. Just as in the solid/gas case, the repulsion can, in one sense, be considered to be of entropic origin. This, however, is far too simplistic a picture.

We again stress that we are not considering polymer chains that are attached to the surface of the particle. Rather, we are considering adsorption in its broader manifestation relating to the presence of Gibbs surface layers at the particle/solution interface. Closer approach results in greater negative adsorption of the polymer chains, which is accompanied by a repulsion.

Depletion flocculation

The interpretation of the origin of depletion flocculation using the generalized adsorption model is more subtle than that for depletion stabilization. The layer of essentially pure dispersion medium that exists between colloidal particles when quite close together must be regarded as a region of *negative* adsorption. This is, of course, precisely what it is: the polymer molecules are almost completely excluded from the space between the particles because their entry therein would result in an unacceptably large loss of configurational entropy. Closer approach of the particles destroys regions of negative adsorption which can be considered to be equivalent mathematically to creating positive adsorption on closer approach. Thus $\partial \Gamma_2^{(1)}/\partial h_{\mu_2}$ is negative and gives rise to an attractive interaction between the particles, which, if sufficiently large, may result in depletion flocculation.

The origin of the phenomenon that is responsible for depletion flocculation is somewhat different from the gas/solid situation in that the solvent molecules are squeezed out to a lower potential energy in the bulk rather than a lower potential energy in the adsorbed state. The attraction arises from a free energy of mixing of essentially pure solvent with bulk polymer solution rather than from a simple internal energy change.

Qualitative example

The foregoing thermodynamic concepts can be illustrated in a qualitative, yet revealing, fashion as follows. Consider two parallel flat interfaces separated by an infinite distance. Suppose, arbitrarily, the segment density distribution increases linearly with distance from the interface until it reaches the bulk value at a distance W, as shown in Fig. 17.18a. If c_2^b is the bulk polymer concentration (in appropriately chosen units), the surface excess (in its simplest form) at each interface is simply $Wc_2^b/2 - Wc_2^b = -Wc_2^b/2$. For two interfaces, the surface excess is thus $-Wc_2^b$.

As the particles come closer together, the surface excess at each interface is unchanged until the distance of separation is less than $2W$. Thereafter, the value of the surface excess becomes more negative until at a separation infinitesimally smaller than $h = W$, no polymer remains between the plates (see Fig. 17.18b). The surface excess per plate is now $0 - Wc_2^b = -Wc_2^b$. For two plates it is $-2Wc_2^b$. The surface excess for the pair of plates must thus decrease from $-Wc_2^b$ to $-2Wc_2^b$ as h diminishes from $2W$ to W.

Even closer approach results in the surface excess becoming less negative until at $h=0$, when the plates are in contact, its value is zero.

The qualitative features of the surface excess as a function of the distance of separation of the plates are summarized schematically in Fig. 17.19a. Its derivative gives the qualitative shape of the force curve shown in Fig. 17.19b. It is clear that the shape of the force curve predicted by the thermodynamic approach coincides with that predicted by the alternative approaches to depletion stabilization considered previously.

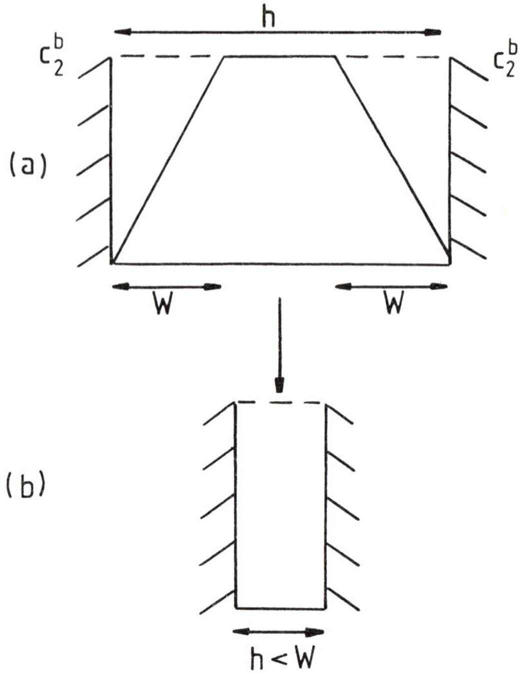

Fig. 17.18. Schematic representation of the change in the segmental density profile between two flat plates in a polymer solution on close approach.

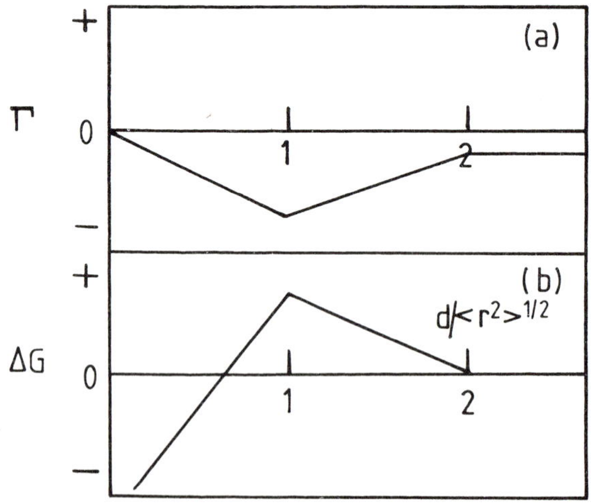

Fig. 17.19. (a) The relative surface excess for two flat plates as a function of the distance of separation; (b) characteristic free energy of interaction arising from depletion effects according to the thermodynamic approach.

17.6. Predictions of the steric layer theory

The steric layer theory of Vincent, Luckham and Waite (1980) also provides reasonable semi-quantitative predictions of the onset of flocculation. According to this theory, the onset of flocculation marks the entrance into the semi-dilute polymer concentration regime (i.e. $v_2^* \simeq c_2^*$). Flocculation arises because there is a net decrease in free energy when two particles come together and displace into the bulk solution some of the free polymer molecules that are interpenetrating the stabilizing moieties. Restabilization is said to be associated with the onset of the so-called concentrated polymer regime (i.e. $v_2^{**} \simeq c_2^{**}$). It should be stressed, however, that according to scaling law theory, neither c_2^* nor c_2^{**} represent well-defined concentrations but rather concentration regions that may be rather broad.

Vincent *et al.* (1980) have argued that v_2^* occurs at the free polymer concentration where the centres of the coils, packed in some close-packed array, are separated by twice the radius of gyration, i.e.

$$v_2^* = M/(bR_g^3 N_A \rho) \qquad (17.25)$$

where M = molecular weight of the free polymer, ρ = density of the polymer and R_g = radius of gyration. The value of $b = 8$ for cubic close packing whereas $b = 5{\cdot}63$ for hexagonal close packing. An identical expression applies to v_2^{**} except that R_g now represents the unperturbed radius of gyration and the value of b is reduced to $b = 2{\cdot}52$ for cubic close packing and $b = 1{\cdot}36$ for hexagonal close packing. Note that Vincent, Luckham and Waite claim that below v_2^*, the presence of the free polymer has no effect on the steric interaction free energy, i.e. all depletion effects are ignored.

A comparison is presented in Table 17.6 of the calculated values of c_2^* and c_2^{**} and some experimental values for v_2^* and v_2^{**} for two polystyrene latices sterically stabilized by poly(oxyethylene) of two different molecular weights (750 and 2 000). Crude agreement is apparent, at least in so far as the general trends are concerned.

It was claimed by Vincent *et al.* (1980) that a more quantitative indication of the predictions of the steric layer theory could be ascertained as follows. The free energy of interaction in the presence of free polymer is given by

$$\Delta G = G^{LL,O} - (n/2)G^{PP} \qquad (17.26)$$

where $G^{LL,O}$ = steric free energy in the absence of free polymer, G^{PP} = coil-coil interaction energy and n = number of free polymer coils that are displaced into the bulk solution. The value of $G^{LL,O}$ varies with the distance of separation between the particles and so ΔG can exhibit a minimum. This minimum value was calculated for a range of polymer concentrations. The theoretical value of

TABLE 17.6

Comparison of the predictions of the steric layer theory with experiment for polystyrene latices flocculated by poly(oxyethylene)

Free polymer molecular weight	Experimental v_2* 750	2 000	c_2*	v_2** 750	c_2**
800	0·47	—	0·22	>0·8	0·78
1 000	0·38	0·60	0·19	>0·6	0·70
1 500	0·28	0·47	0·16	>0·55	0·58
4 000	0·22	0·37	0·09	0·55	0·35
6 000	0·16	0·30	0·07	0·30	0·29
10 000	0·055	0·20	0·05	0·21	0·22
20 000	0·03	0·12	0·03	0·14	0·16
300 000	0·01	0·04	0·004	0·04	0·04

v_2* was that value whose corresponding minimum value of ΔG led to the following equation being satisfied

$$v_3{}^d = v_3{}^{cp}\exp\{z\Delta G_{min}/2kT\} \qquad (17.27)$$

where $v_3{}^d$ = volume fraction of latex particles, $v_3{}^{cp}$ = volume fraction in the floc phase ($v_3{}^{cp}=0{\cdot}62$) and z = floc coordination number (assumed to be $z=8$). Values of ΔG_{min} usually lays in the range (1–2)kT.

This theory has been applied to a range of dispersions, both aqueous (Vincent *et al.*, 1980) and nonaqueous (Clarke and Vincent, 1981a,b) in character. The precise results depend upon the nature of the segment density distribution function assumed for the steric layer but semi-quantitative agreement could be achieved between theory and experiment in this fashion. The results obtained will not be presented in detail because they span the values of c_2* and c_2** tabulated in Table 17.5. Accordingly, these more detailed calculations really provide no better prediction of the values of v_2* and v_2** than do c_2* and c_2** if account is taken of the number of adjustable parameters necessary to perform the calculations.

17.7. Summary of the effects of polymers on colloid stability

Figure 17.20 summarizes schematically the effects of polymers on an idealized colloid. The colloid is idealized to the extent that its surface, chameleon like, can change from being inert to strongly anchoring as is required for heuristic purposes.

At very low polymer concentrations (say several ppm), the predominant effect, if an effect is indeed observed, is bridging flocculation. This can be observed not only with electrostatically stabilized particles but also with certain sterically stabilized particles (Evans and Napper, 1973d). It has been shown that even quite low molecular weight poly(oxyethylene) can

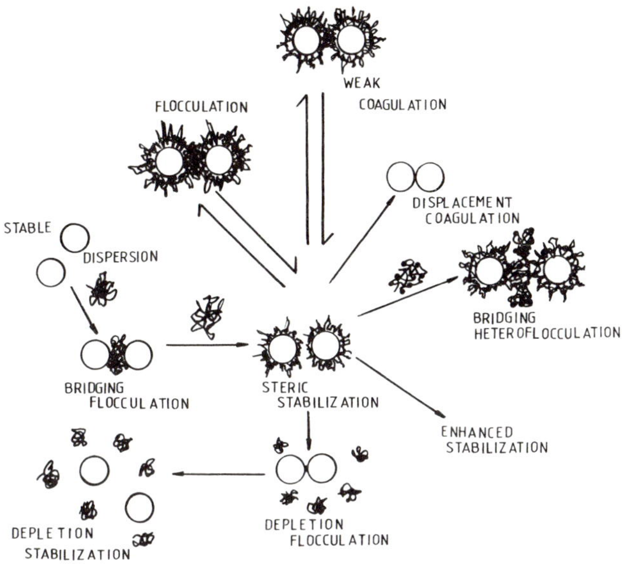

Fig. 17.20. Diagrammatic representation of some of the ways that polymer chains can affect colloid stability.

flocculate poly(acrylic acid) stabilized polystyrene latices at low pH. Flocculation could also be observed when the locations of these stabilizing moieties and the flocculating chains were reversed.

At higher polymer concentrations, steric stabilization can be observed. Sterically stabilized particles can undergo displacement coagulation, weak coagulation and flocculation in worse than θ-solvents.

At even higher polymer concentrations, depletion flocculation is observed and, finally, at the highest polymer concentrations depletion stabilization is observed.

This diagram is able to explain some puzzling observations disclosed by Cowell and Vincent (1982). They report clear differences between the stability behaviour of latex dispersions depending upon whether the particles are naked or pre-coated by terminally anchored chains. The addition of free polymer to the sterically stabilized particles resulted in the transition sequence stability→instability→stability. This is precisely what would be predicted from Fig. 17.20 if bridging flocculation is absent (as it must be if the free polymer and stabilizing moieties are identical in chemical composition).

For naked particles, entry into the sequence of events shown in Fig. 17.20 occurs earlier because the particles are not coated. Addition of polymer causes steric stabilization to occur prior to depletion flocculation at much higher polymer concentrations. It is scarcely surprising that in this instance Cowell and Vincent reported the sequence instability→stability→instability. The puzzling difference in behaviour is accordingly explained.

References

Akcasu, A. Z. and Han, C. C. (1979). *Macromolecules* **12**, 276.

Albers, W. and Overbeek, J. Th. G. (1959). *J. Colloid Sci.* **14**, 510.

Alexander, A. E. and Johnson, P. (1949). "Colloid Science", p. 124. Clarendon, Oxford.

Allen, G. (1976a). *In* "Structural Studies of Macromolecules by Spectroscopic Methods" (K. J. Ivin *ed.*). p. 1. Wiley, London.

Allen, G. (1976b). *Proc. Roy. Soc. London A* **351**, 381.

Andrews, D. M., Manev, E. D. and Haydon, D. A. (1970). *Spec. Disc. Faraday Soc.* **1**, 46.

Anon (1968a). *Encyclopaedia Britannica* **17**, 38.

Anon (1968b). *Encyclopaedia Britannica* **12**, 257.

Arfken, G. (1970). "Mathematical Methods for Physicists", p. 794. Academic Press, New York.

Asakura, S. and Oosawa, F. (1954). *J. Chem. Phys.* **22**, 1255.

Asakura, S. and Oosawa, F. (1958). *J. Polymer Sci.* **33**, 183.

Ash, S. G. (1974). *J. Chem. Soc. Faraday Trans. II* **70**, 895.

Ash, S. G., Everett, D. H. and Radke, C. (1973). *J. Chem. Soc. Faraday Trans. II* **69**, 1256.

Bagchi, P. (1974a). *J. Colloid Interface Sci.* **47**, 86.

Bagchi, P. (1974b). *J. Colloid Interface Sci.* **47**, 100.

Bagchi, P. (1975). *J. Colloid Interface Sci.* **50**, 115.

Bagchi, P. (1976). *Colloid Polymer Sci.* **254**, 890.

Barber, M. N., Guttman, A. J., Middlemiss, K. M., Torrie, G. M. and Whittington, S. G. (1978). *J. Phys. A* **11**, 1833.

Barclay, L., Harrington, A. and Ottewill, R. H. (1972). *Kolloid-Z. Z. Polymere* **250**, 655.

Barker, J. A. (1963). "Lattice Theories of the Liquid State". Pergamon, London.

Barker, J. A. and Fock, W. (1953). *Disc. Faraday Soc.* **15**, 188.

Barnett, K. G., Cosgrove, T., Vincent, B., Burgess, A. N., Crowley, T. L., King, T., Turner, J. D. and Tadros, Th. F. (1981a). *Polymer* **22**, 283.

Barnett, K. G., Cosgrove, T., Vincent, B., Sissons, D. S. and Cohen-Stuart, M. (1981b). *Macromolecules* **14**, 1018.

Barnett, K. G., Cosgrove, T., Crowley, T. L., Tadros, Th. F. and Vincent, B. (1982). *In* "The Effects of Polymers on Dispersion Properties" (Th. F. Tadros *ed.*). p. 183. Academic Press, London.

Barrett, K. E. J. (1975). "Dispersion Polymerization in Organic Media". John Wiley, London.

Bauer, D. R. and Ullman, R. (1980). *Macromolecules* **13**, 392.

Benoit, H., Cotton, J. P., Decker, D., Farnoux, B., Higgins, J. S., Jannick, G., Ober, R. and Picot, C. (1973a). *Nature* **245**, 13.

Benoit, H., Decker, D., Higgins, J. S., Picot, C., Cotton, J. P., Farnoux, B. and Jannick, G. (1973b). *Nature* **246**, 13.

Bluestone, S., Mark, J. E. and Flory, P. J. (1974). *Macromolecules* **7**, 325.

Bondy, C. (1939). *Trans. Faraday Soc.* **35**, 1093.

Boucher, E. A. and Hines, P. M. (1976). *J. Polymer Sci. Polymer Phys.* **14**, 2241.

Boyer, R. F. and Spencer, R. S. (1944). *J. Appl. Phys.* **15**, 398.

Braun, D. (1976). *Angew. Chem. Int. Ed.* **15**, 451.

Bridger, K. (1979). PhD Thesis, University of Bristol.

Bridger, K. and Vincent, B. (1980). *Europ. Polymer J.* **16**, 1017.

Bridger, K., Fairhurst, D. and Vincent, B. (1979). *J. Colloid Interface Sci.* **68**, 190.

Bueche, F. (1972). *J. Colloid Interface Sci.* **41**, 374.

Burchard, W. (1983). *In* "Chemistry and Technology of Water Soluble Polymers" (C. A. Finch *ed.*). Plenum, London.

Buscall, R. (1981). *J. Chem. Soc. Faraday Trans. I* **77**, 909.

Cain, F. W., Ottewill, R. H. and Smitham, J. B. (1978). *Faraday Disc. Chem. Soc.* **65**, 33.

Cairns, R. J. R. and Neustadter, E. L. (1975). "Proceedings of the International Conference on Colloid and Surface Science" (E. Wolfram *ed.*). Vol. 1. p. 323. Elsevier, Amsterdam.

Cairns, R. J. R. and Neustadter, E. L. (1978). *Faraday Disc. Chem. Soc.* **65**, 318.

Cairns, R. J. R., Ottewill, R. H., Osmond, D. W. J. and Wagstaff, I. (1976). *J. Colloid Interface Sci.* **54**, 45.

Cairns, R. J. R., van Megen, W. and Ottewill, R. H. (1981). *J. Colloid Interface Sci.* **79**, 511.

Candau, F., Rempp, P. and Benoit, H. (1972). *Macromolecules* **5**, 627.

Carslaw, H. S. (1921). "The Conduction of Heat", Macmillan, London.

Casassa, E. F. (1970). *J. Polymer Sci. A-2* **8**, 1651.

Chahal, R. S., Kao, W. and Patterson, D. (1973). *J. Chem. Soc. Faraday Trans. I* **69**, 1834.

Chew, B. and Couper, A. (1976). *J. Chem. Soc. Faraday Trans. I* **72**, 389.

Clark, A. T. and Lal, M. (1977). *Bri. Polymer J.* **9**, 92.

Clark, A. T. and Lal, M. (1978). *J. Chem. Soc. Faraday Trans. II* **74**, 1857.

Clark, A. T. and Lal, M. (1981). *J. Chem. Soc. Faraday Trans. II* **77**, 981.

Clarke, J. and Vincent, B. (1981a). *J. Colloid Interface Sci.* **82**, 208.

Clarke, J. and Vincent, B. (1981b). *J. Chem. Soc. Faraday Trans. I* **77**, 1831.

Clayfield, E. J. and Lumb, E. C. (1966a). *J. Colloid Sci.* **22**, 269.

Clayfield, E. J. and Lumb, E. C. (1966b). *J. Colloid Sci.* **22**, 285.

Clayfield, E. J. and Lumb, E. C. (1968). *Macromolecules* **1**, 133.

Clayfield, E. J. and Lumb, E. C. (1974a). *J. Colloid Interface Sci.* **47**, 6.

Clayfield, E. J. and Lumb, E. C. (1974b). *J. Colloid Interface Sci.* **47**, 16.

Clayfield, E. J. and Lumb, E. C. (1974c). *J. Colloid Interface Sci.* **49**, 489.

Cornet, C. F. and van Ballegooijen, H. (1966). *Polymer* **7**, 293.

Cosgrove, T. and Barnett, K. G. (1981). *J. Mag. Resonance* **43**, 15.

Cosgrove, T., Crowley, T. L., Vincent, B., Barnett, K. G. and Tadros, Th. F. (1982). *Faraday Disc. Chem. Soc.* (in press).
Cotton, J. P., Nierlich, M., Boue, F., Daoud, M., Farnoux, B., Jannink, G., Dupplessix, R. and Picot, C. (1976). *J. Chem. Phys.* **65**, 1101.
Cowell, C. and Vincent, B. (1982a). *In* "The Effects of Polymers on Dispersion Properties" (Th. F. Tadros *ed.*). p. 263. Academic Press, London.
Cowell, C., and Vincent, B. (1982b). *J. Colloid Interface Sci.* **87**, 518.
Cowell, C., Li-in-on, R. and Vincent, B. (1978). *J. Chem. Soc. Faraday Trans. I* **74**, 337.
Cowley, A. C., Fuller, N. L., Rand, R. P. and Parsegian, V. A. (1978). *Biochemistry* **17**, 3163.
Cowie, J. M. G. and McEwen, I. J. (1975). *Polymer* **16**, 244.
Croucher, M. D. and Hair, M. L. (1978). *Macromolecules* **11**, 874.
Croucher, M. D. and Hair, M. L. (1979). *J. Phys. Chem.* **83**, 1712.
Croucher, M. D. and Hair, M. L. (1980a). *Colloids Surf.* **1**, 349.
Croucher, M. D. and Hair, M. L. (1980b). *In* "Polymer Colloids II" (R. M. Fitch *ed.*). p. 497. Plenum, New York.
Croucher, M. D. and Hair, M. L. (1981). *J. Colloid Interface Sci.* **81**, 257.
Croucher, M. D., Hair, M. L. and Sundararajan, P. R. (1980). *Colloid Polymer Sci.* **258**, 408.
Daoud, M., Cotton, J. P., Farnoux, B., Jannick, G., Sharma, G., Benoit, H., Duplessix, R., Picot, C. and de Gennes, P.-G. (1975). *Macromolecules* **8**, 804.
Daoud, M. and Jannink, G. (1976). *J. Phys. (Paris)* **37**, 973.
Darling, D. F. (1982). *In* "The Effects of Polymers on Dispersion Properties" (Th. F. Tadros *ed.*). p. 285. Academic Press, London.
Darling, D. F. and Dickson, J. (1979). *J. Dairy Res.* **46**, 329.
Darling, D. F. and van Hooydonk, A. C. M. (1981). *J. Dairy Res.* **48**, 189.
Dawkins, J. V. (1973). *In* "Block Copolymers" (D. C. Allport and W. H. James *eds*). Chapter 9. Applied Science, London.
Dawkins, J. V. and Taylor, G. (1980). *Colloid Polymer Sci.* **258**, 79.
Day, J. C. and Robb, I. D. (1981). *Polymer* **22**, 1530.
de Gennes, P.-G. (1969). *Rep. Prog. Phys.* **32**, 187.
de Gennes, P.-G. (1972). *Phys. Lett. A* **38**, 339.
de Gennes, P.-G. (1975). *J. Phys. Lett.* **36**, L55.
de Gennes, P.-G. (1979). "Scaling Concepts in Polymer Physics". Cornell University Press, Ithaca.
de Gennes, P.-G. (1980). *Macromolecules* **13**, 1069.
de Gennes, P.-G. (1981). *Macromolecules* **14**, 1637.
de Gennes, P.-G. (1982). *Macromolecules* **15**, 492.
de Hek, H. and Vrij, A. (1979). *J. Colloid Interface Sci.* **70**, 592.
de Hek, H. and Vrij, A. (1981). *J. Colloid Interface Sci.* **84**, 409.
Derjaguin, B. V. (1934). *Kolloid-Z.* **69**, 155.
Derjaguin, B. V. and Muller, V. M. (1967). *Doklady Phys. Chem.* **176**, 738.
des Cloizeaux, J. (1970). *J. Physique* **31**, 715.
DiMarzio, E. A. (1965). *J. Chem. Phys.* **42**, 2101.
DiMarzio, E. A. and McCrackin, F. L. (1965). *J. Chem. Phys.* **43**, 539.
Dobbie, J. W., Evans, R., Gibson, D. V., Smitham, J. B. and Napper, D. H. (1973). *J. Colloid Interface Sci.* **45**, 557.
Dodd, J. (1980). BSc Honours Thesis, University of Sydney.
Dodd, J. and Napper, D. H. (1980). Unpublished results.
Dolan, A. K. and Edwards, S. F. (1974). *Proc. Roy. Soc. London A* **337**, 509.
Dolan, A. K. and Edwards, S. F. (1975). *Proc. Roy. Soc. London A* **343**, 427.

Domb, C. (1974). *Polymer* **15**, 259.
Doroszkowski, A. and Lambourne, R. (1968). *J. Colloid Interface Sci.* **26**, 214.
Doroszkowski, A. and Lambourne, R. (1971). *J. Polymer Sci. C* **34**, 253.
Doroszkowski, A. and Lambourne, R. (1973). *J. Colloid Interface Sci.* **43**, 97.
Doroszkowski, A. and Monk, C. J. H. (1969). *J. Sci. Instr.* **2**, 536.
Duckworth, D. S., Lips, A. and Staples, E. S. (1978). *Faraday Disc. Chem. Soc.* **65**, 288.
Dzyaloshinskii, I. E., Lifshitz, E. M. and Pitaevskii, L. P. (1961). *Advan. Phys.* **10**, 165.
Eagland, D. (1975). *In* "Water" (F. Franks *ed.*). Vol. 5. Chapter 1. Plenum, New York.
Edwards, S. F. (1965). *Proc. Phys. Soc.* **85**, 613.
Edwards, S. F. (1966). *Proc. Phys. Soc.* **88**, 265.
Edwards, S. F. and Jeffers, E. F. (1979). *J. Chem. Soc. Faraday Trans. II* **75**, 1020.
Eichinger, B. E. (1970). *J. Chem. Phys.* **53**, 561.
Eighinger, B. E. and Flory, P. J. (1968). *Trans. Faraday Soc.* **64**, 2053.
Eisenlauer, J., Killmann, E. and Korn, M. (1977). *J. Polymer Sci. C* **61**, 413.
Elias, H.-G. (1959). *Makromol. Chem.* **33**, 140.
Elias, H.-G. (1961). *Makromol. Chem.* **50**, 1.
Elias, H.-G. (1977). *In* "Fractionation of Synthetic Polymers" (L. H. Tung *ed.*). Chapter 4. Dekker, New York.
Erlander, S. R. (1970). *J. Colloid Interface Sci.* **34**, 53.
Evans, R. (1971). BSc Honours Thesis, University of Sydney.
Evans, R. (1976). PhD Thesis, University of Sydney.
Evans, R. and Napper, D. H. (1973a). *Kolloid-Z. Z. Polymere* **251**, 329.
Evans, R. and Napper, D. H. (1973b). *Kolloid-Z. Z. Polymere* **251**, 409.
Evans, R. and Napper, D. H. (1973c). *J. Colloid Interface Sci.* **45**, 138.
Evans, R. and Napper, D. H. (1973d). *Nature* **246**, 34.
Evans, R. and Napper, D. H. (1975). *J. Colloid Interface Sci.* **52**, 260.
Evans, R. and Napper, D. H. (1977). *J. Chem. Soc. Faraday Trans. I* **73**, 1377.
Evans, R. and Napper, D. H. (1978). *J. Colloid Interface Sci.* **63**, 43.
Evans, R., Napper, D. H. and Ewald, A. H. (1975). *J. Colloid Interface Sci.* **51**, 552.
Evans, R., Smitham, J. B. and Napper, D. H. (1977). *Colloid Polymer Sci.* **255**, 161.
Evans, R., Davison, J. B. and Napper, D. H. (1972). *J. Polymer Sci. B* **10**, 449.
Everett, D. H. (1976). *Pure Appl. Chem.* **48**, 419.
Everett, D. H. (1978). *Faraday Disc. Chem. Soc.* **65**, 215.
Everett, D. H. (1980). *Proc. Roy. Inst.* **52**, 209.
Everett, D. H. (1981). Private communication.
Everett, D. H. and Stageman, J. F. (1977). *Colloid Polymer Sci.* **255**, 293.
Everett, D. H. and Stageman, J. F. (1978a). *Faraday Disc. Chem. Soc.* **65**, 230.
Everett, D. H. and Stageman, J. F. (1978b). *Faraday Disc. Chem. Soc.* **65**, 314.
Faraday, M. (1857). *Phil. Trans.* **147**, 145.
Feat, G. R. and Levine, S. (1976). *J. Colloid Interface Sci.* **54**, 34.
Feigin, R. I. (1980). PhD Thesis, University of Sydney.
Feigin, R. I. and Napper, D. H. (1979). *J. Colloid Interface Sci.* **71**, 117.
Feigin, R. I. and Napper, D. H. (1980a). *J. Colloid Interface Sci.* **74**, 567.
Feigin, R. I. and Napper, D. H. (1980b). *J. Colloid Interface Sci.* **75**, 525.
Feigin, R. I. and Napper, D. H. (1980c). *Colloid Polymer Sci.* **258**, 1153.
Feigin, R. I., Dodd, J. and Napper, D. H. (1981). *Colloid Polymer Sci.* **259**, 1027.
Feller, W. (1968). "An Introduction to Probability Theory and its Applications", 3rd edn, Chapter XIV. Wiley, New York.
Findlay, A. (1951). "The Phase Rule", 9th edn, pp. 7–8. Dover, New York.
Fischer, E. W. (1958). *Kolloid-Z.* **160**, 120.
Fleer, G. J., Koopal, L. K. and Lyklema, J. (1972). *Kolloid-Z. Z. Polymere* **250**, 689.

Flory, P. J. (1941). *J. Chem. Phys.* **9**, 660.
Flory, P. J. (1942). *J. Chem. Phys.* **10**, 51.
Flory, P. J. (1949). *J. Chem. Phys.* **17**, 1347.
Flory, P. J. (1953). "Principles of Polymer Chemistry". Cornell University Press, Ithaca.
Flory, P. J. (1965). *J. Am. Chem. Soc.* **87**, 1833.
Flory, P. J. (1969). "Statistical Mechanics of Chain Molecules". Interscience, New York.
Flory, P. J. (1970). *Disc. Faraday Soc.* **49**, 7.
Flory, P. J. (1971). *Pure Appl. Chem.* **26**, 309.
Flory, P. J. (1974). *Macromolecules* **7**, 381.
Flory, P. J. (1976). *J. Polymer Sci. C* **54**, 19.
Flory, P. J. (1978a). *Macromolecules* **11**, 1138.
Flory, P. J. (1978b). *Macromolecules* **11**, 1141.
Flory, P. J. and Krigbaum, W. R. (1950). *J. Chem. Phys.* **18**, 1086.
Flory, P. J. and Osterheld, J. E. (1954). *J. Phys. Chem.* **58**, 653.
Fock, V. (1930). *Z. Phys.* **61**, 126.
Fredenslund, Aa., Gymehling, J. and Rasmussen, P. (1977). "Vapour-Liquid Equilibria Using UNIFAC". Elsevier, Amsterdam.
Freeman, P. I. and Rowlinson, J. S. (1960). *Polymer* **1**, 20.
Freundlich, H. (1926). "Colloid and Capillary Chemistry", pp. 589–593. Methuen, London.
Fuchs, N. (1934). *Z. Phys.* **89**, 736.
Gaeckle, D. and Patterson, D. (1972). *Macromolecules* **5**, 136.
Gaeckle, D., Kao, W., Patterson, D. and Rinfret, M. (1973). *J. Chem. Soc. Faraday Trans. I* **69**, 1849.
Garvey, M. J. (1982). *In* "The Effects of Polymers on Dispersion Properties" (Th. Tadros *ed.*). p. 203. Academic Press, London.
Garvey, M. J. and Robb, I. D. (1979). *J. Chem. Soc. Faraday Trans. I* **75**, 993.
Garvey, M. J., Tadros, Th. F. and Vincent, B. (1974). *J. Colloid Interface Sci.* **49**, 57.
Garvey, M. J., Tadros, Th. F. and Vincent, B. (1976). *J. Colloid Interface Sci.* **55**, 440.
Garvey, M. J., Mitchell, D. and Smith, A. L. (1979). *Colloid Polymer Sci.* **257**, 70.
Geoghegan, M. J. and Brian, R. C. (1948). *Biochem. J.* **43**, 5 and 13.
Gerber, P. R. and Moore, M. A. (1976). *Macromolecules* **9**, 478.
Gerber, P. R. and Moore, M. A. (1977). *Macromolecules* **10**, 476.
Gottlieb, M. and Herskowitz, M. (1981). *Macromolecules* **14**, 1468.
Gradshtein, I. S. and Ryznik, I. M. (1965). "Tables of Integrals, Series and Products". Academic Press, New York.
Gregory, J. (1970). *Adv. Colloid Interface Sci.* **2**, 396.
Guttman, C. M. and DiMarzio, E. A. (1982). *Macromolecules* **15**, 525.
Guttman, A. J. and Whittington, S. G. (1978). *J. Phys. A* **11**, L107.
Guttman, A. J., Middlemiss, K. G., Torrie, G. M. and Whittington, S. G. (1978). *J. Chem. Phys.* **69**, 5375.
Guttman, C. M., DiMarzio, E. A. and Hoffman, J. D. (1981). *Polymer* **22**, 1466.
Hall, D. G. (1972). *J. Chem. Soc. Faraday Trans. II* **68**, 2169.
Hamaker, H. C. (1937). *Physica* **4**, 1058.
Hardgraft, J., Hyde, A. I. and Richards, R. W. (1979). *J. Chem. Soc. Faraday Trans. II* **75**, 1495.
Hartree, D. R. (1928). *Camb. Phil. Soc.* **24**, 89, 111, 426.
Hatschek, E. (1925). "The Foundations of Colloid Science", pp. 89–91. Ernest Benn, London.

Hedges, E. S. (1931). "Colloids", p. 164. Arnold, London.
Heller, W. (1966). *Pure Appl. Chem.* **12**, 249.
Heller, W. and Pugh, T. L. (1954). *J. Chem. Phys.* **22**, 1778.
Heller, W. and Pugh, T. L. (1956). *J. Chem. Phys.* **24**, 1107.
Heller, W. and Tanaka, W. (1951). *Phys. Rev.* **82**, 301.
Hertz, H. (1888). *Mathematik* **92**, 155.
Hesselink, F. Th. (1969). *J. Phys. Chem.* **73**, 3488.
Hesselink, F. Th. (1971). *J. Phys. Chem.* **75**, 65.
Hesselink, F. Th. (1975). *J. Colloid Interface Sci.* **50**, 606.
Hesselink, F. Th. (1977). *J. Polymer Sci. Polymer Symposia* **61**, 439.
Hesselink, F. Th., Vrij, A. and Overbeek, J. Th. G. (1971). *J. Phys. Chem.* **75**, 2094.
Hirschfelder, J., Stevenson, D. and Eyring, H. (1937). *J. Chem. Phys.* **5**, 896.
Hoeve, C. A. J. (1965). *J. Chem. Phys.* **43**, 3007.
Hoeve, C. A. J., DiMarzio, E. A. and Peyser, P. (1965). *J. Chem. Phys.* **42**, 2558.
Homola, A. M. and Robertson, A. A. (1975a). *J. Colloid Interface Sci.* **51**, 202.
Homola, A. M. and Robertson, A. A. (1975b). *Can. J. Chem.* **53**, 389.
Homola, A. M. and Robertson, A. A. (1976). *J. Colloid Interface Sci.* **54**, 286.
Homola, A. M., Snook, I. and van Megen, W. (1977). *J. Colloid Interface Sci.* **61**, 493.
Honig, E. P., Robersen, G. J. and Wiersema, P. H. (1971). *J. Colloid Interface Sci.* **36**, 97.
Huggins, M. L. (1941). *J. Chem. Phys.* **9**, 440.
Huggins, M. L. (1942a). *J. Phys. Chem.* **46**, 151.
Huggins, M. L. (1942b). *J. Am. Chem. Soc.* **64**, 1712.
Hunter, R. J., Neville, P. C. and Firth, B. A. (1975). *ACS Symposium Series* **9**, 193.
Hyde, A. J. and Tanner, A. G. (1968). *J. Colloid Interface Sci.* **28**, 179.
Inokuchi, K. (1951). *Bull. Chem. Soc. Japan* **24**, 78.
Israelachvili, J. N. (1982). *In* "Principles of Colloid and Surface Chemistry" (R. J. Hunter *ed.*). Royal Australian Chemical Institute, Sydney.
Israelachvili, J. N. and Adams, G. E. (1976). *Nature* **262**, 774.
Israelachvili, J. N. and Adams, G. E. (1978). *J. Chem. Soc. Faraday Trans. I* **74**, 975.
Israelachvili, J. N. and Tabor, D. (1972). *Proc. Roy. Soc. London A* **331**, 19.
Israelachvili, J. N., Tandon, R. K. and White, L. R. (1979). *Nature* **277**, 120.
Israelachvili, J. N., Tandon, R. K. and White, L. R. (1980). *J. Colloid Interface Sci.* **78**, 430.
Jäckel, K. (1964). *Kolloid-Z. Z. Polymere* **197**, 143.
Joanny, J. F., Liebler, L. and de Gennes, P.-G. (1979). *J. Polymer Sci. Polymer. Phys. Ed.* **17**, 1073.
Kerker, M. (1969). "The Scattering of Light and Other Electromagnetic Radiation". Academic Press, New York.
Khokhlov, A. R. (1978). *Polymer* **19**, 1387.
Khokhlov, A. R. (1981). *Polymer* **22**, 447.
Kitahara, A. and Hasumuma, M. (1972). *J. Colloid Interface Sci.* **41**, 383.
Klein, J. (1980). *Nature* **288**, 248.
Klein, J. (1982). *Adv. Colloid Interface Sci.* **16**, 101.
Knapschinsky, L., Ehmke, B., Sonntag, H. and Katz, W. (1982). *Colloid Polymer Sci.* (in press).
Koelmans, H. and Overbeek, J. Th. G. (1954). *Disc Faraday Soc.* **18**, 52.
Koningsveld, P. and Staverman, A. J. (1968). *J. Polymer Sci. A2* **6**, 325.
Koningsveld, P., Kleintjens, L. A. and Schultz, A. R. (1970). *J. Polymer Sci. A2* **8**, 1261.
Laible, R. and Hamann, K. (1980). *Adv. Colloid Interface Sci.* **13**, 65.

Lambe, R., Tadros, Th. F. and Vincent, B. (1978). *J. Colloid Interface Sci.* **66**, 77.
La Mer, V. K. (1964). *J. Colloid Sci.* **19**, 291.
Landau, L. D. and Lifshitz, E. M. (1977). "Quantum Mechanics", 3rd edn, p. 60. Pergamon Press, Oxford.
Langmuir, I. (1926). *In* "Colloid Chemistry" (J. Alexander *ed.*). Chapter 29. Chemical Catalog Co., New York.
Langmuir, I. (1938). *J. Chem. Phys.* **6**, 873.
Lax, M. (1974a). *J. Chem. Phys.* **61**, 4133.
Lax, M. (1974b). *Macromolecules* **7**, 660.
Lax, M. (1975). *Macromolecules* **8**, 947.
Lax, M., Barr, R. and Brender, C. (1981). *J. Chem. Phys.* **75**, 460.
Le Neveu, D. M., Rand, R. P., Parsegian, V. A. and Gingell, D. (1977). *Biophys. J.* **18**, 209.
Levine, S., Thomlinson, M. L. and Robinson, K. (1978). *Faraday Disc. Chem. Soc.* **65**, 202.
Lewis, G. and Johnson, A. F. (1969). *J. Chem. Soc. A* 1816.
Lichtenbelt, J. W. Th., Pathmamanoharan, C. and Wiersema, P. H. J. (1974). *J. Colloid Interface Sci.* **49**, 281.
Litshitz, E. M. (1956). *Sov. Phys. JETP* **2**, 73.
Lifshitz, I. M., Grosberg, A. Yu. and Khokhlov, A. (1978). *Rev. Mod. Phys.* **50**, 683.
Li-in-on, F. K., Vincent, B. and Waite, F. A. (1975). *ACS Symposium Series* **9**, 165.
Lipatov, Yu. S. and Sergeeva, L. M. (1974). "Adsorption of Polymers". John Wiley, New York.
Lips, A., Smart, C. and Willis, E. (1971). *Trans. Faraday Soc.* **67**, 2979.
London, F. (1934). *Z. Phys.* **89**, 736.
Long, J. A., Osmond, D. W. J. and Vincent, B. (1973). *J. Colloid Interface Sci.* **42**, 545.
Lottermoser, A. and von Meyer, E. (1897). *J. Prakt. Chemie* **56**, 242.
Lyklema, J. and van Vliet, T. (1978). *Faraday Disc. Chem. Soc.* **65**, 26.
Mackor, E. L. (1951). *J. Colloid Sci.* **6**, 492.
Mackor, E. L. and van der Waals, J. H. (1952). *J. Colloid Sci.* **7**, 535.
Mahanty, J. and Ninham, B. W. (1976). "Dispersion Forces". Academic, London.
Malcolm, G. N. and Rowlinson, J. S. (1957). *Trans. Faraday Soc.* **53**, 921.
March, G. C. and Napper, D. H. (1977). *J. Colloid Interface Sci.* **61**, 383.
Mark, P. and Windwer, S. (1974). *Macromolecules* **7**, 690.
Mark, P., Windwer, S. and Lax, M. (1975). *Macromolecules* **8**, 946.
Mattice, W. L. (1981). *Macromolecules* **14**, 1491.
Mattice, W. L. and Napper, D. H. (1981). *Macromolecules* **14**, 1066.
Mayer, K., Hahnel, E. and Feiner, R. (1945). *Proc. Soc. Exp. Biol. Med.* **58**, 36.
McCrackin, F. L., Mazur, J. and Guttman, C. M. (1974). *Macromolecules* **6**, 859.
Meier, D. J. (1967). *J. Phys. Chem.* **71**, 1861.
Medalia, A. I. (1971). *In* "Surface and Colloid Science" (E. Matijevic *ed.*). Vol. 4. p. 1. Wiley, New York.
Middlemiss, K. M. and Whittington, S. G. (1976). *J. Chem. Phys.* **64**, 4684.
Middlemiss, K. M., Torrie, G. M. and Whittington, S. G. (1977). *J. Chem. Phys.* **66**, 3227.
Molyneux, P. (1975). *In* "Water" (F. Franks *ed.*). Vol. 4. Chapter 7. Plenum, New York.
Monaghan, B. R. and White, H. L. (1935). *J. Gen. Physiol.* **19**, 715.
Moore, W. J. (1972). "Physical Chemistry", 5th edn, p. 273. Longman, London.
Morawetz, H. (1965). "Macromolecules in Solution", 2nd edn, p. 192. John Wiley, New York.

Napper, D. H. (1968a). *Makromol. Chem.* **120**, 231.
Napper, D. H. (1968b). *Trans. Faraday Soc.* **64**, 1701.
Napper, D. H. (1969a). *Polymer* **10**, 181.
Napper, D. H. (1969b). *J. Colloid Interface Sci.* **29**, 168.
Napper, D. H. (1969c). *Kolloid-Z. Z. Polymere* **234**, 1149.
Napper, D. H. (1970a). *J. Colloid Interface Sci.* **32**, 106.
Napper, D. H. (1970b). *J. Colloid Interface Sci.* **33**, 384.
Napper, D. H. (1970c). *Ind. Eng. Chem. Prod. Res. Develop.* **9**, 467.
Napper, D. H. (1977). *J. Colloid Interface Sci.* **58**, 390.
Napper, D. H. (1981). *Chem. Aust.* **48**, 384.
Napper, D. H. (1982). *In* "Colloidal Dispersions" (J. W. Goodwin *ed.*). Royal Chemical Society, London.
Napper, D. H. and Hunter, R. J. (1972). *MTP Int. Rev. Sci. Series 1* **7**, 241.
Napper, D. H. and Hunter, R. J. (1975). *MTP Int. Rev. Sci. Series 2* **7**, 161.
Napper, D. H. and Netschey, A. (1971). *J. Colloid Interface Sci.* **37**, 528.
Nierlich, M., Cotton, J. P. and Farnoux, B. (1978). *J. Chem. Phys.* **69**, 1379.
Nose, T. and Chu, B. (1979a). *Macromolecules* **12**, 590.
Nose, T. and Chu, B. (1979b). *Macromolecules* **12**, 1122.
Oishi, T. and Prausnitz, J. (1978). *J. Ind. Eng. Chem. Process Res. Develop.* **17**, 333.
Orofino, T. A. and Flory, P. J. (1957). *J. Chem. Phys.* **26**, 1067.
Osmond, D. W. J., Vincent, B. and Waite, F. A. (1975). *Colloid Polymer Sci.* **253**, 676.
Ostwald, Wo. (1919). *Kolloidchem. Beihefte* **10**, 234.
Ottewill, R. H. (1969). *Ann. Rep. A* **66**, 183.
Ottewill, R. H. (1973). *In* "Colloid Science" (D. H. Everett *ed.*). Vol. 1. p. 173. The Chemical Society, London.
Ottewill, R. H. (1976). *Prog. Colloid Polymer Sci.* **59**, 14.
Ottewill, R. H. (1980). *Colloid Polymer Sci.* **67**, 71.
Ottewill, R. H. (1982). *In* "Colloidal Dispersions" (J. W. Goodwin *ed.*). Royal Chemical Society, London.
Ottewill, R. H. and Walker, T. (1968). *Kolloid-Z. Z. Polymere* **227**, 108.
Ottewill, R. H. and Walker, T. (1974). *J. Chem. Soc. Faraday Trans. I* **70**, 917.
Overbeek, J. Th. G. (1977). *J. Colloid Interface Sci.* **58**, 408.
Overbeek, J. Th. G. (1982). *In* "Colloidal Dispersions" (J. W. Goodwin *ed.*). Royal Chemical Society, London.
Parfitt, G. D. and Peacock, J. (1978). *Surface Colloid Sci.* **10**, 163.
Parsegian, V. A. and Ninham, B. W. (1971). *J. Colloid Interface Sci.* **37**, 332.
Parsegian, V. A., Fuller, N. and Rand, R. P. (1979). *Proc. Natn. Acad. Sci.* **76**, 2750.
Pashley, R. M. and Israelachvili, J. N. (1981). *Colloids Surfaces* **2**, 169.
Patterson, D. (1968). *J. Polymer Sci. C* **16**, 3379.
Patterson, D. (1969). *Macromolecules* **2**, 672.
Patterson, D. and Delmas, G. (1969). *Trans. Faraday Soc.* **65**, 708.
Pathmamanoharan, C., de Hek, H. and Vrij, A. (1981). *Colloid Polymer Sci.* **259**, 769.
Payens, T. A. J. (1979). *J. Dairy Res.* **46**, 291.
Pearson, K. (1905). *Nature* **72**, 294.
Percus, J. K. and Yevick, G. J. (1958). *Phys. Rev.* **110**, 1.
Peyser, P. and Stromberg, R. R. (1967). *J. Phys. Chem.* **71**, 2066.
Peyser, P., Tutas, D. and Stromberg, R. R. (1967). *J. Polymer Sci. A* **5**, 653.
Phillips, M. C. (1975). *In* "Water: A Comprehensive Treatise" (F. Franks *ed.*). Vol. 5. Chapter 3. Plenum, New York.
Pimentel, G. C. and McClellan, A. L. (1960). "The Hydrogen Bond", pp. 195–196. Freeman, San Francisco.

Pritchard, M. J. and Caroline, D. (1980). *Macromolecules* **13**, 957.
Pritchard, M. J. and Caroline, D. (1981). *Macromolecules* **14**, 424.
Ptitsyn, O. B., Kron, A. K. and Eizner, Y. Y. (1968). *J. Polymer Sci. C* **16**, 3509.
Pusey, P. N. (1982). *In* "Colloidal Dispersions" (J. W. Goodwin *ed.*). Royal Chemical Society, London.
Rayleigh, Lord (1880). *Phil. Mag.* **10**, 73.
Rayleigh, Lord (1905). *Nature* **72**, 318.
Rayleigh, Lord (1919). *Phil. Mag.* **37**, 321.
Rehbinder, P., Lagutina, L. and Wenstrom, E. (1930). *Z. Phys. Chem. A* **196**, 63.
Richards, R. W., Maconnachie, A. and Allen, G. (1981a). *Polymer* **22**, 147.
Richards, R. W., Maconnachie, A. and Allen, G. (1981b). *Polymer* **22**, 153.
Richards, R. W., Maconnachie, A. and Allen, G. (1981c). *Polymer* **22**, 158.
Richmond, P. and Lal, M. (1974). *Chem. Phys. Lett.* **24**, 594.
Roberts, A. D. and Tabor, D. (1971). *Proc. Roy. Soc. London A* **325**, 323.
Robb, I. D. (1982). *In* "The Chemistry and Technology of Water Soluble Polymers" (C. A. Finch *ed.*). Plenum, London.
Robb, I. D. and Smith, R. (1977). *Polymer* **18**, 500.
Roe, R.-J. and Zin, W.-C. (1980). *Macromolecules* **13**, 122.
Rossi, C., Bianchi, E. and Conio, G. (1963). *Chem. Ind. (Milan)* **45**, 1498.
Rubin, R. J., Mazur, J. and Weiss, G. H. (1976). *Pure Appl. Chem.* **46**, 143.
Saeki, S., Kuwahara, N., Konno, S. and Kaneko, M. (1973). *Macromolecules* **6**, 246
Saeki, S., Kuwahara, N., Nakata, M. and Kaneko, M. (1976). *Polymer* **17**, 685.
Sanchez, I. C. (1979). *Macromolecules* **12**, 980.
Sanchez, I. C. (1980). *J. Macromol. Sci. Phys. B* **17**, 565.
Sanchez, I. C. and Lacombe, R. H. (1978). *Macromolecules* **11**, 1145.
Sato, T. (1971). *J. Appl. Polymer Sci.* **15**, 1053.
Sato, T. (1979). *J. Appl. Polymer Sci.* **23**, 1693.
Sato, T. and Ruch, R. (1980). "Stabilization of Colloidal Dispersions by Polymeric Adsorption". Dekker, New York.
Sato, T. and Sieglaff, C. L. (1980). *J. Appl. Polymer Sci.* **25**, 1781.
Scheutjens, J. M. H. M. and Fleer, G. J. (1979). *J. Phys. Chem.* **83**, 1619.
Scheutjens, J. M. H. M. and Fleer, G. J. (1980). *J. Phys. Chem.* **84**, 178.
Scheutjens, J. M. H. M. and Fleer, G. J. (1982). *Adv. Colloid Interface Sci.* **16**, 361.
Sieglaff, C. L. (1959). *J. Polymer Sci.* **41**, 319.
Silberberg, A. (1962a). *J. Phys. Chem.* **66**, 1872.
Silberberg, A. (1962b). *J. Phys. Chem.* **66**, 1884.
Silberberg, A. (1967). *J. Chem. Phys.* **46**, 1105.
Silberberg, A. (1968). *J. Chem. Phys.* **48**, 2835.
Silberberg, A. (1976). *Prog. Colloid Polymer Sci.* **59**, 33.
Silberberg, A., Eliassaf, J. and Katchalsky, A. (1957). *J. Polymer Sci.* **23**, 259.
Slagowski, E., Tsai, B. and McIntyre, D. (1976). *Macromolecules* **9**, 687.
Smitham, J. B. (1976). PhD Thesis, University of Sydney.
Smitham, J. B. and Napper, D. H. (1976a). *J. Chem. Soc. Faraday Trans. I* **72**, 2429.
Smitham, J. B. and Napper, D. H. (1976b). *J. Colloid Interface Sci.* **54**, 467.
Smitham, J. B. and Napper, D. H. (1979). *Colloid Polymer Sci.* **257**, 748.
Smitham, J. B., Gibson, D. V. and Napper, D. H. (1973). *J. Colloid Interface Sci.* **45**, 211.
Smitham, J. B., Evans, R. and Napper, D. H. (1975). *J. Chem. Soc. Faraday Trans. I* **71**, 285.
Smoluchowski, M. (1917). *Z. Phys. Chem.* **92**, 129.
Snook, I. and van Megen, W. (1975). *Chem. Phys. Lett.* **33**, 156.

Snook, I. and van Megen, W. (1976). *J. Chem. Soc. Faraday Trans. II* **72**, 216.
Sonntag, H. (1974). *Abh. AdW. DDR Klasse Chemie N1* **2**, 517.
Sonntag, H., Ehmke, B., Miller, R. and Knapschinsky, L. (1982). *In* "The Effects of Polymers on Dispersion Properties" (Th. F. Tadros *ed.*). p. 207. Academic Press, London.
Sperry, P. R. (1982). *J. Colloid Interface Sci.* **87**, 375.
Sperry, P. R., Hopfenberg, H. B. and Thomas, N. L. (1981). *J. Colloid Interface Sci.* **82**, 62.
Spielman, L. A. (1970). *J. Colloid Interface Sci.* **33**, 562.
Spring, M. W. (1900). *Rec. Trav. Chim.* **19**, 204.
Stockmayer, W. H. (1955). *J. Polymer Sci.* **15**, 595.
Stockmayer, W. H. (1960). *Makromol. Chem.* **35**, 54.
Stockmayer, W. H. and Albrecht, A. C. (1958). *J. Polymer Sci.* **32**, 215.
Stockmayer, W. H., Moore, L. D., Fixman, M. and Epstein, B. M. (1955). *J. Polymer Sci.* **16**, 517.
Sundararajan, R. R., Hamer, G. K. and Croucher, M. D. (1980). *Macromolecules* **13**, 971.
Sun, S.-T., Nishio, I., Swislow, G. and Tanaka, T. (1980). *J. Chem. Phys.* **73**, 5971.
Swislow, G., Sun, S.-T., Nishio, I. and Tanaka, T. (1980). *Phys. Rev. Lett.* **44**, 796.
Tabor, D. and Winterton, R. H. S. (1969). *Proc. Roy. Soc. London A* **312**, 435.
Tadros, Th. F. (1980). *Adv. Colloid Interface Sci.* **12**, 141.
Tadros, Th. F. (1982). *In* "The Effects of Polymers on Dispersion Properties" (Th. F. Tadros *ed.*). p. 1. Academic Press, London.
Talamini, G. and Vidotto, G. (1967). *Makromol. Chem.* **110**, 111.
Tanaka, T. (1977). *Macomolecules* **10**, 51.
Thompson, L. and Pryde, D. N. (1981). *J. Chem. Soc. Faraday Trans. I* **77**, 2405.
Tompa, H. (1956). "Polymer Solutions". Butterworth, London.
Torrie, G. M. and Whittington, S. G. (1975). *J. Phys. A* **8**, 1178.
Torrie, G. M., Middlemiss, K. M., Bly, S. H. P. and Whittington, S. G. (1976). *J. Chem. Phys.* **65**, 1867.
Traube, I. (1925). *Gummi Ztg.* **39**, 434; 1647.
Twiss, D. F. and Carpenter, A. S. (1938). *Proc. Rubber Technol. Conf. London II* 81.
van der Schee, H. A. and Lyklema, J. (1982). *In* "The Effects of Polymers on Dispersion Properties" (Th. F. Tadros *ed.*). p. 81. Academic, London.
van der Waarden, M. (1950). *J. Colloid Sci.* **5**, 317.
van der Waarden, M. (1951). *J. Colloid Sci.* **6**, 443.
van Megen, W. and Snook, I. (1975). *J. Colloid Interface Sci.* **53**, 172.
van Megen, W. and Snook, I. (1976a). *J. Colloid Interface Sci.* **57**, 40.
van Megen, W. and Snook, I. (1976b). *J. Colloid Interface Sci.* **57**, 47.
Vester, C. F. (1938a). *Proc. Rubber Technol. Conf. London* 126.
Vester, C. F. (1938b). *Kolloid-Z.* **84**, 63.
Verwey, E. J. and de Boer, J. H. (1938). *Rec. Trav. Chim.* **57**, 383.
Vincent, B. (1973a). *In* "Colloid Science" (D. H. Everett *ed.*). Vol. I. p. 220. The Chemical Society, London.
Vincent, B. (1973b). *J. Colloid Interface Sci.* **42**, 270.
Vincent, B. (1973c). *In* "Colloid Science". (D. H. Everett *ed.*). Vol. I. p. 238. The Chemical Society, London.
Vincent, B. (1974). *Adv. Colloid Interface Sci.* **4**, 193.
Vincent, B. (1978). *Faraday Disc. Chem. Soc.* **65**, 313.
Vincent, B. (1982). *In* "Water Soluble Polymers" (C. A. Finch *ed.*). Plenum, London.
Vincent, B. and Whittington, S. (1981). *In* "Surface and Colloid Science". Vol. 12. p. 1. Plenum Press, New York.

Vincent, B., Luckham, P. F. and Waite, F. A. (1980). *J. Colloid Interface Sci.* **73**, 508.
Visser, J. (1972). *Adv. Colloid Interface Sci.* **3**, 331.
Visser, J. (1981). *Adv. Colloid Interface Sci.* **15**, 157.
Volkov, V. A. (1973). *Colloid J. USSR* **35**, 980.
Volkov, V. A. (1974). *Colloid J. USSR* **36**, 855.
Volkov, V. A. (1976). *Colloid J. USSR* **38**, 117.
Vollmert, B. (1975). *Pure Appl. Chem.* **43**, 183.
Vrij, A. (1974). *J. Polymer Sci. C* **44**, 169.
Vrij, A. (1976). *Pure Appl. Chem.* **48**, 471.
Wall, F. T. and Seitz, W. A. (1977). *J. Chem. Phys.* **67**, 3722.
Wellington, S. L. (1981). *Polymer Preprints* **22**, 63.
Wheeler, J. C. (1975). *J. Chem. Phys.* **62**, 433.
Whittington, S. G. (1975). *J. Chem. Phys.* **63**, 779.
Wolter, K. D. (1978). *J. Colloid Interface Sci.* **64**, 592.
Yamakawa, H. (1971). "Modern Theory of Polymer Solutions". Harper and Row, New York.
Yamakawa, H. (1976). *Pure Appl. Chem.* **46**, 135.
Zeman, L. and Patterson, D. (1972). *Macromolecules* **5**, 513.
Zsigmondy, R. (1901). *Z. Anal. Chem.* **40**, 697.
Zsigmondy, R. and Schultz, N. (1902). *Z. Chem. Physiol. Path.* **3**, 1902.

Subject Index